Current Advances in Oxytocin Research

Current Advances in Oxytocin Research

Guest Editor

Claudia Camerino

MDPI

Basel • Beijing • Wuhan • Barcelona • Belgrade • Novi Sad • Cluj • Manchester

Guest Editor
Claudia Camerino
Department of Precision and
Regenerative Medicine
University of Bari Aldo Moro
Bari
Italy

Editorial Office
MDPI AG
Grosspeteranlage 5
4052 Basel, Switzerland

This is a reprint of the Special Issue, published open access by the journal *Current Issues in Molecular Biology* (ISSN 1467-3045), freely accessible at: https://www.mdpi.com/journal/cimb/special_issues/3NPET33FKL.

For citation purposes, cite each article independently as indicated on the article page online and as indicated below:

Lastname, A.A.; Lastname, B.B. Article Title. *Journal Name* **Year**, *Volume Number*, Page Range.

ISBN 978-3-7258-5357-1 (Hbk)
ISBN 978-3-7258-5358-8 (PDF)
https://doi.org/10.3390/books978-3-7258-5358-8

Contents

About the Editor

Claudia Camerino

Claudia Camerino is a Physiologist, appointed as Assistant Professor in the Department of Precision and Regenerative Medicine, School of Medicine, University of Bari Aldo Moro, Italy. She has continuously collaborated with American Universities, where she worked for more than a decade and has established a fervid collaboration that is still ongoing. She has long been a member of the American Society for Bone and Mineral Research and the European Calcified Tissue Society. Her research focuses on the effects of hormones and neurotransmitters on bone metabolism and energy homeostasis, with a special interest in oxytocin and brain-derived neurotrophic factor. Currently, she is studying the expression of oxytocin and oxytocin receptor in different skeletal muscle phenotypes after thermogenic challenges and the gene relationship between the neurotrophin nerve growth factor, brain-derived neurotrophic factor, and osteocalcin/oxytocin in bone, brain, and energy regulation and reproductive organs after cold stress challenges in mice. Her research also focuses on diseases related to oxytocin dysfunctions such as Prader–Willi syndrome and autism spectrum disorder. She currently collaborates with several journals as an Assistant Editor and reviewer.

Preface

This Special Issue Reprint, curated under the name "Current advance in oxytocin research", brings together recent advances exploring the roles of oxytocin in the etiology of obesity, skeletal muscle disorders, polycystic ovary syndrome, Prader–Willi syndrome, the reaction to stress and autism spectrum disorder. The aim of this collection is to deepen our understanding of the physiological effects of oxytocin, beyond its role in pregnancy and lactation, and its role in several diseases. The contributions herein—from basic research to translational applications—are intended to support both early-career and established researchers, clinicians, and neurobiologists working in the intersecting fields of basic science and translational research. As a Guest Editor, I was also motivated to compile this focused issue in light of the special anniversary for oxytocin this year, in 2025, as it has been 70 years since oxytocin's discovery in 1955. The discovery of the pituitary neurohormone oxytocin led to the Nobel Prize in Chemistry in 1955 being awarded to Vincent du Vigneaud. This represented the culmination of a research programme dating back to 1895, when Professors Oliver and Schafer reported that a substance extracted from the pituitary gland elevates blood pressure when intravenously injected into dogs. Professor Dale later reported on a neurohypophysial substance that triggers uterine contraction, stimulates lactation and functions as an antidiuretic. In 1950, the amino acid sequences of vasopressin and oxytocin were determined, and both peptides were chemically synthesised. This characterisation led to the Nobel Prize being awarded to Vincent du Vigneaud. I extend my sincere appreciation to the contributing authors of this Special Issue for their high-quality submissions, to the peer reviewers for their valuable input, and to the editorial staff at Current Issues in Molecular Biology for their professional support throughout the process.

Claudia Camerino

Guest Editor

current issues in
molecular biology

MDPI

Editorial

Editorial for Special Issue "Current Advances in Oxytocin Research"

Claudia Camerino [1,2]

1 Department of Precision and Regenerative Medicine, School of Medicine, University of Bari Aldo Moro, P.za G. Cesare 11, 70100 Bari, Italy; ccamerino@libero.it
2 Department of Physiology and Pharmacology "V. Erspamer", Sapienza University of Rome, P.le Aldo Moro 5, 00185 Rome, Italy

Very few hormones and neurotransmitters are as fascinating as oxytocin. The discovery of oxytocin dates back to 1955, when Professor Vincent du Vigneaud was awarded the Nobel Prize in Chemistry for sequencing oxytocin [1]. This achievement marked the culmination of a long line of research initiated in 1895. For the following 50 years or so, oxytocin was primarily studied for its role in uterine contractility and milk ejection during lactation. However, in the early 2000s, its physiological functions were reconsidered. Researchers discovered that oxytocin also plays a role in glucose metabolism and obesity. Notably, a lack of oxytocin was found to contribute to the onset of obesity and metabolic syndrome, even with no changes in food intake [2,3]. Since then, thousands of studies have explored the wide range of beneficial physiological effects associated with oxytocin. The oxytocin receptor, a G-protein coupled receptor, has been shown to exhibit sexual dimorphisms, indicating that its role is not exclusive to the female gender and, by extension, not limited to uterine contraction. But what is the real physiological identity of oxytocin? The answer begins with the early evidence showing that oxytocin is involved in thermoregulation [4–6] and muscle contraction [7,8]. Indeed, like many hypothalamic hormones, oxytocin regulates thermogenesis, food intake, and reproduction—three fundamental processes essential to life. Oxytocin plays a distinctive role in regulating muscle contraction and thermoregulation. This collection of articles was conceived to provide readers with a comprehensive overview of the physiological functions of oxytocin in both health and disease. A total of eight articles are featured in this Special Issue, including two original research articles that explore the novel physiological functions of oxytocin. One study found that the hindbrain administration of oxytocin reduced body weight and energy intake, while stimulating the temperature of interscapular brown adipose tissue in male mice, in which obesity had been induced through diet. In contrast, in female rats fed a high-fat diet, the increase in thermogenesis and weight loss elicited by oxytocin occurred independently of the sympathetic innervation to brown adipose tissue. Moreover, in an in vitro model of oxygen-glucose deprivation, oxytocin was found to confer a degree of neuroprotection, specifically in cases of severe lesions in hippocampal neurons after 7 days in vitro. This findings offer new therapeutic possibilities for the prevention of perinatal asphyxia and hypoxic-ischemic encephalopathy. The remaining six articles in this Special Issue are reviews. Three of them discuss the role of oxytocin in Prader–Willi syndrome and autism spectrum disorder, highlighting the importance of oxytocin's pharmacological properties for the treatment of several diseases, including obesity, Prader–Willi syndrome, and autism spectrum disorder [9–11]. The fourth review article examines the role of oxytocin in mesodermal stem cell-derived lineages. The fifth review investigates the presence of possible alterations in oxytocin levels in polycystic ovary syndrome, noting that serum oxytocin levels are reported to be lower in affected individuals. Finally, the sixth review article introduces the innovative concept of hormesis and the role of oxytocin and vasopressin in the physiological adaptation to stressor events. In conclusion, it was most gratifying to receive such enthusiastic responses from all the contributors, who submitted excellent work in both basic and translational research, as well as from the reviewers and the editorial team. The completion of this collection paves the way for new opportunities and collaborations. It was an honour to serve as Guest Editor, and I extend my sincere congratulations to all the contributors for finalizing this collection of articles, published in commemoration of 70th anniversary of oxytocin's discovery (1955–2025).

Conflicts of Interest: The author declares no conflict of interest.

References

1. Odekunle, E.A.; Elphick, M.R. Comparative and Evolutionary Physiology of Vasopressin/Oxytocin-Type Neuropeptide Signaling in Invertebrates. *Front. Endocrinol.* **2020**, *11*, 225. [CrossRef] [PubMed]
2. Camerino, C. Low sympathetic tone and obese phenotype in oxytocin-deficient mice. *Obesity* **2009**, *17*, 980–984. [CrossRef] [PubMed]
3. Takayanagi, Y.; Kasahara, Y.; Onaka, T.; Takahashi, N.; Kawada, T.; Nishimori, K. Oxytocin receptor-deficient mice developed late-onset obesity. *Neuroreport* **2008**, *19*, 951–955. [CrossRef] [PubMed]
4. Conte, E.; Romano, A.; De Bellis, M.; de Ceglia, M.; Rosaria Carratù, M.; Gaetani, S.; Maqoud, F.; Tricarico, D.; Camerino, C. Oxtr/TRPV1 expression and acclimation of skeletal muscle to cold-stress in male mice. *J. Endocrinol.* **2021**, *249*, 135–148. [CrossRef]
5. Slattery, J.D.; Rambousek, J.R.; Tsui, E.; Honeycutt, M.K.; Goldberg, M.; Graham, J.L.; Wietecha, T.A.; Wolden-Hanson, T.; Williams, A.L.; O'Brien, K.D.; et al. Effects of systemic oxytocin and beta-3 receptor agonist (CL 316243) treatment on body weight and adiposity in male diet-induced obese rats. *Front. Endocrinol.* **2025**, *16*, 1503096. [CrossRef] [PubMed]
6. Blevins, J.E.; Olszewski, P.K. Editorial: Oxytocin and metabolic dysregulation: From pathophysiology to pharmacotherapy. *Front. Endocrinol.* **2025**, *16*, 1579937. [CrossRef] [PubMed]
7. Asla, Q.; Garrido, M.; Urgell, E.; Terzan, S.; Santos, A.; Varghese, N.; Atila, C.; Biagetti, B.; Plessow, F.; Christ-Crain, M.; et al. Melatonin as a possible stimulus to unmask an oxytocin deficient state in hypopituitarism and hypothalamic damage. *J. Clin. Endocrinol. Metab.* **2025**, dgaf201. [CrossRef] [PubMed]
8. Aulinas, A.; Lawson, E.A. The Oxytocin System and Implications for Oxytocin Deficiency in Hypothalamic-Pituitary Disease. *Endocr. Rev.* **2025**, bnaf008. [CrossRef] [PubMed]
9. Camerino, C. The Pivotal Role of Oxytocin's Mechanism of Thermoregulation in Prader-Willi Syndrome, Schaaf-Yang Syndrome, and Autism Spectrum Disorder. *Int. J. Mol. Sci.* **2024**, *25*, 2066. [CrossRef] [PubMed]
10. Lawson, E.A. Understanding oxytocin in human physiology and pathophysiology: A path towards therapeutics. *Compr. Psychoneuroendocrinol.* **2024**, *19*, 100242. [CrossRef] [PubMed]
11. Rice, L.J.; Agu, J.; Carter, C.S.; Cho, Y.H.; Harris, J.; Heilman, K.; Nazarloo, H.P.; Naanai, H.; Porges, S.; Einfeld, S.L. The relationship between cardiac activity, behaviour and endogenous oxytocin and vasopressin in Prader-Willi Syndrome: An exploratory study. *Int. J. Psychophysiol.* **2024**, *205*, 112429. [CrossRef] [PubMed]

current issues in
molecular biology

MDPI

Article

Sympathetic Innervation of Interscapular Brown Adipose Tissue Is Not a Predominant Mediator of Oxytocin-Induced Brown Adipose Tissue Thermogenesis in Female High Fat Diet-Fed Rats

Andrew D. Dodson [1], Adam J. Herbertson [1], Mackenzie K. Honeycutt [1], Ron Vered [1], Jared D. Slattery [1], Matvey Goldberg [1], Edison Tsui [1], Tami Wolden-Hanson [1], James L. Graham [2], Tomasz A. Wietecha [3,4], Kevin D. O'Brien [4,5], Peter J. Havel [2,6], Carl L. Sikkema [1,7], Elaine R. Peskind [1,7], Thomas O. Mundinger [3], Gerald J. Taborsky, Jr. [1,3] and James E. Blevins [1,3,*]

1 VA Puget Sound Health Care System, Office of Research and Development Medical Research Service, Department of Veterans Affairs Medical Center, Seattle, WA 98108, USA; anddodson@gmail.com (A.D.D.); adamherb12@gmail.com (A.J.H.); machoney@ucdavis.edu (M.K.H.); rvered@live.com (R.V.); jared.slattery@va.gov (J.D.S.); matveyg@uw.edu (M.G.); edisontsui2000@gmail.com (E.T.); tami.woldenhanson@gmail.com (T.W.-H.); sikkema@uw.edu (C.L.S.); peskind@uw.edu (E.R.P.); taborsky@uw.edu (G.J.T.J.)

2 Department of Nutrition, University of California, Davis, CA 95616, USA; jlgraham@ucdavis.edu (J.L.G.); pjhavel@ucdavis.edu (P.J.H.)

3 Division of Metabolism, Endocrinology and Nutrition, Department of Medicine, University of Washington School of Medicine, Seattle, WA 98195, USA; wietechatomasz@gmail.com (T.A.W.); mundin714@gmail.com (T.O.M.)

4 UW Medicine Diabetes Institute, University of Washington School of Medicine, Seattle, WA 98109, USA; cardiac@cardiology.washington.edu

5 Division of Cardiology, Department of Medicine, University of Washington School of Medicine, Seattle, WA 98195, USA

6 Department of Molecular Biosciences, School of Veterinary Medicine, University of California, Davis, CA 95616, USA

7 Department of Psychiatry and Behavioral Sciences, University of Washington School of Medicine, Seattle, WA 98195, USA

* Correspondence: jeblevin@uw.edu

Abstract: Recent studies have indicated that hindbrain [fourth ventricle (4V)] administration of the neurohypophyseal hormone, oxytocin (OT), reduces body weight, energy intake and stimulates interscapular brown adipose tissue temperature (T_{IBAT}) in male diet-induced obese (DIO) rats. What remains unclear is whether chronic hindbrain (4V) OT can impact body weight in female high fat diet-fed (HFD) rodents and whether this involves activation of brown adipose tissue (BAT). We hypothesized that OT-elicited stimulation of sympathetic nervous system (SNS) activation of interscapular brown adipose tissue (IBAT) contributes to its ability to activate BAT and reduce body weight in female high HFD-fed rats. To test this hypothesis, we determined the effect of disrupting SNS activation of IBAT on OT-elicited stimulation of T_{IBAT} and reduction of body weight in DIO rats. We first measured the impact of bilateral surgical SNS denervation to IBAT on the ability of acute 4V OT (0.5, 1, and 5 μg ≈ 0.5, 0.99, and 4.96 nmol) to stimulate T_{IBAT} in female HFD-fed rats. We found that the high dose of 4V OT (5 μg ≈ 4.96 nmol) stimulated T_{IBAT} similarly between sham rats and denervated rats (p = NS). We subsequently measured the effect of bilateral surgical denervation of IBAT on the effect of chronic 4V OT (16 nmol/day ≈ 16.1 μg/day) or vehicle infusion to reduce body weight, adiposity and energy intake in female HFD-fed rats (N = 7–8/group). Chronic 4V OT reduced body weight gain (sham: -18.0 ± 4.9 g; denervation: -15.9 ± 3.7 g) and adiposity (sham: -13.9 ± 3.7 g; denervation: -13.6 ± 2.4 g) relative to vehicle treatment ($p < 0.05$) and these effects were similar between groups (p = NS). These effects were attributed, in part, to reduced energy intake evident during weeks 2 ($p < 0.05$) and 3 ($p < 0.05$). To test whether these results translate to other female rodent species, we also examined the effect of chronic 4V infusion of OT on body weight and adiposity in two strains of female HFD-fed mice. Similar to what we found in the HFD-fed rat model, we also found that chronic 4V OT (16 nmol/day) infusion resulted in reduced body weight gain, adiposity and energy intake in female DIO C57BL/6J and DBA/2J mice ($p < 0.05$ vs. vehicle). Together, these findings suggest that (1) sympathetic innervation of IBAT is not necessary for OT-elicited increases in BAT thermogenesis and weight loss in female HFD-fed rats and (2) the effects of OT to reduce weight gain and adiposity translate to other female mouse models of diet-induced obesity (DIO).

Keywords: obesity; brown adipose tissue; oxytocin; female rodents

1. Introduction

The neuropeptide, oxytocin (OT), has been largely associated with eliciting prosocial (i.e., pair bonds and increased trust) [1,2] and reproductive behavior (i.e., uterine contraction, milk ejection reflex) [3,4], but the role of OT in the regulation of body weight, particularly in female rodents [5–8], is not entirely clear. Recent studies have shown that acute intracerebroventricular (ICV) OT administration reduces food intake in female single-minded 1 (SIM1) haploinsufficient mice [9] and female rats [10]. However, few long-term treatment studies exploring the mechanism (s) by which chronic central nervous system (CNS) administration of OT reduces body weight and adiposity in female rodents have been reported.

Although suppression of food intake is thought to contribute, at least in part, to the effects of hindbrain (fourth ventricle (4V)) OT-elicited weight loss in male rodents, the findings from pair-feeding studies from male rodents suggest that OT-elicited reductions of food intake cannot fully explain OT elicited weight loss [11–13]. In addition to OT's well established effects on food intake, previous studies in rodents and nonhuman primates have shown that OT may also evoke weight loss, in part, by stimulating energy expenditure (EE) [14–17] and lipolysis [12,14,18]. While it is clear that brown adipose tissue thermogenesis (BAT) plays an important role in the regulation of EE (see [19,20] for review), less is known about whether OT's effects on EE result from (1) non-shivering BAT thermogenesis, (2) spontaneous physical activity-induced thermogenesis [21], (3) non-shivering and shivering thermogenesis in skeletal muscle [22,23] and/or through the anabolic effects of OT on muscle [24,25], (5) white adipose tissue thermogenesis or (6) hormonal mediators (e.g., fibroblast growth factor-21 [26], irisin [27], leptin [28], thyroid hormone [29] or secretin [30,31] (see [32,33] for review). We and others have found that acute injections of OT into either the forebrain (third (3V)) or hindbrain (4V) elevate interscapular BAT temperature (T_{IBAT}) (surrogate measure of BAT thermogenesis) [34,35] and/or core temperature [36] in male rats or mice. Furthermore, the onset of OT-elicited weight loss coincides with OT-elicited elevations of T_{IBAT} in male diet-induced obese (DIO) rats [34]. In addition, an earlier study found that chemogenetic excitation of hypothalamic paraventricular nucleus (PVN) OT neurons increases both subcutaneous BAT temperature and EE in *Oxytocin-Ires Cre* mice [37]. In addition, a recent study reported that chronic subcutaneous infusion of OT increases core temperature, IBAT thermogenic gene expression and differentiation of BAT in vitro in male DIO mice [38]. On the other hand, genetic knockdown or pharmacological blockade of OT signaling reduces cold-induced BAT thermogenesis [39–42], decreases EE [16,17,39,43] and promotes obesity [17,43–45] in mice. We recently determined the impact of bilateral surgical sympathetic nervous system (SNS) denervation to IBAT on the ability of chronic hindbrain (4V) OT infusion to reduce body weight and adiposity in male DIO mice. We found that chronic 4V OT produced similar reductions of body weight and adiposity between groups suggesting that SNS innervation of IBAT is not required for OT to reduce body weight and adiposity in male DIO mice [35]. This finding raised the question as to whether OT stimulates IBAT thermogenesis and evokes weight loss through a mechanism that requires increased SNS outflow to IBAT in female high fat diet (HFD)-fed rats and whether the effect of OT on BAT thermogenesis may involve hindbrain oxytocin receptors (OTR). Here, we aimed to determine the role of SNS outflow to IBAT in contributing to the effect of chronic hindbrain (4V) OT to stimulate BAT thermogenesis and evoke weight loss in a female HFD-fed rat model.

Based on our previous findings that linked 4V OT to increases in BAT thermogenesis in male DIO rats, we hypothesized that OT-induced stimulation of SNS outflow to IBAT contributes to its ability to stimulate non-shivering BAT thermogenesis and evoke weight loss in female HFD-fed rats. To assess if SNS innervation of BAT is required for OT to stimulate non-shivering thermogenesis in IBAT (as surrogate measure of energy expenditure), we determined the effects of acute 4V injections of OT (0.5, 1, and 5 µg) on T_{IBAT} in female HFD-fed rats following bilateral surgical SNS denervation to IBAT. To determine whether SNS innervation of IBAT is required for OT to elicit weight loss, we measured the ability of chronic 4V OT (16 nmol/day over 29 days) to decrease body weight and adiposity in female HFD-fed rats following bilateral surgical or sham denervation of IBAT. We subsequently determined if these effects were associated with a reduction of adipocyte size and energy intake. To test whether the effects of chronic 4V OT to reduce body weight and adiposity could translate to other female rodent models of diet-induced obesity (DIO), we also examined the effect of chronic 4V infusion of OT on body weight and adiposity in two different strains of female HFD-fed mice (C57BL/6J and DBA/2J). Our findings suggest that (1) sympathetic innervation of IBAT is not necessary for OT-elicited increases in BAT thermogenesis and weight loss in female HFD-fed rats and (2) the effects of OT to elicit weight loss translate to other mouse models of diet-induced obesity (DIO).

2. Materials and Methods

2.1. Animals

Adult female Long-Evans rats and C57BL/6J (strain 000664) and DBA/2J (strain 000671) mice were initially obtained from Envigo [Indianapolis, IN (rats)] or [The Jackson Laboratory; Bar Harbor, ME (mice)] and maintained for at least 4 months on a HFD prior to study onset. All animals were housed individually in Plexiglas cages in a temperature-controlled room (22 ± 2 °C) under a 12:12-h light-dark cycle. All rats and mice were maintained on a 6 a.m. (lights on)/6 p.m. (lights off) light cycle. Rats and mice had *ad libitum* access to water and a HFD providing 60% kcal from fat [Research Diets, D12492 (rats) or D12492i (mice), New Brunswick, NJ, USA]. The research protocols were approved both by the Institutional Animal Care and Use Committee of the Veterans Affairs Puget Sound Health Care System (VAPSHCS) and the University of Washington in accordance with NIH Guidelines for the Care and Use of Animals.

2.2. Drug Preparation

The beta-3 adrenergic receptor (β3-AR) agonist, CL 316243 (Tocris/Bio-Techne Corporation, Minneapolis, MN), was solubilized in sterile water on each day of each experiment (Study 1). OT acetate salt (Bachem Americas, Inc., Torrance, CA, USA) was solubilized in sterile water on each day of each study (Study 2). For Studies, 3–5, OT acetate salt (Bachem Americas, Inc., Torrance, CA, USA) was dissolved in sterile water and subsequently added to Alzet® minipumps (model 2004; DURECT Corporation, Cupertino, CA, USA) and primed in sterile vehicle (0.9% saline) at 37 °C for approximately 40 h.

2.3. SNS Denervation Procedure

The procedure for SNS denervation of IBAT has been described previously [35]. Briefly, a dissecting microscope (Leica M60/M80; Leica Microsystems, Buffalo Grove, IL, USA) was used for the denervation/sham surgeries. Rats were treated pre-operatively with the analgesic ketoprofen (2 mg/kg; Fort Dodge Animal Health, Overland Park, KS, USA) prior to the completion of the denervation or sham procedure. This IBAT denervation procedure was combined with transponder implantations for studies that involved IBAT temperature measurements in response to acute (Studies 1–2) IP or 4V administration. Animals were allowed to recover for approximately 5–7 days prior to implantation of 4V cannulas.

2.4. 4V Cannulations for Acute Injections in Rats

The procedure for 4V cannulations for acute injections in rats has been described previously [34,46]. Briefly, rats were implanted with a cannula (P1 Technologies, Roanoke, VA, USA) that was directed towards the 4V [47–49]. Rats were initially anesthetized with isoflurane and subsequently positioned on a stereotaxic device [Digital Lab Standard Stereotaxic, Rat, (Item 51900), Stoelting Co., Wood Dale, IL, USA] with the incisor bar fixed at 3.3 mm below the interaural line. A 26-gauge cannula (P1 Technologies) was stereotaxically implanted into the 4V [−3.5 mm caudal to the interaural line; 1.4 mm lateral to the midline, and 6.2 mm ventral to the skull surface [50] and fastened to the surface of the skull with dental acrylic and stainless-steel screws. Rats were treated with the analgesic ketoprofen (2 mg/kg; Fort Dodge Animal Health) and the antibiotic enrofloxacin (5 mg/kg; Bayer Healthcare LLC., Animal Health Division, Shawnee Mission, KS, USA) at the completion of the 4V cannulations. Animals were allowed to recover for at least 10 days prior to study onset.

2.5. 4V Cannulations for Chronic Infusions in Rats

The procedure for 4V cannulations for chronic infusions in rats has been described previously [34,46]. Briefly, rats were implanted with a cannula within the 4V with a side port that was connected to an osmotic minipump (model 2004, DURECT Corporation) [34,46,51]. Rats were initially anesthetized with isoflurane anesthesia and subsequently placed in a stereotaxic device [Digital Lab Standard Stereotaxic, Rat, (Item 51900), Stoelting Co.] with the incisor bar fixed at 3.3 mm below the interaural line. A 30-gauge cannula (P1 Technologies) was stereotaxically directed into the 4V [−3.5 mm caudal to the interaural line; 1.4 mm lateral to the midline, and 7.2 mm ventral to the skull surface [50]] and secured to the surface of the skull with stainless steel screws and dental acrylic. Rats were given the analgesic ketoprofen (2 mg/kg; Fort Dodge Animal Health) and the antibiotic enrofloxacin (5 mg/kg; Bayer Healthcare LLC., Animal Health Division, Shawnee Mission, KS, USA) at the completion of the 4V cannulations and were allowed to recover at least 10 days prior to implantation of osmotic minipumps.

2.6. 4V Cannulations for Chronic Infusions in Female C57BL/6J and DBA/2J Mice

The procedure for 4V cannulations for chronic infusions in mice has been described previously [52]. Mice were implanted with a cannula within the 4V with a side port that was connected to an osmotic minipump (model 2004, DURECT Corporation) as previously described [52]. Mice were initially anesthetized with isoflurane anesthesia and subsequently positioned on a stereotaxic device [Digital Just for Mouse Stereotaxic, (Item 51730D), Stoelting Co.] with the incisor bar positioned 4.5 mm below the interaural line. A 30-gauge cannula (P1 Technologies) was stereotaxically positioned into the 4V of either female C57BL/6J or DBA/2J mice (−5.9 mm caudal to bregma; 0.4 mm lateral to the midline, and 3.7 mm ventral to the skull surface) [53] and secured to the surface of the skull with dental cement and stainless steel screws. Mice were treated with the analgesic ketoprofen (5 mg/kg; Fort Dodge Animal Health) and the antibiotic enrofloxacin (5 mg/kg; Bayer Healthcare LLC., Animal Health Division Shawnee Mission, KS, USA) at the completion of the 4V cannulations and were allowed to recover for at least 10 days prior to implantation of osmotic minipumps.

2.7. Implantation of Temperature Transponders Underneath IBAT

The procedure for temperature transponder implantations underneath IBAT in rats and mice has been described previously [35,52,54]. Animals were initially anesthetized with isoflurane and had the dorsal surface along the upper midline of the back shaved and the area was scrubbed with 70% ethanol followed by betadine swabs. A one-inch incision was made at the midline of the interscapular area. The temperature transponder (14 mm long/2 mm wide) (HTEC IPTT-300; Bio Medic Data Systems, Inc., Seaford, DE, USA) was implanted underneath the left IBAT pad as previously described [34,35,52,54–56] and secured in place by suturing it to the brown fat pad with sterile silk suture. The interscapular incision was closed with Nylon sutures (5-0), which were removed in awake animals 10–14 days after surgery. Rats or mice were treated pre-operatively with the analgesic ketoprofen [rats: 2 mg/kg; mice: 5 mg/kg (Fort Dodge Animal Health)] prior to the completion of the temperature transponder implantation procedure.

2.8. Acute IP or 4V Injections and Measurements of T_{IBAT}

On an experimental day, 4-h fasted animals received either IP (CL 316243 or sterile water vehicle; 0.1 mL/kg injection volume) or 4V injections (OT or saline vehicle; 1 µL injection volume) during the early part of the light cycle. Injections were completed in a crossover design over approximately 7-day (CL 316243) or 48-h (OT) intervals such that each animal served as its own control. Animals remained fasted for an additional 4 h (Study 3–4) during the course of the T_{IBAT} measurements. A handheld reader (DAS-8007-IUS Reader System; Bio Medic Data Systems, Inc.) was used to collect measurements of T_{IBAT}. Rats underwent all treatments in a randomized order separated by at least 48-h (OT) or 7–8 days (CL 316243) between treatments.

2.9. Body Composition

The procedure for measuring body composition in mice and rats has been described previously [35,54]. Briefly, determinations of lean body mass and fat mass were made on un-anesthetized mice and rats by quantitative magnetic resonance using an EchoMRI 4-in-1-700™ instrument (Echo Medical Systems, Houston, TX, USA) at the VAPSHCS Rodent Metabolic Phenotyping Core. Measurements were taken prior to 4V cannulations and/or minipump implantations as well as at the end of the infusion period.

2.10. Tissue Collection for Norepinephrine (NE) Content Measurements

Rats were euthanized by rapid conscious decapitation at 8 weeks (Study 1–2) or 7–12 weeks (Study 3) post-sham or denervation procedure. Trunk blood and tissues (IBAT, EWAT, IWAT, liver and/or pancreas) were collected from 4-h fasted rats. Tissue was rapidly removed, wrapped in foil and frozen in liquid N2. Samples were stored frozen at −80 °C until analysis. Note that rapid conscious decapitation was used in place of anesthesia when collecting tissue for NE content as anesthesia can cause the release of NE from SNS terminals within the tissue [57].

2.11. NE Content Measurements (Biochemical Confirmation of IBAT Denervation Procedure)

The procedure for measuring NE content has been described previously [58]. Specifically, NE content was measured in IBAT, EWAT, IWAT, liver and/or pancreas using previously established techniques [35,58]. Successful denervation was noted by ≥60% reduction in IBAT NE content as previously noted [35,59]. Experimental animals that did not meet this criterion were excluded from the data analysis.

2.12. Study Protocols

2.12.1. Study 1: Determine if Surgical Denervation of IBAT Changes the Ability of the β-3R Agonist, CL 316243, to Increase T_{IBAT} in DIO Rats

Rats (N = 13 at study onset) from Study 2 were used in these studies. Rats were fed *ad libitum* and maintained on HFD for approximately 4.5 months prior to underdoing sham or SNS denervation procedures and implantation of temperature transponders underneath the left IBAT depot. Animals were subsequently implanted with 4V cannulas approximately 1 week following sham/denervation procedures and implantation of temperature transponders. Rats were allowed to recover for at least 2 weeks during which time they were adapted to a daily 4-h fast, handling and mock injections. On an experimental day, 4-h fasted rats received Cl 316243 (0.1 or 1 mg/kg) or vehicle (sterile water) during the early part of the light cycle in a crossover design at approximately 7-day intervals such that each animal served as its own control (approximately 1–3 weeks post-sham or denervation procedures). T_{IBAT} was measured at baseline (−2 h; 9:00 a.m.), immediately prior to IP injections (0 h; 9:45–10:00 a.m.), and at 0.25, 0.5, 0.75, 1, 1.25, 1.5, 2, 3, 4, and 24-h post-injection (10:00 a.m.). Food intake and body weight were measured daily. Daily food intake was determined by measuring the difference in weight of the high fat diet pre- vs. post-intervention of a 24-h period and converting grams/day to units of energy intake/day (kcal/day; 5.24 kcal/gram [60]). This dose range was based on doses of CL 316243 found to be effective at reducing food intake and weight gain in rats [52,61]. Animals were euthanized by rapid conscious decapitation at 13 weeks post-sham or denervation procedure.

2.12.2. Study 2: Determine the Extent to Which OT-Induced Activation of Sympathetic Outflow to IBAT Contributes to Its Ability to Increase T_{IBAT} in DIO Rats

Rats (N = 16 at study onset) from Study 1 were used in these studies. On an experimental day, 4-h fasted rats received OT (1 or 5 μg/μL) or vehicle during the early part of the light cycle in order to maximize the effects of OT [17,52] during a time when circulating NE levels [62] and IBAT catecholamine levels are lower [63]. Injections were completed in a crossover design at approximately 48-h to 72-h intervals such that each animal served as its own control (approximately 4-weeks post-sham or denervation procedures). T_{IBAT} was measured at baseline (−2 h; 9:00 a.m.), immediately prior to 4V injections (0 h; 9:45–10:00 a.m.), and at 0.25, 0.5, 0.75, 1, 1.25, 1.5, 2, 3, 4, and 24-h post-injection (10:00 a.m.). Food intake and body weight were measured daily. This dose range was based on doses of 4V OT found to be effective at stimulating T_{IBAT} in male DIO rats in previous studies [34].

In addition, we examined the impact of a lower dose of OT (0.5 μg/μL) in an identical manner following the completion of the initial studies in Study 2.

2.12.3. Study 3A: Determine the Extent to Which OT-Induced Activation of Sympathetic Outflow to IBAT Contributes to Its Ability to Reduce Weight Gain in Female HFD-Fed Rats

Rats (N = 30 at study onset) were used for these studies. Animals were fed *ad libitum* and maintained on HFD for approximately 5.25 months prior to receiving implantations of temperature transponders underneath IBAT, 4V cannulas and subcutaneous minipumps to infuse vehicle or OT (16 nmol/day) over 29 days as previously described [34]. This dose was selected based on a dose of 4V OT found to be effective at reducing body weight in male DIO rats [34]. Daily food intake and body weight were also tracked for 29 days. Animals were euthanized by rapid conscious decapitation at 7 weeks post-sham or denervation procedure. Trunk blood and tissues [IBAT, epididymal white adipose tissue (EWAT), inguinal white adipose tissue (IWAT), liver and pancreas)] were collected from 4-h fasted rats and tissues were subsequently analyzed for IBAT NE content to confirm success of denervation procedure relative to sham operated animals and other tissues (EWAT, IWAT, liver and pancreas).

2.12.4. Study 3B: Determine the Extent to Which 4V OT Impacts Thermogenic Gene Expression in IBAT and IWAT in Female HFD-Fed Rats

Rats from Study 3A were used for these studies. All rats received chronic infusions of 4V vehicle or OT (16 nmol/day) and were euthanized by rapid conscious decapitation following a 4-h fast.

2.12.5. Study 4A: Determine the Effects of Chronic 4V OT Treatment on Body Weight, Adiposity and Energy Intake in Female HFD-Fed C57BL/6J Mice

Female mice (N = 20 at study onset) were fed *ad libitum* and maintained on HFD for approximately 4.5 months prior to being implanted with temperature transponders underneath IBAT. Mice were allowed up to 1-week post-op recovery prior to receiving 4V cannulas. Mice were allowed up to 2 weeks post-op recovery prior to being implanted with minipumps as previously described [34]. T_{IBAT} was measured daily at baseline (−2 h; 9:00 a.m.) and immediately prior to access to food (10:00 a.m.). Daily food intake, body weight and T_{IBAT} were tracked for 28 days.

2.12.6. Study 4B: Determine the Effects of Chronic 4V OT Treatment on Body Weight, Adiposity and Energy Intake in Female DIO DBA2J Mice

Female mice (N = 20 at study onset) were used for these studies. Animals were fed *ad libitum* and maintained on HFD for approximately 4.5 months prior to being implanted with temperature transponders underneath IBAT. Mice were allowed up to 2 weeks post-op recovery prior to receiving 4V cannulas. Mice were allowed up to 4 weeks post-op recovery prior to being implanted with minipumps as previously described [34]. T_{IBAT} was measured daily at baseline (−2 h; 9:00 a.m.) and immediately prior to access to food (10:00 a.m.). Daily food intake, body weight and T_{IBAT} were tracked for 28 days.

2.12.7. Study 5: Determine the Effects of Chronic Systemic OT Treatment (16 and 50 nmol/day) on Body Weight, Adiposity and Energy Intake in Female DIO DBA/2J Mice

Female mice (N = 22 mice at study onset) were fed *ad libitum* and maintained on HFD for approximately 4.5 months prior to being implanted with temperature transponders underneath IBAT. Mice were allowed up to 4 weeks post-op recovery prior to being implanted with subcutaneous (SC) minipumps as previously described [34]. T_{IBAT} was measured daily at baseline (−2 h; 9:00 a.m.) and immediately prior to access to food (10:00 a.m.). Daily food intake, body weight and T_{IBAT} were tracked for 27 days.

2.13. *Blood Collection*

Trunk blood (Study 1–2) or blood from cardiac stick (Study 3–5) was collected from 4-h fasted rats or mice within a 2-h window towards the beginning of the light cycle (10:00 a.m.–12:00 p.m.) as previously described in DIO CD®IGS and Long-Evans rats and C57BL/6J mice [34,48]. Treatment groups were counterbalanced at time of euthanasia to avoid time of day bias. Blood samples [up to 1 mL (mice) or 3 mL (rats)] were collected from trunk or via cardiac puncture in chilled K2 EDTA Microtainer Tubes (Becton-Dickinson, Franklin Lakes, NJ, USA). Whole blood was centrifuged at 6000 rpm for 1.5-min at 4 °C; plasma was removed, aliquoted and stored at −80 °C for subsequent analysis.

2.14. *Plasma Hormone Measurements*

Plasma leptin and insulin were measured using electrochemiluminescence detection [Meso Scale Discovery (MSD®), Rockville, MD, USA] using established procedures [34,64]. Intra-assay coefficient of variation (CV) for leptin was 2.8% and 2.4% for rat and mouse, respectively. Intra-assay CV for insulin was 2.7% and 2.4% for rat and mouse, respectively. The range of detectability for the leptin assay is 0.07–51.9 ng/mL and 0.069–50 ng/mL for insulin. Plasma glucagon (Mercodia, Winston Salem, NC, USA), fibroblast growth factor-21 (FGF-21) (R&D Systems, Minneapolis, MN, USA) and irisin (AdipoGen, San Diego, CA, USA) levels were determined by ELISA. The intra-assay CV for glucagon was 1.6% and 1.9% for rat and mouse, respectively, and the range of detection was 2–182 pmol/L. The intra-assay CV for FGF-21 was 2.7% and 2.3% for rat and mouse, respectively. The intra-assay CV for irisin was 6.9% for mouse (not obtained for rat). The ranges of detectability were 31.3–2000 pg/mL (FGF-21) and 0.078–5 μg/mL (irisin). Plasma adiponectin was also measured using ELISA (Alpco, Salem, NH, USA) using established procedures [34,64]. Intra-assay CV for adiponectin was 1.7% and 1.6% for rat and mouse, respectively. The range of detectability for the adiponectin assay is 0.25–10 ng/mL (rat) and 0.025–1 ng/mL (mice). The data were normalized to historical values using a pooled plasma quality control sample that was assayed in each plate.

2.15. *Blood Glucose and Lipid Measurements*

Blood was collected for glucose measurements by tail vein nick in 4-h fasted rats and measured with a glucometer using the AlphaTRAK 2 blood glucose monitoring system (Abbott Laboratories, Abbott Park, IL, USA) [34,65]. Total cholesterol (TC) [Fisher Diagnostics (Middletown, VA, USA)] and free fatty acids (FFAs) [Wako Chemicals USA, Inc., Richmond, VA, USA)] were measured using an enzymatic-based kits. Intra-assay CVs for TC were 3.5% and 3.4% for rat and mouse, respectively. Intra-assay CV for FFA were 1.4% and 2.5% for rat and mouse, respectively. These assay procedures have been validated for rodents [66].

2.16. Adipose Tissue Processing for Adipocyte Size

Adipose tissue depots were collected at the end of the infusion period in DIO rats from Study 3B (EWAT). EWAT was processed as previously described [35,46,52,54]. EWAT was dissected and placed in 4% paraformaldehyde-PBS for 24 h and then placed in 70% ethanol (EtOH) prior to paraffin embedding. Sections (5 μm) sampled were obtained using a rotary microtome, slide-mounted using a floatation water bath (37 °C), and baked for 30 min at 60 °C to give approximately 15–16 slides/fat depot with two sections/slide.

2.17. Adipocyte Size Analysis

Adipocyte size analysis was performed as previously described [35,46,52,54]. Analysis was completed on deparaffinized and digitized EWAT sections. The average cell area from two randomized photomicrographs was determined using the built-in particle counting method of ImageJ software (https://imagej.net/ij/, assessed date 30 September 2024) (National Institutes of Health, Bethesda, MD, USA). Slides were visualized using bright field on an Olympus BX51 microscope (Olympus Corporation of the Americas; Center Valley, PA, USA) and photographed using a Canon EOS 5D SR DSLR (Canon U.S.A., Inc., Melville, NY, USA) camera at ×10 magnification. Values for each tissue within a treatment were averaged to obtain the mean of the treatment group.

2.18. Tissue Collection for Quantitative Real-Time PCR (qPCR)

Tissue (IBAT and IWAT) was collected from a subset of 4-h (Study 3B). IBAT and IWAT were collected within a 2-h window towards the start of the light cycle (10:00 a.m.–12:00 p.m.) as previously described in DIO CD®IGS/Long-Evans rats and C57BL/6J mice [34,48,52]. Tissue was rapidly removed, wrapped in foil and frozen in liquid N2. Samples were stored frozen at −80 °C until analysis.

2.19. qPCR

RNA extracted from samples of IBAT and IWAT (Study 3B) were analyzed using the RNeasy Lipid Mini Kit (Qiagen Sciences Inc., Germantown, MD, USA) followed by reverse transcription into cDNA using a high-capacity cDNA archive kit (Applied Biosystems, Foster City, CA, USA). Quantitative analysis for relative levels of mRNA in the RNA extracts was measured in duplicate by qPCR on an Applied Biosystems 7500 Real-Time PCR system (Thermo Fisher Scientific, Waltham, MA, USA) and normalized to the cycle threshold value of Nono mRNA in each sample. The TaqMan® probes used in the study were Thermo Fisher Scientific Gene Expression Assay probes. The probe for rat *Nono* (Rn01418995_g1), uncoupling protein-1 (UCP-1) (*Ucp1*; catalog no. Rn00562126_m1), β1-adrenergic receptor (β1-AR) (*Adrb1*; catalog no. Rn00824536_s1), β3-adrenergic receptor (β3-AR) (*Adrb3*; catalog no. Rn01478698_g1), type 2 deiodinase (D2) (*Dio2*; catalog no. Rn00581867_m1), PR domain containing 16 (*Prdm16*; catalog no. Rn01516224_m1), G-protein coupled receptor 120 (*Gpr120*; catalog no. Rn01759772_m1), cell death-inducing DNA fragmentation factor alpha-like effector A (*Cidea*; catalog no. Rn04181355_m1), and peroxisome proliferator-activated receptor gamma coactivator 1 alpha (*Ppargc1a*; catalog no. Rn00580241_m1) were acquired from Thermo Fisher Scientific. Relative amounts of target mRNA were determined using the Comparative C_T or $2^{-\Delta\Delta CT}$ method [67] following adjustment for the housekeeping gene, Nono. Specific mRNA levels of all genes of interest were normalized to the cycle threshold value of *Nono* mRNA in each sample and expressed as changes normalized to controls (vehicle/sham treatment).

2.20. Statistical Analyses

All results are expressed as means ± SE. Comparisons between multiple groups involving between-subjects designs were made using one-way ANOVA as appropriate, followed by a post-hoc Fisher's least significant difference test. Comparisons involving within-subjects designs were made using a one-way repeated-measures ANOVA followed by a post-hoc Fisher's least significant difference test. Analyses were performed using the statistical program SYSTAT (https://grafiti.com/systat/, accessed date 30 September 2024) (Systat Software, Point Richmond, CA, USA). Differences were considered significant at $p < 0.05$, 2-tailed. Non-statistical trends ($0.05 < p < 0.1$) have been included in the analysis where appropriate.

3. Results

3.1. Study 1: Determine if Surgical Denervation of IBAT Changes the Ability of the β3-AR Agonist, CL 316243, to Increase T_{IBAT} in Female HFD-Fed Rats

Our objective was to extend recently published results in a mouse model [35] and verify there was no functional impairment in the ability of IBAT to respond to direct β3-AR stimulation as a result of the denervation procedure relative to female sham operated rats. As expected, female HFD-fed rats were borderline obese as confirmed by both body weight (336.8 ± 7.8 g) and adiposity (99.0 ± 6.7 g fat mass; 28.9 ± 1.3% body fat) after being maintained on the HFD for approximately 4 months prior undergoing sham/IBAT denervation surgeries.

All IBAT tissues from Study 1/Study 2 animals were analyzed for IBAT NE content and only 1 out of 5 animals was removed on account of having a failed IBAT denervation procedure. Surgical IBAT denervation resulted in a 76.9 ± 2.7% reduction of IBAT NE content relative to sham-operated control rats [($F(1,10) = 18.975$, $p = 0.001$). Similar to what we reported in a mouse model [35], NE content was unaltered in IWAT, EWAT, liver or pancreas in IBAT denervated rats relative to sham-operated rats (p = NS). As expected [35], there was no significant difference in body weight between female sham-operated and IBAT denervated rats at the conclusion of the study (p = NS).

In sham-operated rats, CL 316243 (1 mg/kg) stimulated T_{IBAT} throughout the post-injection measurement period (0.25, 0.5, 0.75, 1, 1.25, 1.5, 1.75, 2, 3 and 4-h post-injection). In addition, the lower dose of CL 316243 (0.1 mg/kg) also elevated T_{IBAT} throughout the post-injection measurement period (0.25, 0.5, 0.75, 1, 1.25, 1.5, 1.75, 2, 3 and 4-h post-injection) ($p < 0.05$; Figure 1A).

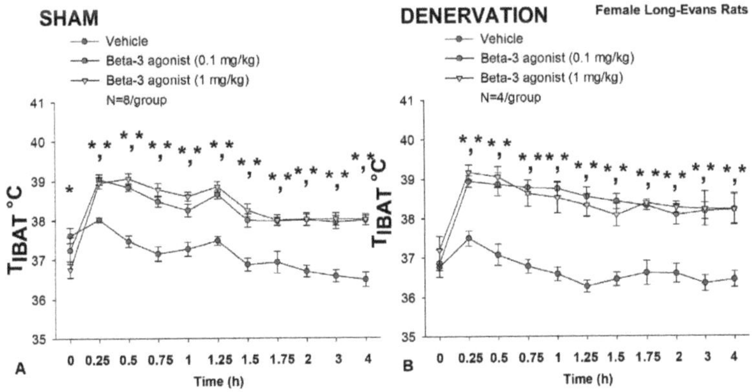

Figure 1. *Cont.*

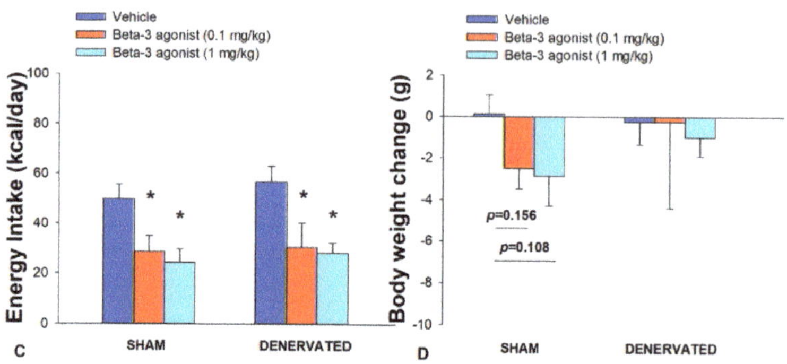

Figure 1. (**A–D**) **Effect of systemic β3-AR agonist (CL 316243) administration (0.1 and 1 mg/kg) on IBAT temperature (T_{IBAT}), energy intake and body weight post-sham or IBAT denervation in female HFD-fed rats.** Rats were maintained on HFD (60% kcal from fat; N = 4–8/group) for approximately 4.5 months prior to undergoing a sham or bilateral surgical IBAT denervation and implantation of temperature transponders underneath IBAT. Animals were subsequently adapted to a 4-h fast prior to receiving IP injections of CL 316243 (0.1 or 1 mg/kg, IP) or vehicle (sterile water) where each animal received each treatment at approximately 7-day intervals. (**A**,**B**) Effect of CL 316243 on T_{IBAT} in (**A**) sham operated or (**B**) IBAT denervated DIO rats; (**C**) Effect of CL 316243 on change in energy intake in sham or IBAT denervated DIO rats; (**D**) Effect of CL 316243 on change in body weight in sham or IBAT denervated DIO rats. Data are expressed as mean ± SEM. * $p < 0.05$ CL 316243 vs. vehicle.

Likewise, in IBAT denervated animals, CL 316243 (1 mg/kg) also stimulated T_{IBAT} at 0.25, 0.5, 0.75, 1, 1.25, 1.5, 1.75, 2, 3 and 4-h post-injection. CL 316243 (0.1 mg/kg) also stimulated T_{IBAT} at 0.25, 0.5, 0.75, 1, 1.25, 1.5, 1.75, 2, 3 and 4-h post-injection ($p < 0.05$; Figure 1B).

Notably, there was no significant difference in the ability of CL 316243 to (0.1 or 1 mg/kg) stimulate T_{IBAT} when averaged over the 1-h or 4-h post-injection period between sham and IBAT denervated animals (p = NS).

Collectively, these findings indicate that IBAT denervation did not result in a functional change in the effectiveness of CL 316243 to increase BAT thermogenesis relative to sham-operated animals.

3.1.1. Energy Intake

In sham-operated rats, CL 316243 decreased daily energy intake at both the low (0.1 mg/kg) and high dose (1 mg/kg) by 42.3 and 51.4% ($p < 0.05$). Likewise, in IBAT denervated animals, CL 316243 also decreased daily energy intake at both the low (0.1 mg/kg) and high (1 mg/kg) doses ($p < 0.05$) by 46.2 and 50.4%, respectively (Figure 1C).

3.1.2. Body Weight

CL 316243 had no effect on either body weight or body weight gain in either sham or IBAT denervated rats (p = NS; Figure 1D). CL 316243 tended to reduce body weight gain at the high dose (1 mg/kg) in the sham-operated group but this did not reach significance ($p = 0.111$).

Similar to what we observed on T_{IBAT}, there was no significant difference in the ability of CL 316243 (0.1 or 1 mg/kg) to reduce energy intake between sham and IBAT denervated rats (p = NS).

Collectively, these results indicate that IBAT denervation did not result in a functional change in the effectiveness of CL 316243 to reduce energy intake relative to sham-operated animals.

3.2. Study 2: Determine the Extent to Which OT-Induced Activation of Sympathetic Outflow to IBAT Contributes to Its Ability to Increase T_{IBAT} in Female HFD-Fed Rats

Following confirmation that there was no functional defect in the effectiveness of IBAT to respond to CL 316243-elicited stimulation of β3-AR (Study 1), the objective here was to examine whether OT-induced stimulation of T_{IBAT} requires intact SNS innervation of IBAT. Three of the sixteen rats available at study onset were euthanized during the course of the study and were excluded from the data analysis.

In sham-operated rats, OT (5 μg) stimulated T_{IBAT} throughout the post-injection measurement period (0.75, 1, 1.25, 1.5, 1.75, and 2-h post-injection) ($p < 0.05$). In addition, the lower dose (1 μg) also stimulated T_{IBAT} throughout the post-injection measurement period (0.5, 0.75, 1, 1.25, and 1.5-h post-injection) ($p < 0.05$; Figure 2A).

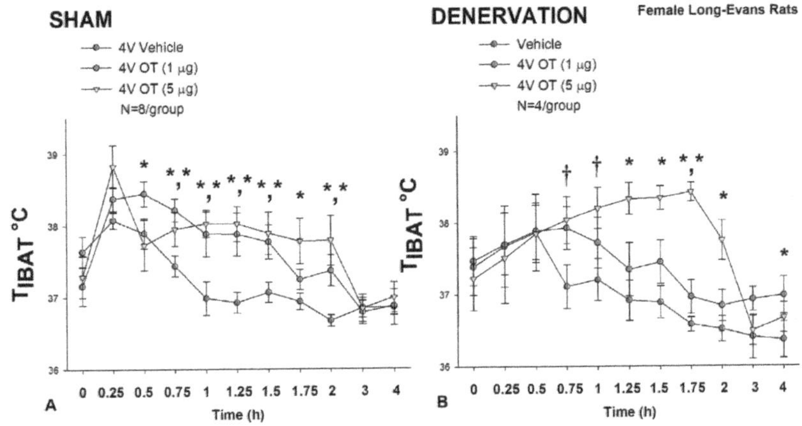

Figure 2. (A,B) Effect of acute 4V OT administration (1 and 5 μg) on T_{IBAT} post-sham or IBAT denervation in female HFD-fed rats. Rats were maintained on HFD (60% kcal from fat; N = 4–8/group) for approximately 4.5 months prior to undergoing a sham or bilateral surgical IBAT denervation and implantation of temperature transponders underneath IBAT. Rats were subsequently implanted with 4V cannulas and allowed to recover for 2 weeks prior to receiving acute 4V injections of OT or vehicle. Animals were subsequently adapted to a 4-h fast prior to receiving acute 4V injections of OT or vehicle (**A**,**B**) Effect of acute 4V OT on T_{IBAT} in (**A**) sham operated or (**B**) IBAT denervated DIO rats. Data are expressed as mean ± SEM. * $p < 0.05$, † $0.05 < p < 0.1$ OT vs. vehicle.

Likewise, in IBAT denervated animals, OT (5 μg) stimulated T_{IBAT} throughout the post-injection measurement period (1.25, 1.5, 1.75, and 2-h post-injection) ($p < 0.05$) and also tended to increase T_{IBAT} at 0.75, and 1-h post-injection ($0.05 < p < 0.1$). The low dose (1 μg) increased T_{IBAT} at both 1.75 and 4-h post-injection ($0.05 < p < 0.1$; Figure 2B).

Notably, there was no significant change in the ability of 4V OT (5 μg) to increase T_{IBAT} when averaged over the 4-h post-injection period between sham and IBAT denervated animals (p = NS).

There were, however, seizures, barrel-rolling and unexpected deaths that occurred in three out of the sixteen rats (1 sham, 2 denervated) shortly after 4V administration of OT at the high dose (5 μg). Rinaman also reported that acute ICV administration of a higher dose (10 μg) also resulted in seizure-like activity and barrel-rolling in a subset of adult male Sprague-Dawley rats [68]. These findings raise the possibility that females may be more sensitive to the effects of acute injections of 4V OT compared to what we have observed previously at similar doses in males in the absence of such effects [34,46]. Thus, following the completion of these studies, we also examined the effectiveness of a lower dose of 4V OT (0.5 μg/μL) on T_{IBAT} in an identical manner.

In sham-operated animals, 4V administration of OT (0.5 μg/μL) increased T_{IBAT} at 1.25-h post-injection ($p < 0.05$; Figure 3A) and it also appeared to increase T_{IBAT} during the post-injection period (0.5, 0.75, and 1-h post-injection) ($0.05 < p < 0.1$; Figure 3A). Acute 4V administration of OT also tended to decrease T_{IBAT} at 2-h post-injection ($0.05 < p < 0.1$; Figure 3A).

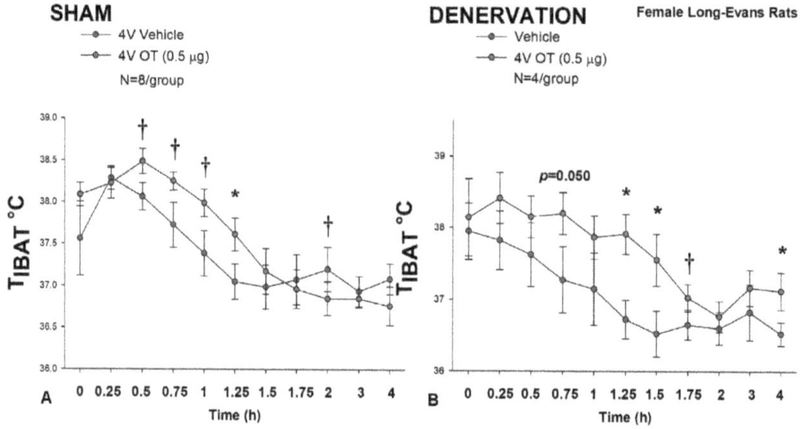

Figure 3. (**A**,**B**) **Effect of acute 4V OT administration (0.5 μg) on T_{IBAT} post-sham or IBAT denervation in female HFD-fed rats.** Rats were maintained on HFD (60% kcal from fat; N = 4–8/group) for approximately 4.5 months prior to undergoing a sham or bilateral surgical IBAT denervation and implantation of temperature transponders underneath IBAT. Rats were subsequently implanted with 4V cannulas and allowed to recover for 2 weeks prior to receiving acute 4V injections of OT or vehicle. Animals were subsequently adapted to a 4-h fast prior to receiving acute 4V injections of OT or vehicle (**A**,**B**) Effect of acute 4V OT on T_{IBAT} in (**A**) sham operated or (**B**) IBAT denervated DIO rats. Data are expressed as mean ± SEM. * $p < 0.05$, † $0.05 < p < 0.1$ OT vs. vehicle.

In IBAT denervated animals, 4V OT (0.5 μg/μL) administration increased T_{IBAT} throughout the post-injection period (1.25, 1.5, and 4-h post-injection) ($p < 0.05$; Figure 3B). Acute 4V OT administration also tended to increase T_{IBAT} at both 0.75 ($p = 0.050$) and 1.75-h ($p = 0.053$) post-injection (Figure 3B). Acute 4V OT also stimulated T_{IBAT} at 24-h post-injection ($p < 0.05$).

Notably, there was no significant change in the ability of 4V OT (0.5 μg) to stimulate T_{IBAT} when averaged over the 1-h post-injection or at 1.25-h post-injection between sham and denervated rats (p = NS).

Collectively, these results indicate that IBAT denervation did not result in a functional change in the effectiveness of 4V OT administration to stimulate BAT thermogenesis in between denervated rats and sham-operated animals.

Plasma Hormone Concentrations

Here, we determined the effects of acute 4V OT (5 μg/μL) on plasma hormones in sham HFD-fed rats (Table 1). Samples from the denervated HFD-fed rats were excluded due to the lack of samples/group for valid comparisons (N = 1–2/group). There were no significant differences in any of the plasma measurements between vehicle and 4V OT-treated rats in the sham-operated groups.

Table 1. Plasma measurements following acute injections of 4V OT (5 µg/µL) or vehicle in female sham and IBAT denervated DIO rats. Data are expressed as mean ± SEM. (N = 3–4/group).

4V Treatment	Vehicle	OT
	Sham	Sham
Leptin (ng/mL)	22.2 ± 6.0 [a]	20.7 ± 5.7 [a]
Insulin (ng/mL)	2.6 ± 1.3 [a]	2.4 ± 0.7 [a]
Glucagon (pmol/L)	6.0 ± 0.3 [a]	18.1 ± 8.9 [a]
FGF-21 (pg/mL)	218.3 ± 71.3 [a]	392.6 ± 172.6 [a]
Irisin (mg/mL)	3.1 ± 0.7 [a]	4.6 ± 0.9 [a]
Adiponectin (mg/mL)	6.9 ± 0.7 [a]	8.3 ± 0.8 [a]
Blood Glucose (mg/dL)	149.3 ± 3.5 [a]	138 ± 6.8 [a]
FFA (mEq/L)	0.4 ± 0.1 [a]	0.6 ± 0.1 [a]
Total Cholesterol (mg/dL)	88.3 ± 12.4 [a]	85.0 ± 8.8 [a]

Blood was collected by tail vein nick (blood glucose) or from the trunk following a 6-h fast. Different letters denote significant differences between treatments. Shared letters are not significantly different from one another.

3.3. Study 3A: Determine the Extent to Which OT-Induced Activation of Sympathetic Outflow to IBAT Contributes to Its Ability to Impact Body Weight in Female HFD-Fed Rats

The objective of Study 3A was to determine whether OT-evoked weight loss requires intact SNS outflow to IBAT. Initially, female rats were lean as defined by body weight (230 ± 2.1 g). As was the case with Study 1, HFD-fed rats were borderline obese as determined by both body weight (380 ± 8.3 g) and adiposity (126.4 ± 6.4 g fat mass; 32.8 ± 1.0% adiposity) after having been maintened on the HFD for at least 4.5 months prior to sham/denervation procedures.

Note that a subset of rats from Study 3 have been analyzed (9 out of 15) for IBAT NE content and all had successful IBAT denervation procedures. All 15 animals were included in the subsequent analyses. IBAT NE content was decreased by 83.1 ± 3.0% in a subset of denervated (9 out of 15) rats relative to a subset of sham-operated control rats (11 out of 15) [(F(1,18) = 64.663, p = 0.000)]. On the other hand, NE content was unchanged in other tissues, including IWAT, EWAT, liver or pancreas in denervated rats relative to sham rats (p = NS). As expected [35], there was no significant change in body weight between female sham-operated and IBAT denervated rats at the start of the study prior to minipump implantation (p = NS).

Chronic 4V administration of vehicle resulted in a 4.5 ± 1.2% weight gain compared to vehicle pre-treatment [(F(1,6) = 14.125, p = 0.009)] in female sham-operated rats. In contrast, chronic 4V OT treatment resulted in a 5.5 ± 1.3% reduction of body weight compared to 4V OT pre-treatment [(F(1,7) = 8.169, p = 0.024)] (Figure 4A). Chronic 4V OT treatment also decreased weight gain throughout the 29-day infusion period, specifically over treatment days 9–29 ($p < 0.05$; Figure 4B). At the conclusion of the study (infusion day 29), OT had decreased body weight by −18 ± 5.0 g in comparison with vehicle-treated rats (15.0 ± 3.8 g; $p < 0.05$). Chronic 4V OT treatment also decreased relative fat mass (pre- vs. post-intervention) (Figure 4C; $p < 0.05$), fat mass and relative lean mass (pre- vs. post-intervention) but had no impact on total lean body mass (p = NS). These effects that were associated with a modest decrease in energy intake that was apparent over treatment weeks 2 and 3 (Figure 4D; $p < 0.05$).

Chronic 4V vehicle treatment resulted in 8.9 ± 1.2% weight gain relative to vehicle pre-treatment in female IBAT denervated rats [(F(1,6) = 65.633, p = 0.000)]. In addition, 4V OT treatment also resulted in a 4.1 ± 1.0% reduction of body weight relative to 4V OT pre-treatment [(F(1,7) = 13.723, p = 0.008)]. (Figure 4A). Chronic 4V OT also reduced weight gain throughout the 29-day infusion period, specifically over treatment days 3–29 ($p < 0.05$; Figure 4B). At the conclusion of the study (infusion day 29), OT had decreased body weight by −15.9 ± 3.7 g in comparison with vehicle-treated rats (30 ± 2.9 g; $p < 0.05$). Chronic 4V OT treatment also decreased relative fat mass (pre- vs. post-intervention) (Figure 4C; $p < 0.05$) and fat mass ($p < 0.05$) but had no impact on total lean body mass (p = NS). These effects that were associated with a modest decrease in energy intake that was apparent during treatment weeks 2 and 3 (Figure 4D; $p < 0.05$). Chronic 4V OT treatment also tended to decrease energy intake during week 1 ($p = 0.052$) and 4 ($p = 0.075$) of the infusion period. There was also no significant effect of chronic 4V OT to increase kaolin intake over the course of the treatment period (p = NS).

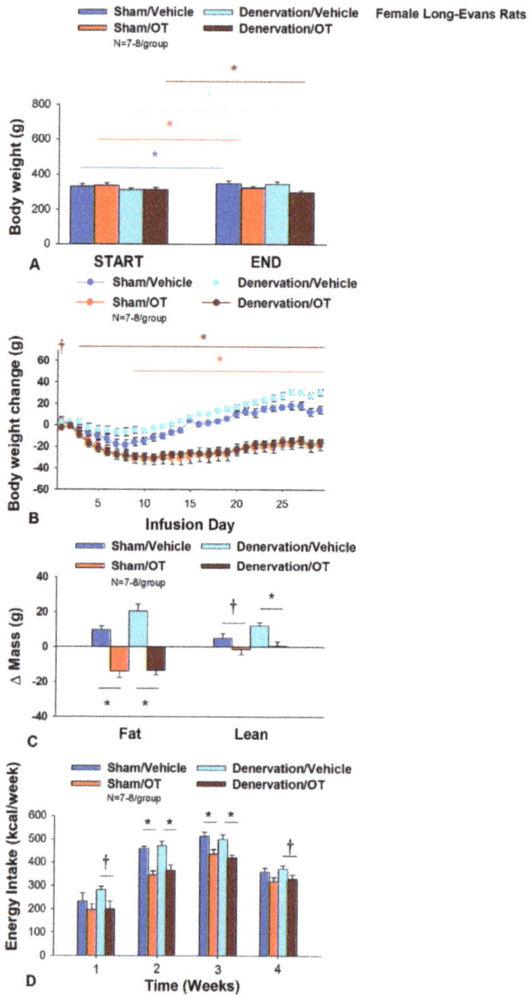

Figure 4. (A–D) Effect of chronic 4V OT infusions (16 nmol/day) on body weight, adiposity and energy intake post-sham or IBAT denervation in female HFD-fed rats. (**A**) Rats were maintained on HFD (60% kcal from fat; N = 7–8/group) for approximately 4.75–5.25 months prior to undergoing a sham or bilateral surgical IBAT denervation. Rats were subsequently implanted with 4V cannulas and allowed to recover for 2 weeks prior to being implanted with subcutaneous minipumps that were subsequently attached to the 4V cannula. (**A**) Effect of chronic 4V OT or vehicle on body weight in sham operated or IBAT denervated DIO rats; (**B**) Effect of chronic 4V OT or vehicle on body weight change in sham operated or IBAT denervated DIO rats; (**C**) Effect of chronic 4V OT or vehicle on adiposity in sham operated or IBAT denervated DIO rats; (**D**) Effect of chronic 4V OT or vehicle on adiposity in sham operated or IBAT denervated DIO rats. Data are expressed as mean ± SEM. * $p < 0.05$, † $0.05 < p < 0.1$ OT vs. vehicle.

Notably, there was no change in the ability of chronic 4V OT treatment to decrease body weight, energy intake, and relative fat mass (pre- vs. post-intervention) or fat mass between female sham-operated and IBAT denervated rats (p = NS).

T_{IBAT}

Similar to what we have previously reported following chronic third ventricular (3V) [34] and 4V [52] treatment in male rats, chronic 4V administration of OT appeared to increase T_{IBAT} (at onset of light cycle) relative to vehicle treatment in female sham-operated rats. This was evident when the data were averaged over week 2 of the infusion period (Table 2A; $p = 0.081$) and throughout the infusion period in ad libitum fed rats on days 6 ($p = 0.068$), 12 ($p < 0.05$) and 21 ($p < 0.05$).

Table 2. Changes in T_{IBAT} following 4V infusions of OT or vehicle in female sham or IBAT denervated DIO rats. (**A**), Changes in T_{IBAT} following 4V infusions of OT or vehicle in ad libitum fed female sham or IBAT denervated DIO rats; (**B**), Changes in T_{IBAT} following 4V infusions of OT or vehicle in 4-h fasted female sham or IBAT denervated DIO rats. (**C**) Changes in TIBAT following 4V infusions of OT or vehicle in ad libitum fed female HFD-fed IBAT denervated rats. (**D**) Changes in TIBAT following 4V infusions of OT or vehicle in 4-h fasted female HFD-fed IBAT denervated rats. Data are expressed as mean ± SEM. * $p < 0.05$ OT, † $0.05 < p < 0.1$ OT vs. vehicle (N = 7–8/group).

Table 2A Changes in T_{IBAT} following 4V infusions of OT or vehicle in ad libitum fed female HFD-fed sham rats				
4V	**Week 1**	**Week 2**	**Week 3**	**Week4**
SHAM	**Temp (°C)**	**Temp (°C)**	**Temp (°C)**	**Temp (°C)**
Vehicle	37.5 ± 0.3	37.5 ± 0.3	37.7 ± 0.3	37.5 ± 0.3
OT	37.9 ± 0.2	38.0 ± 0.2 †	38.0 ± 0.1	37.9 ± 0.2
Table 2B Changes in T_{IBAT} following 4V infusions of OT or vehicle in 4-h fasted fasted female HFD-fed sham rats				
4V	**Week 1**	**Week 2**	**Week 3**	**Week4**
SHAM	**Temp (°C)**	**Temp (°C)**	**Temp (°C)**	**Temp (°C)**
Vehicle	37.7 ± 0.3	37.8 ± 0.2	37.6 ± 0.2	37.7 ± 0.3
OT	38.2 ± 0.2 †	37.56 ± 0.3	37.5 ± 0.3	37.7 ± 0.1
Table 2C Changes in T_{IBAT} following 4V infusions of OT or vehicle in ad libitum fed female HFD-fed IBAT denervated rats				
4V	**Week 1**	**Week 2**	**Week 3**	**Week4**
DENERVATION	**Temp (°C)**	**Temp (°C)**	**Temp (°C)**	**Temp (°C)**
Vehicle	37.4 ± 0.1	37.3 ± 0.2	37.3 ± 0.1	37.1 ± 0.3
OT	37.6 ± 0.2	37.5 ± 0.1	37.9 ± 0.2 †	37.8 ± 0.2 *
Table 2D Changes in T_{IBAT} following 4V infusions of OT or vehicle in 4-h fasted female HFD-fed IBAT denervated rats				
4V	**Week 1**	**Week 2**	**Week 3**	**Week4**
DENERVATION	**Temp (°C)**	**Temp (°C)**	**Temp (°C)**	**Temp (°C)**
Vehicle	38.1 ± 0.1	37.4 ± 0.3	37.7 ± 0.2	37.9 ± 0.2
OT	38.0 ± 0.1	37.5 ± 0.2	37.6 ± 0.2	37.6 ± 0.3

In order to minimize the confound of diet-induced thermogenesis, we collected T_{IBAT} from the same sham operated rats following a 4-h fast. Chronic 4V OT elevated T_{IBAT} when the data were averaged over week 1 of the infusion period (Table 2B; $p = 0.066$) and on infusion days 3 ($p < 0.05$), 5 ($p < 0.05$), and 7 ($p < 0.05$).

In addition, chronic 4V administration of OT also appeared to stimulate T_{IBAT} in denervated rats when the data were averaged over weeks 3 (Table 2C; $p = 0.054$) and 4 (Table 2C; $p < 0.05$) and throughout the infusion period on days 15 ($p < 0.05$), 18 ($p = 0.063$), 20 ($p = 0.057$), 21 ($p = 0.050$), 23 ($p < 0.05$), 24 ($p < 0.05$) and 26 ($p < 0.05$). In contrast, chronic

4V OT failed to elicit a change in T_{IBAT} in IBAT denervated rats that underwent a 4-h fast (Table 2D; p = NS).

Based on these findings as a whole, we conclude that SNS innervation of IBAT does not appear to be a predominant contributor of OT-elicited reduction of weight gain and adiposity in female HFD-fed rats.

3.4. Study 3B: Determine the Extent to Which 4V OT Impacts Thermogenic Gene Expression in IBAT and IWAT in Female HFD-Fed Rats

The objective of this study was to determine whether chronic 4V OT treatment stimulates thermogenic gene expression in IBAT and EWAT from female sham-operated rats.

3.4.1. IBAT

We found that chronic 4V OT treatment elicited a near significant increase in β3-AR mRNA expression (*Adrb3*; $p = 0.063$) and a near significant reduction of Dio2 mRNA expression ($p = 0.077$; Table 3A).

Table 3. (**A**,**B**). Changes in IBAT and IWAT gene expression following 4V infusions of OT or vehicle in female sham or IBAT denervated DIO rats. (**A**), Changes in IBAT mRNA expression 4V infusions of OT or vehicle in female sham or IBAT denervated DIO rats; (**B**), Changes in IWAT mRNA expression 4V infusions of OT or vehicle in female sham or IBAT denervated DIO rats. Shared letters are not significantly different from one another. Data are expressed as mean ± SEM (N = 7–8/group).

Table 3A. Changes in IBAT mRNA Expression Following Chronic 4V Infusions of OT or Vehicle in Female HFD-Fed Rats

4V Treatment **IBAT**	**Vehicle** **Sham**	**OT** **Sham**
Adrb1	1.0 ± 0.4 [a]	1.5 ± 0.4 [a]
Adrb3	1.0 ± 0.3 [a]	1.9 ± 0.3 [a]
Ucp1	1.0 ± 0.3 [a]	1.2 ± 0.2 [a]
Cidea	1.0 ± 0.3 [a]	1.3 ± 0.3 [a]
Dio2	1.0 ± 0.4 [a]	0.3 ± 0.2 [a]
Gpr120	1.0 ± 0.5 [a]	1.2 ± 0.4 [a]
Prdm16	1.0 ± 0.3 [a]	1.2 ± 0.2 [a]
Ppargc1a	1.0 ± 0.3 [a]	1.2 ± 0.2 [a]

Table 3B. Changes in IWAT mRNA Expression Following Chronic 4V Infusions of OT or Vehicle in Female HFD-Fed Rats

4V Treatment **IWAT**	**Vehicle** **Sham**	**OT** **Sham**
Adrb1	1.0 ± 0.2 [a]	2.7 ± 0.6 [b]
Adrb3	1.0 ± 0.4 [a]	2.0 ± 0.5 [a]
Ucp1	1.0 ± 0.3 [a]	0.5 ± 0.1 [a]
Cidea	1.0 ± 0.2 [a]	2.8 ± 0.5 [b]
DIio2	1.0 ± 0.3 [a]	1.0 ± 0.2 [a]
Gpr120	1.0 ± 0.3 [a]	2.0 ± 0.4 [a]
Prdm16	1.0 ± 0.1 [a]	1.6 ± 0.3 [a]
Ppargc1a	1.0 ± 0.2 [a]	0.9 ± 0.1 [a]

IBAT was collected following a 4-h fast. Different letters denote significant differences between treatments. Shared letters are not significantly different from one another. N = 7–8/group. IWAT was collected following a 4-h fast. Different letters denote significant differences between treatments. Shared letters are not significantly different from one another.

3.4.2. IWAT

4V OT treatment was associated with a significant increase of the thermogenic markers, beta 1 adrenergic receptor (β1-AR) (*Adrb1*; $p < 0.05$; Table 3B) and Cidea ($p < 0.05$) mRNA expression. 4V OT treatment also elicited a near significant increase in Gpr120 mRNA expression ($p = 0.060$) as well as a near significant reduction of Dio2 mRNA expression ($p = 0.097$) in IWAT from sham-operated rats.

Together, these findings raise the possibility that different thermogenic markers in IBAT and IWAT may contribute, in part, to the ability of chronic 4V OT to reduce body weight and adiposity in sham-operated rats.

3.5. Study 4A: Determine the Effects of Chronic 4V OT Treatment (16 nmol/day) on Body Weight, Adiposity and Energy Intake in Female DIO C57BL/6J Mice

The objective of this study was to determine the susceptibility of female C57BL/6J mice to DIO and whether the effects of chronic hindbrain (4V) administration to reduce body weight and adiposity could translate to another female rodent model (female C57BL/6J mice). Initially, female C57BL/6J mice were lean as determined by both body weight (17.4 ± 0.3 g) and adiposity (2.2 ± 0.2 g fat mass; 12.7 ± 0.8% adiposity). HFD-fed C57BL/6J mice became borderline DIO as demonstrated by both body weight (31.2 ± 1.4 g) and adiposity (11.7 ± 1.2 g fat mass; 36 ± 2.2% adiposity) after maintenance on the HFD for at least 4.5 months prior to being IBAT temperature transponder and minipump implantations as described earlier. By design, there was no significant difference in initial body weight or adiposity between vehicle and OT treatment groups prior to minipump implantation (p = NS). Three of the twenty mice available at study onset were euthanized during the course of the study and were excluded from the data analysis (including one whose head cap had become detached).

Chronic 4V vehicle treatment in female C57BL/6J mice resulted in modest amount of weight relative to vehicle pre-treatment ($p = 0.114$). While chronic 4V OT treatment failed to evoke weight loss (p = NS; Figure 5A), it reduced weight gain on treatment day 8 ($p < 0.05$) and it also tended to reduce weight gain on treatment days 5, 7, 9–12, and 23 ($0.05 < p < 0.1$; Figure 5B). These effects were associated with a reduction of relative fat mass (pre- vs. post-intervention) (Figure 5C; $p < 0.05$). Chronic 4V OT treatment also tended to reduce total lean mass ($p = 0.131$) but had no effect on relative lean mass (pre- vs. post-intervention) or total lean mass (p = NS). These effects were not associated with significant reductions in energy intake (Figure 5D; p = NS) or increased kaolin consumption (p = NS) during the course of the treatment period.

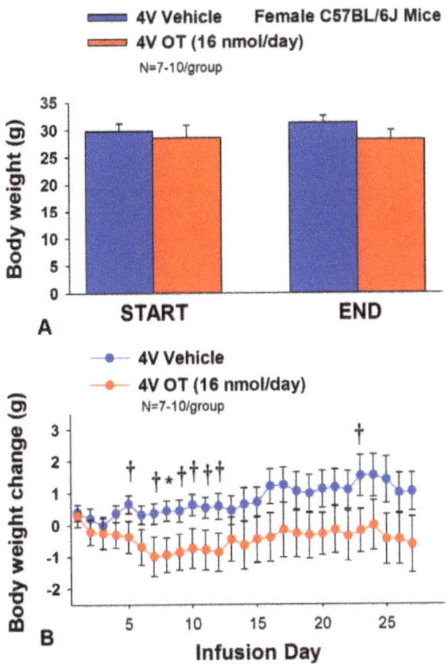

Figure 5. *Cont.*

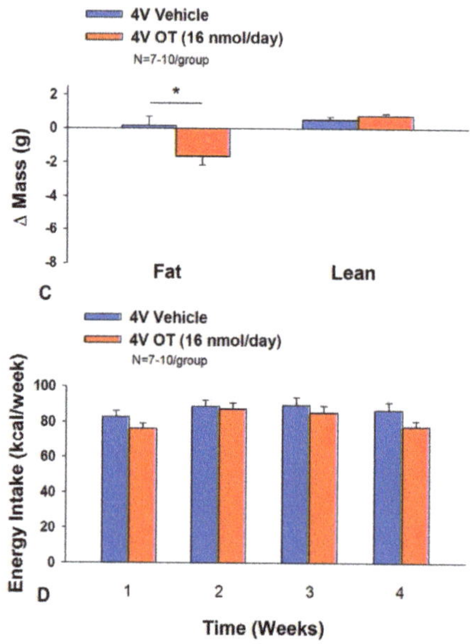

Figure 5. (**A–D**) **Effect of chronic 4V OT infusions (16 nmol/day) on body weight, adiposity and energy intake in female HFD-fed C57BL/6J mice.** (**A**) Mice were maintained on HFD (60% kcal from fat; N = 7–10/group) for approximately 4.5 months prior to implantation of temperature transponders underneath IBAT. Mice were subsequently implanted with 4V cannulas and allowed to recover for 2 weeks prior to being implanted with subcutaneous minipumps that were subsequently attached to the 4V cannula. (**A**) Effect of chronic 4V OT or vehicle on body weight in female C57BL/6J mice rats; (**B**) Effect of chronic 4V OT or vehicle on body weight change in female C57BL/6J mice; (**C**) Effect of chronic 4V OT or vehicle on adiposity in female C57BL/6J mice; (**D**) Effect of chronic 4V OT or vehicle on adiposity in female C57BL/6J mice. Data are expressed as mean ± SEM. * $p < 0.05$, † $0.05 < p < 0.1$ OT vs. vehicle.

3.5.1. T_{IBAT}

In contrast to the effects of observed following chronic 4V infusions of OT (16 nmol/day) in male [54] and female DIO rats, we found that chronic 4V infusions of OT at the same dose (16 nmol/day) in female C57BL/6J mice largely had no effect significant effects on T_{IBAT} in ad libitum fed mice when the data were averaged over weeks 1, 2, 3 and 4 of the infusion period (p = NS).

While largely similar results were obtained following a 4-h fast over weeks 1–3 (p = NS), there was a tendency for chronic 4V OT to reduce T_{IBAT} over week 4 ($p = 0.076$) in 4-h fasted mice. Specifically, chronic 4V OT reduced T_{IBAT} on infusion days 17, 20, and 25 ($p < 0.05$) and tended to reduce T_{IBAT} on infusion days 13, 23, and 24 ($0.05 < p < 0.1$).

3.5.2. Plasma Hormone Concentrations

Here, we determined the effects of chronic 4V OT (16 nmol/day) on plasma hormones in female C57BL/6J mice (Table 4). We did not find any significant differences in any of the plasma measurements in female C57BL/6J mice that received chronic 4V infusions of vehicle or OT.

Table 4. Plasma measurements following chronic 4V infusions of OT (16 nmol/day) or vehicle in female HFD-fed C57BL/6J mice. Data are expressed as mean ± SEM (N = 7–10/group).

4V Treatment	Vehicle	OT
Leptin (ng/mL)	10.0 ± 1.3 [a]	6.7 ± 1.3 [a]
Insulin (ng/mL)	0.5 ± 0.1 [a]	0.3 ± 0.04 [a]
Glucagon (pmol/L)	23.3 ± 3.7 [a]	34.1 ± 13.2 [a]
FGF-21 (pg/mL)	467.5 ± 129.7 [a]	513.6 ± 179.6 [a]
Irisin (mg/mL)	4.1 ± 0.2 [a]	3.9 ± 0.2 [a]
Adiponectin (mg/mL)	19.3 ± 1.0 [a]	20.1 ± 1.8 [a]
Blood Glucose (mg/dL)	155 ± 2.7 [a]	152.6 ± 8.5 [a]
FFA (mEq/L)	0.13 ± 0.01 [a]	0.11 ± 0.01 [a]
Total Cholesterol (mg/dL)	105.9 ± 2.9 [a]	95.3 ± 9.5 [a]

Blood was collected by tail vein nick (blood glucose) or from the trunk following a 6-h fast. Different letters denote significant differences between treatments. Shared letters are not significantly different from one another.

3.6. Study 4B: Determine the Effects of Chronic 4V OT Treatment (16 nmol/day) on Body Weight, Adiposity and Energy Intake in Female DIO DBA/2J Mice

The objective of this study was to determine the susceptibility of female DMA/2J mice to DIO and whether the effects of chronic hindbrain (4V) administration to reduce body weight and adiposity could translate to another rodent model (female DBA/2J mice). This particular strain of female mice was previously found to be susceptible to becoming DIO [69,70]. Initially, female DBA/2J mice were lean as determined by both body weight (20.3 ± 0.3 g) and adiposity (3.4 ± 0.3 g fat mass; 16.8 ± 1.1% adiposity). DBA/2J mice became DIO as determined by both body weight (35.8 ± 0.9 g) and adiposity (15.1 ± 0.8 g fat mass; 41.9 ± 1.2% adiposity) after maintenance on the HFD for at least 4.5 months prior to IBAT temperature transponder and minipump implantations as described earlier. There was no significant difference in initial body weight or adiposity between vehicle and OT treatment groups prior to minipump implantation (p = NS). Four of the twenty mice available at study onset were euthanized during the course of the study and were excluded from the data analysis (including two whose head caps had become detached).

Unexpectedly, chronic 4V vehicle resulted in 6.4 ± 2.0% weight loss relative to chronic 4V vehicle pre-treatment [($F(1,7) = 12.781$, $p = 0.009$)] whereas chronic 4V OT reduced body weight by 11.9 ± 2.1% relative to chronic 4V OT pre-treatment [($F(1,6) = 24.802$, $p = 0.003$)] (Figure 6A). Furthermore, chronic 4V OT treatment reduced weight gain on treatment days 8–10, 12–13, and 15 ($p < 0.05$) and it also tended to reduce weight gain on treatment days 4, 5 ($p = 0.050$), 6–7, 11, 14, 19–21, and 23–25 ($0.05 < p < 0.1$; Figure 6B). Chronic 4V OT also tended to reduce relative fat mass (pre- vs. post-intervention) (Figure 6C; $0.05 < p < 0.1$) and total fat mass ($p < 0.05$) but had no effect on adipocyte size or relative lean mass (pre- vs. post-intervention) or total lean mass (p = NS). These effects were associated with a modest reduction of energy intake that was apparent during weeks 1 and 2 of the treatment period (Figure 6D; $p < 0.05$). There was no effect of chronic 4V OT to increase kaolin consumption during the treatment period (p = NS).

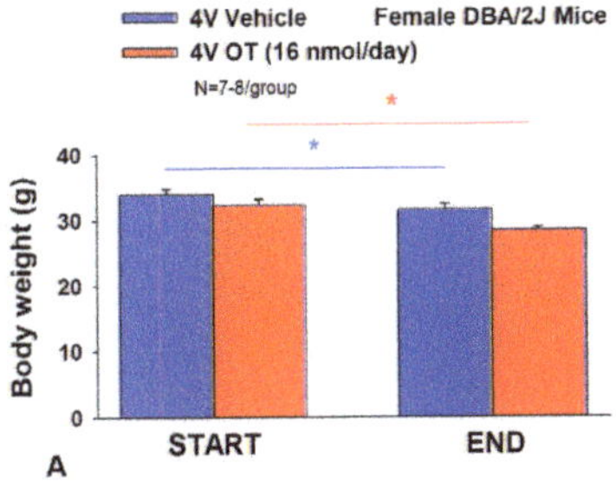

Figure 6. *Cont.*

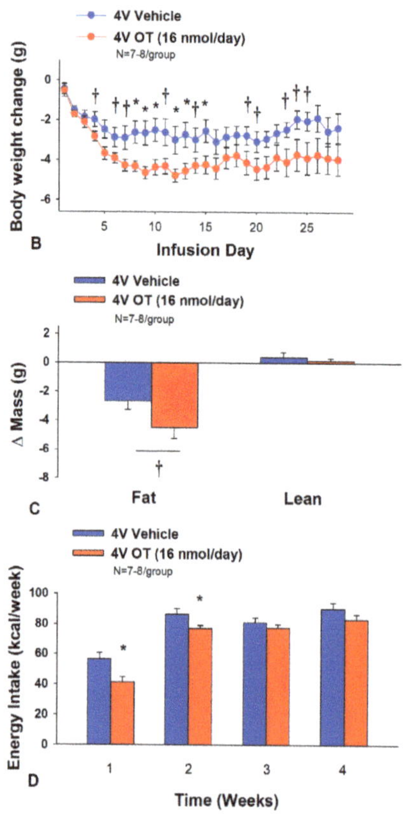

Figure 6. (A–D) Effect of chronic 4V OT infusions (16 nmol/day) on body weight, adiposity and energy intake in female HFD-fed DBA/2J mice. (**A**) Mice were maintained on HFD (60% kcal from fat; N = 7–8/group) for approximately 4.5 months prior to implantation of temperature transponders underneath IBAT. Mice were subsequently implanted with 4V cannulas and allowed to recover for 2 weeks prior to being implanted with subcutaneous minipumps that were subsequently attached to the 4V cannula. (**A**) Effect of chronic 4V OT or vehicle on body weight in female DBA/2J mice; (**B**) Effect of chronic 4V OT or vehicle on body weight change in female DBA/2J mice; (**C**) Effect of chronic 4V OT or vehicle on adiposity in female DBA/2J mice; (**D**) Effect of chronic 4V OT or vehicle on adiposity in female DBA/2J mice. Data are expressed as mean ± SEM. * $p < 0.05$, † $0.05 < p < 0.1$ OT vs. vehicle.

3.6.1. T_{IBAT}

Chronic 4V infusions of OT (16 nmol/day) had no significant effect on T_{IBAT} in ad libitum fed mice when the data were averaged over weeks 1, 2, 3 and 4 (p = NS).

While similar results were obtained following a 4-h fast over weeks 1–4 (p = NS), chronic 4V OT reduced T_{IBAT} on infusion day 8 ($p < 0.05$) and tended to reduce T_{IBAT} on infusion day 26 ($0.05 < p < 0.1$).

3.6.2. Plasma Hormone Concentrations

Here, we determined the effects of chronic SC OT (16 and 50 nmol/day) on plasma hormones in female DBA/2J mice (Table 5). We did not find any significant differences in any of the plasma measurements in female DBA/2J mice that received chronic 4V infusions of vehicle or OT.

Table 5. Plasma Measurements Following 4V Infusions of OT or Vehicle in Female HFD-Fed DBA/2J Mice.

4V Treatment	Vehicle	OT
Leptin (ng/mL)	5.1 ± 1.4 [a]	3.1 ± 0.5 [a]
Insulin (ng/mL)	2.6 ± 0.4 [a]	1.4 ± 0.2 [b]
Glucagon (pmol/L)	46.1 ± 11.2 [a]	46.9 ± 8.7 [a]
FGF-21 (pg/mL)	1419.1 ± 251.3 [a]	1297.8 ± 282 [a]
Irisin (mg/mL)	5.0 ± 0.2 [a]	4.2 ± 0.3 [a]
Adiponectin (mg/mL)	8.1 ± 0.2 [a]	8.4 ± 0.3 [a]
Blood Glucose (mg/dL)	140 ± 4.3 [a]	145.4 ± 4.2 [a]
FFA (mEq/L)	0.2 ± 0.02 [a]	0.3 ± 0.1 [a]
Total Cholesterol (mg/dL)	96.4 ± 4.1 [a]	103.4 ± 5.2 [a]

Blood was collected by tail vein nick (blood glucose) or from the trunk following a 6-h fast. Different letters denote significant differences between treatments. Shared letters are not significantly different from one another. Data are expressed as mean ± SEM (N = 7–8/group).

3.7. Study 5: Determine the Effects of Chronic Systemic OT Treatment (16 and 50 nmol/day) on Body Weight, Adiposity and Energy Intake in Female DIO DBA/2J Mice

The objective of this study was to extend previous findings from Study 4B and determine whether SC infusion of a centrally effective dose of OT (16 nmol/day) can decrease both body weight and adiposity in female DIO DBA/2J mice (same strain used in Study 4B). DBA/2J mice became DIO as determined by the increased body weight (32.2 ± 0.9 g) and adiposity (12.7 ± 0.7 g fat mass; 38.7 ± 1.2% adiposity) after being maintained on the HFD for at least 4.5 months prior to IBAT temperature transponder implantations. By design, there was no significant difference in body weight or adiposity between vehicle and OT treatment groups at the start of the study prior to minipump implantation (p = NS).

In contrast to chronic 4V OT treatment in female DBA/2J mice (Study 4B), chronic SC OT treatment did not result in a significant reduction of body weight (Figure 7A). SC OT treatment (16 nmol/day) reduced weight gain on treatment days 2–3 ($p < 0.05$) and tended to reduce weight gain on treatment day 1 ($p = 0.138$; Figure 7B). The higher dose (50 nmol/day) also reduced weight gain on treatment days 1–3 and tended to increase body weight gain on treatment days 4 ($p = 0.055$), 5 ($p = 0.106$), 7 ($p = 0.077$), 8 ($p = 0.051$) and 9 ($p = 0.092$) (Figure 7B). There was no effect of SC OT at either dose on relative fat mass or lean mass (pre- vs. post-intervention) (Figure 7C). OT (50 nmol/day) produced a transient reduction of energy intake during week 2 (Figure 7D; $p < 0.05$) but OT failed to impact energy intake at any other time. There was also no effect of chronic SC OT to significant increase kaolin consumption during weeks 2–4 of the treatment period (p = NS) but a slight reduction of kaolin intake during week 1 in response to the higher dose 50 nmol/day ($p = 0.016$).

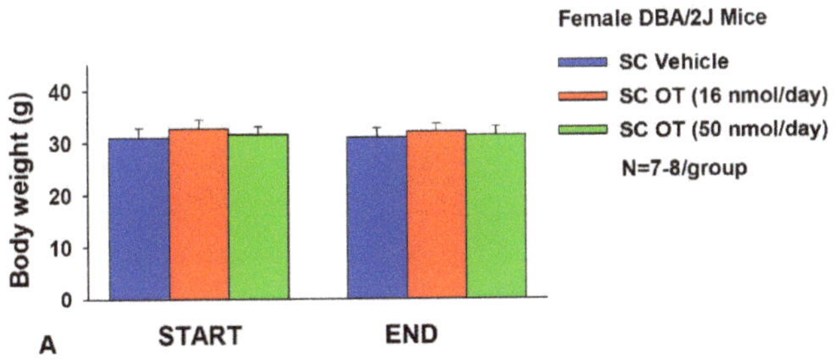

Figure 7. *Cont.*

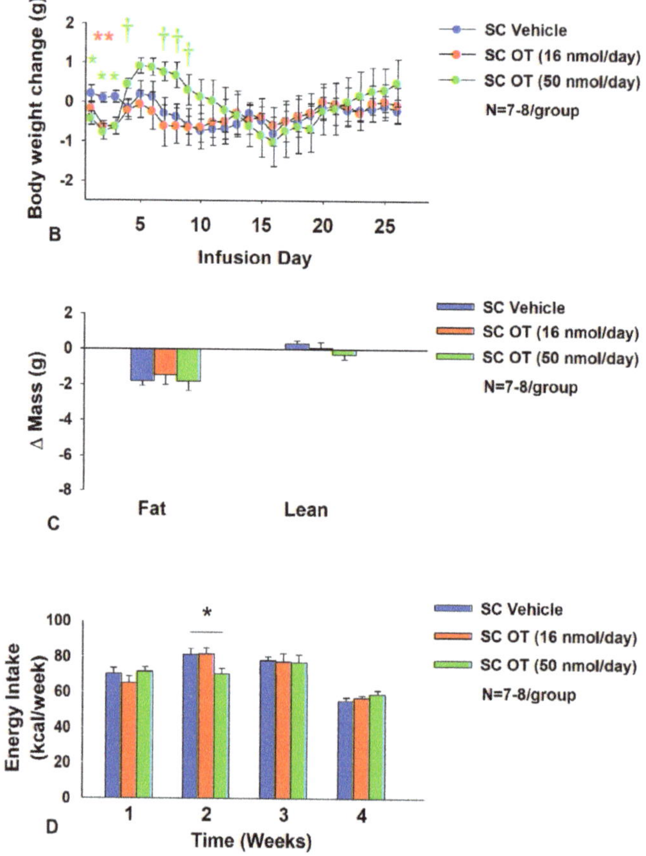

Figure 7. (**A–D**) Effect of chronic systemic OT infusions (16 and 50 nmol/day) on body weight, adiposity and energy intake in female HFD-fed DBA/2J mice. (**A**) Mice were maintained on HFD (60% kcal from fat; N = 7–8/group) for approximately 4.5 months prior to implantation of temperature transponders underneath IBAT. Mice were subsequently implanted with 4V cannulas and allowed to recover for 2 weeks prior to being implanted with SC minipumps that were subsequently attached to the 4V cannula. (**A**) Effect of chronic 4V OT or vehicle on body weight in female DBA/2J mice; (**B**) Effect of chronic 4V OT or vehicle on body weight change in female DBA/2J mice; (**C**) Effect of chronic 4V OT or vehicle on adiposity in female DBA/2J mice; (**D**) Effect of chronic 4V OT or vehicle on adiposity in female DBA/2J mice. Data are expressed as mean ± SEM. * $p < 0.05$, † $0.05 < p < 0.1$ OT vs. vehicle.

3.7.1. T_{IBAT}

In contrast to what we found following chronic 4V administration, chronic SC administration of OT (16 and 50 nmol/day) reduced T_{IBAT} relative to vehicle in ad libitum fed mice when the T_{IBAT} data were averaged over weeks 3 and 4 (Table 6A; $p < 0.05$) of the infusion period. Similar results were obtained from the same mice following a 4-h fast over the same period (Table 6B; $p < 0.05$).

Table 6. Changes in T_{IBAT} following chronic systemic infusions of OT (16 and 50 nmol/day) or vehicle in female DBA/2J mice. A, Changes in T_{IBAT} following chronic systemic infusions of OT (16 and 50 nmol/day) or vehicle in ad libitum fed female DBA/2J mice; B, Changes in T_{IBAT} following chronic systemic infusions of OT (16 and 50 nmol/day) or vehicle in 4-h fasted female DBA/2J mice. Shared letters are not significantly different from one another. Data are expressed as mean ± SEM. * $p < 0.05$ OT, vs. vehicle (N = 7–8/group).

Table 6A Changes in T_{IBAT} following SC infusions of OT or vehicle in ad libitum fed female DIO DBA/2J mice

SC	Week 1	Week 2	Week 3	Week4
	Temp (°C)	Temp (°C)	Temp (°C)	Temp (°C)
Vehicle	37.5 ± 0.2	37.1 ± 0.2	37.0 ± 0.1	37.2 ± 0.1
OT (16 nmol/day)	37.2 ± 0.2	36.7 ± 0.2	36.5 ± 0.2 *	36.4 ± 0.2 *
OT (50 nmol/day)	37.2 ± 0.2	36.7 ± 0.1	36.4 ± 0.2 *	36.5 ± 0.2 *

Table 6B Changes in T_{IBAT} following SC infusions of OT or vehicle in 4-h fasted female DIO DBA/2J mice

SC	Week 1	Week 2	Week 3	Week4
	Temp (°C)	Temp (°C)	Temp (°C)	Temp (°C)
Vehicle	37.3 ± 0.2	37.1 ± 0.2	37.2 ± 0.1	37.3 ± 0.1
OT (16 nmol/day)	37.1 ± 0.1	36.8 ± 0.2	36.6 ± 0.1 *	36.6 ± 0.2 *
OT (50 nmol/day)	37.2 ± 0.2	36.9 ± 0.2	36.7 ± 0.2 *	36.6 ± 0.2 *

In addition to the findings in DIO female DBA/2J mice, we found that there appeared to be a very modest effect of chronic systemic OT (16 nmol/day) to reduce T_{IBAT} in 4-h fasted male DIO (C57Bl/6J) mice on infusion day 9 ($p < 0.05$) and tended to reduce T_{IBAT} on days 10 ($0.05 < p < 0.1$), 14 ($0.05 < p < 0.1$), and 21 ($p = 0.05$) (unpublished findings). Likewise, OT (50 nmol/day) tended to reduce T_{IBAT} in 4-h fasted C57Bl/6J mice on infusion day 4 ($p < 0.05$) and tended to reduce T_{IBAT} on infusion day 21 ($0.05 < p < 0.1$). In contrast, we found that OT (100 nmol/day) tended to increase T_{IBAT} on infusion day 13 ($0.05 < p < 0.1$), but this was only evident in ad libitum fed C57Bl/6J mice but not in 4-h fasted mice.

3.7.2. Plasma Hormone Concentrations

Here, we examined the effects of chronic systemic OT (16 and 50 nmol/day) on plasma hormones in female DBA/2J mice (Table 7). Chronic SC OT (16 nmol/day) treatment was associated with a significant increase in plasma glucagon in female DBA/2J mice. Chronic SC OT treatment also tended to increase total cholesterol at the low (16 nmol/day; $p = 0.057$) and high dose (50 nmol/day; $p = 0.062$). In addition, chronic SC Ot at the high dose (50 nmol/day) also tended to produce an increase in plasma leptin ($p = 0.092$). and FGF-21 ($p = 0.051$).

Table 7. Plasma measurements following chronic systemic infusions of OT (16 and 50 nmol/day) or vehicle in female HFD-fed DBA/2J mice. Data are expressed as mean ± SEM. Different letters denote significant differences between treatments. Shared letters are not significantly different from one another (N = 7/group).

SC Treatment	Vehicle	OT (16 nmol/day)	OT (50 nmol/day)
Leptin (ng/mL)	5.6 ± 1.1 [a]	7.2 ± 1.1 [a]	8.6 ± 1.3 [a]
Insulin (ng/mL)	6.3 ± 1.1 [a]	8.1 ± 0.4 [a]	9.5 ± 2.2 [a]
Glucagon (pmol/L)	12.2 ± 2.2 [a]	21.0 ± 2.2 [b]	14.2 ± 3.3 [ab]
FGF-21 (pg/mL)	1158.6 ± 112 [a]	1505.2 ± 174.1 [a]	1577.1 ± 131.6 [a]
Irisin (mg/mL)	3.7 ± 0.2 [a]	4.0 ± 0.3 [a]	3.8 ± 0.5 [a]
Adiponectin (mg/mL)	8.6 ± 0.4 [a]	9.5 ± 0.5 [a]	8.1 ± 0.6 [a]
Blood Glucose (mg/dL)	159.58 ± 5.7 [a]	153.6 ± 5.2 [a]	151.4 ± 4.0 [a]
FFA (mEq/L)	0.2 ± 0.02 [a]	0.2 ± 0.02 [a]	0.2 ± 0.03 [a]
Total Cholesterol (mg/dL)	98.5 ± 2.8 [a]	112.4 ± 4.0 [a]	112.0 ± 6.7 [a]

4. Discussion

The objectives of the current set of studies were to (1) establish whether sympathetic innervation of IBAT is required for 4V (hindbrain) administration of OT to stimulate BAT thermogenesis and decrease body weight and adiposity in female HFD-fed rats and (2) establish whether the ability of hindbrain (4V) infusion of OT to elicit weight loss translates to other rodent species. To accomplish these goals, we examined the effect of disrupting SNS activation of IBAT on OT-induced stimulation of T_{IBAT} and reduction of body weight in HFD-fed rats. We initially determined the impact of bilateral surgical SNS denervation to IBAT on the ability of acute 4V OT (0.5, 1, and 5 μg) to stimulate T_{IBAT} in female HFD-fed rats. We found that the high dose of 4V OT (5 μg) stimulated T_{IBAT} similarly between sham rats and denervated rats. We subsequently determined if OT-elicited reductions of body weight and adiposity require intact SNS outflow to IBAT. To accomplish this, we determined the effect of bilateral surgical or sham denervation of IBAT on the ability of chronic 4V OT (16 nmol/day) or vehicle administration to reduce body weight, adiposity and food intake in female HFD-fed rats. Chronic 4V OT reduced body weight gain (sham: -18.0 ± 4.9 g; denervation: -15.9 ± 3.7 g) and adiposity (sham: -13.9 ± 3.7 g; denervation: -13.6 ± 2.4 g) relative to vehicle treatment and these effects were similar between groups. These effects were attributed, in part, to reduced energy intake evident during weeks 2 and 3. To test whether the effects of 4V OT to elicit weight loss translate to other female rodent species, we also examined the effect of chronic 4V infusion of OT on body weight in two separate strains of female HFD-fed mice. Similar to what we found in the HFD-fed rat model, we also found that chronic 4V OT (16 nmol/day) infusion resulted in reduced body weight gain, adiposity and/or energy intake in female HFD-fed C57BL/6J and DBA/2J mice. Together, these findings suggest that (1) sympathetic innervation of IBAT is not necessary for OT-elicited increases in BAT thermogenesis and weight loss in female HFD-fed rats and (2) the effects of OT to reduce weight gain and adiposity translate to other female mouse models of diet-induced obesity (DIO).

We have now determined that chronic 4V OT-elicited reduction of body weight loss does not require SNS innervation of IBAT in multiple animal models (female HFD-fed rats and male DIO mice) [35]. These data suggest that 4V administration of OT increases BAT thermogenesis and evokes weight loss through a mechanism that does not require SNS innervation of IBAT in male and female rodent models. As mentioned in [35], we have not addressed what mechanism might be required for hindbrain (4V) OT to stimulate BAT thermogenesis if not through SNS innervation of IBAT. We have largely ruled out the possibility that, in mice, 4V OT might be leaking into the periphery to act at peripheral OTRs by showing that systemic administration of OT, at a centrally effective dose was unable to replicate the effects of hindbrain (4V) OT to reduce body weight and stimulate BAT thermogenesis in DIO mice [35]. One mechanism that we did not address in this body of work is whether 4V OT-elicited activation of hindbrain and/or spinal cord OTRs might elicit the release of epinephrine from the adrenal medulla and activate BAT thermogenesis through direct activation of β-adrenergic receptors. However, we recently determined that systemic administration of the β3-AR antagonist, SR 59230A, failed to block the effects of acute 4V OT to increase T_{IBAT} (unpublished findings) in male DIO Long-Evans rats, suggesting that signaling through the β3-AR is not required for OT-elicited BAT thermogenesis. Other potential mediators of 4V-OT elicited BAT thermogenesis include other beta-receptor subtypes, namely the β1-AR and β2-AR, both of which are expressed in IBAT in both mice [71] and rats [72,73]. While we did not find a significant increase in β1-AR mRNA expression in response to 4V OT in IBAT in this study, we did see an increase in β1-AR mRNA expression in IWAT (see discussion below). It is well appreciated that the β1-AR is important in the control of thermogenesis in rodents [74,75] but the β2-AR may be more important in the control of thermogenesis in humans than rodents [76–78]. Furthermore, β1-AR and β2-AR have nearly equal affinity for L-epinephrine in Chinese hamster ovary cells [79] and epinephrine administration to brown adipocytes stimulates fatty acids and respiration [80]. However, only 1% of parvocellular or magnocellular PVN

OT neurons have poly-synaptic projections to the adrenal gland [81], despite the hindbrain and spinal cord being relay sites in outgoing poly-synaptic projections to the adrenal gland [81,82]. Future studies should address whether adrenal demedulation impairs the ability of hindbrain (4V) OT to stimulate BAT thermogenesis and elicit weight loss in DIO rodents.

Our finding that 4V OT treatment elicited an increase in both β1-AR and Cidea mRNA expression in IWAT raises the possibility that WAT browning or beiging may also contribute, in part, to the metabolic effects of 4V OT in female rodents. Beige depots within WAT may account for up to 5% of total UCP-1 [83,84]. It is possible that hindbrain OTRs could also be a component of descending projections that originate in the PVN and are important in the regulation of SNS outflow to IWAT [85]. In fact, there are well established poly-synaptic circuits that link parvocellular PVN OT neurons to IWAT [86,87]. Thus, OT neurons within the parvocellular PVN are anatomically situated to control WAT thermogenesis. One outstanding question is whether these effects are mediated by parvocellular PVN OT neurons that project directly to the hindbrain (nucleus tractus solitarius [88,89]) and/or spinal cord [89]. Further studies that determine the extent to which 4V OT treatment (1) elicits more functional changes in IWAT thermogenesis (increased temperature of IWAT) [85,90] and (2) reduces body weight and adiposity in animals following IWAT denervation will be helpful in assessing the role of WAT in contributing to the effects of 4V OT to reduce body weight and adiposity.

We acknowledge the possibility that the effects of 4V OT on BAT thermogenesis in female rats could be due, in part, to increased activity-induced thermogenesis [21] as well as skeletal muscle thermogenesis [22,23]. While we did not assess the effects of 4V OT on non-shivering and shivering thermogenesis in skeletal muscle [22,23] in this study, we recently determined that acute 4V administration of OT (5 μg) stimulated T_{IBAT}, core temperature and gross motor activity in male DIO rats (unpublished observations). However, we found that 4V OT-associated elevations of T_{IBAT} and core temperature occurred before significant increases in gross motor activity suggesting that changes in gross motor activity are not likely tied to the changes in T_{IBAT} and core temperature that preceded changes in activity. Our findings are similar to what others have reported following ICV (0.5 μg) [91] and ventromedial hypothalamic administration (1 nmol $\approx$ 1.0072 μg) [15]. Taken together, acute CNS administration of OT can increase activity in rodents, but, based on our unpublished findings, these activity related increases do not appear to contribute to the effects of 4V OT on BAT thermogenesis in male DIO rats. It remains to be determined if this holds true in female HFD-fed rats.

One limitation to this study is that we did not account for the contribution of other BAT depots in contributing to the ability of 4V OT to stimulate BAT thermogenesis and reduce body weight in IBAT denervated rats. We chose to make IBAT the focus of our studies given that it contains up to 45% of total UCP-1 [92] and represents $\geq$70% of total BAT mass [93]. In addition, this particular depot is the best characterized of BAT depots [94]. However, other BAT depots [axillary (subscapular), cervical, mediastinal and perirenal depots] show cold-induced elevations of UCP-1 [83]. In particular, the axillary (subscapular), cervical, periaortic and perirenal BAT depots [19,84] may provide up to 50% of total UCP-1 mRNA. Fischer reported that the axillary (subscapular) depot, also showed a significant 2-fold increase of total UCP-1 (UCP-1/scBAT) in response to HFD (diet-induced thermogenesis) in IBAT denervated mice [92]. There also appeared to be an increase of axillary UCP-1 in response to HFD in sham mice but it was not significant and there were no significant differences in UCP-1 between sham vs. denervation groups in response to HFD [92]. Moreover, Nugyen reported that is potential crosstalk between SNS circuits that innervate IBAT and WAT [85]. Nguyen found that there is increased NE turnover and IWAT UCP-1 mRNA expression in hamsters following SNS denervation of IBAT [85]. It will be helpful to selectively denervate other BAT and WAT depots in order to determine whether these depots may contribute, in part, to the effects of 4V OT to reduce body weight gain in female rodents.

While a recent study reported that systemic infusions of OT (100 nmol/day) result in an elevation of core temperature and increased IBAT gene expression in male HFD-fed mice (C57BL6/J) [38], we found that systemic infusion (16 and 50 nmol/day) resulted in a reduction of T_{IBAT} temperature in female DBA/2J mice. Similarly, we found that acute peripheral administration of OT (5 and 10 μg/μL) elicited an initial reduction of T_{IBAT} prior to a subsequent elevation of T_{IBAT} [35]. Furthermore, others have found that systemic injections of higher doses (1 mg/kg) have also resulted in hypothermic effects [95], which is thought to be mediated, in part, by activation of arginine vasopressin receptor 1A (AVPR1A) [96]. It is possible that differences between our study and Yuan's study are due, in part, to strain, sex, age, length of time that the mice were maintained on the HFD prior to study onset (8 weeks rather than 18 weeks in our study) and/or time of day that the core temperature vs. T_{IBAT} measurements were taken. Being able to include measurements of core temperature and T_{IBAT} from the same animal will help enable more direct comparisons with other studies.

Based on recent findings [10], it is possible that differences in estrus cycle might have impaired the effectiveness of OT to reduce food intake during the measurement period. The authors found that there was an impaired ability of ICV OT to reduce food intake during the pro-estrus stage of the estrus cycle, during which time there is an increase in estrogen [10]. Despite this, we still found an effect of 4V OT to reduce weight gain suggesting that other mechanisms (i.e., lipolysis, energy expenditure) may also contribute to 4V OT-elicited changes in body weight in female rodents. Future studies, however, should take into account estrus cycle when measuring energy intake in response to OT treatment.

Our findings showing that chronic 4V administration of OT reduced energy intake in female DIO DBA/2J mice recapitulated the effects an earlier study that found following chronic systemic administration in female DIO C57BL/6J mice [97]. However, we failed to find an effect of chronic 4V OT to reduce food intake in female DIO C57BL/6J mice. In addition, we found that systemic OT (16 or 50 nmol/day) produced transient reductions of body weight gain in female DIO DBA/2J mice at doses that the authors (≈27.6 and 55.1 nmol/day) found to reduce body weight in female DIO C57BL/6J mice [97]. However, the authors in that study used a different strain of mice (C57BL/6J) that were younger (18 weeks vs. 31 weeks at onset of minipump infusions in our study), heavier (34.20 g vs. 31.2 ± 1.4 g in our study) and had been on the HFD diet for a shorter period of time (12 weeks vs. 24 weeks at onset of minipump infusions in our study). Thus, there are several differences between studies that might account for the contradictory effects.

In conclusion, our findings indicate that there is no significant difference in the effectiveness of the β3-AR agonist, CL 316243, to stimulate IBAT in female IBAT denervated rats relative to female sham-operated rats with intact SNS innervation of IBAT. In addition, we found that acute 4V administration of OT at both the low (0.5 μg) and high dose (5 μg) resulted in similar increases in T_{IBAT} at in female sham and IBAT denervated rats. Furthermore, we also found that there was no difference in the effectiveness of chronic 4V OT (16 nmol/day) to reduce body weight gain and adiposity in female sham and IBAT denervated rats. Consistent with what we found in the HFD-fed rat model, we found that chronic 4V OT (16 nmol/day) treatment reduced body weight gain, adiposity and/or energy intake in female DIO C57BL/6J and DBA/2J mice relative to chronic 4V vehicle treatment in control mice. Together, these findings suggest that (1) sympathetic innervation of IBAT is not required for OT to increase BAT thermogenesis and reduce body weight in female HFD-fed rats and (2) chronic hindbrain (4V) administration of OT reduces weight gain and adiposity in two different strains of female HFD-fed mice.

Author Contributions: Conceptualization, G.J.T.J. and J.E.B.; Methodology, A.D.D., A.J.H., M.K.H., R.V., J.D.S., M.G., E.T., T.W.-H., J.L.G., T.A.W., K.D.O., P.J.H., C.L.S., E.R.P., T.O.M. and J.E.B.; Formal analysis, T.A.W. and J.E.B.; Investigation, A.D.D., A.J.H., M.K.H., R.V., J.D.S., M.G., E.T., T.W.-H., J.L.G., T.A.W., P.J.H., C.L.S., E.R.P., G.J.T.J. and J.E.B.; Resources, A.D.D., K.D.O., E.R.P. and J.E.B.; Data curation, J.E.B.; Writing—original draft, J.E.B.; Writing—review & editing, A.D.D., A.J.H., M.K.H., R.V., J.D.S., M.G., E.T., T.W.-H., J.L.G., T.A.W., K.D.O., P.J.H., C.L.S., E.R.P., T.O.M. and G.J.T.J.; Supervision, A.D.D., M.K.H., P.J.H. and J.E.B.; Funding acquisition, J.E.B. All authors have read and agreed to the published version of the manuscript.

Funding: This material was based upon work supported by the Office of Research and Development, Medical Research Service, Department of Veterans Affairs (VA) and the VA Puget Sound Health Care System Rodent Metabolic Phenotyping Core and the Cellular and Molecular Imaging Core of the Diabetes Research Center at the University of Washington and supported by National Institutes of Health (NIH) grant P30DK017047. This work was also supported by the VA Merit Review Award 5 I01BX004102, from the United States (U.S.) Department of Veterans Affairs Biomedical Laboratory Research and Development Service and NIH 5R01DK115976 grant to James Blevins. PJH's research program also received research support during the project period from NIH/National Institute of Aging (NIA) grant AG085572, NIH grants DK-095980, HL-091333, HL-107256 and a multi-campus grant from the University of California Office of the President. The contents do not represent the views of the U.S. Department of Veterans Affairs or the United States Government.

Institutional Review Board Statement: The animal study protocols were approved by the VA Puget Sound Health Care System IAUC [Protocol #s 0928 (rat) and 0927 (mouse)]. The original approval dates for 0928 and 0927 were both approved on 10/5/2017.

Data Availability Statement: All relevant data is contained within the article: The original contributions presented in the study are included in the article. Further inquiries can be directed to the corresponding author.

Acknowledgments: The authors thank the technical support of Nishi Ivanov, Hailey Chadwick and Alex Vu.

Conflicts of Interest: Disclosures: JEB had a financial interest in OXT Therapeutics, Inc., a company developing highly specific and stable analogs of oxytocin to treat obesity and metabolic disease. The authors' interests were reviewed and are managed by their local institutions in accordance with their conflict-of-interest policies. The other authors have nothing to report.

References

1. Kosfeld, M.; Heinrichs, M.; Zak, P.J.; Fischbacher, U.; Fehr, E. Oxytocin increases trust in humans. *Nature* **2005**, *435*, 673–676. [CrossRef] [PubMed]
2. Striepens, N.; Kendrick, K.M.; Maier, W.; Hurlemann, R. Prosocial effects of oxytocin and clinical evidence for its therapeutic potential. *Front. Neuroendocrinol.* **2011**, *32*, 426–450. [CrossRef] [PubMed]
3. Gimpl, G.; Fahrenholz, F. The oxytocin receptor system: Structure, function, and regulation. *Physiol. Rev.* **2001**, *81*, 629–683. [CrossRef]
4. Veening, J.G.; de Jong, T.R.; Waldinger, M.D.; Korte, S.M.; Olivier, B. The role of oxytocin in male and female reproductive behavior. *Eur. J. Pharmacol.* **2015**, *753*, 209–228. [CrossRef]
5. Blevins, J.E.; Baskin, D.G. Translational and therapeutic potential of oxytocin as an anti-obesity strategy: Insights from rodents, nonhuman primates and humans. *Physiol. Behav.* **2015**, *152*, 438–449. [CrossRef] [PubMed]
6. Lawson, E.A. The effects of oxytocin on eating behaviour and metabolism in humans. *Nat. Rev. Endocrinol.* **2017**, *13*, 700–709. [CrossRef]
7. Lawson, E.A.; Olszewski, P.K.; Weller, A.; Blevins, J.E. The role of oxytocin in regulation of appetitive behaviour, body weight and glucose homeostasis. *J. Neuroendocrinol.* **2020**, *32*, e12805. [CrossRef] [PubMed]
8. McCormack, S.E.; Blevins, J.E.; Lawson, E.A. Metabolic Effects of Oxytocin. *Endocr. Rev.* **2020**, *41*, 121–145. [CrossRef]
9. Kublaoui, B.M.; Gemelli, T.; Tolson, K.P.; Wang, Y.; Zinn, A.R. Oxytocin deficiency mediates hyperphagic obesity of Sim1 haploinsufficient mice. *Mol. Endocrinol.* **2008**, *22*, 1723–1734. [CrossRef]
10. Liu, C.M.; Davis, E.A.; Suarez, A.N.; Wood, R.I.; Noble, E.E.; Kanoski, S.E. Sex Differences and Estrous Influences on Oxytocin Control of Food Intake. *Neuroscience* **2020**, *447*, 63–73. [CrossRef]
11. Altirriba, J.; Poher, A.L.; Rohner-Jeanrenaud, F. Chronic Oxytocin Administration as a Treatment Against Impaired Leptin Signaling or Leptin Resistance in Obesity. *Front. Endocrinol.* **2015**, *6*, 119. [CrossRef] [PubMed]

12. Deblon, N.; Veyrat-Durebex, C.; Bourgoin, L.; Caillon, A.; Bussier, A.L.; Petrosino, S.; Piscitelli, F.; Legros, J.J.; Geenen, V.; Foti, M.; et al. Mechanisms of the anti-obesity effects of oxytocin in diet-induced obese rats. *PLoS ONE* **2011**, *6*, e25565. [CrossRef] [PubMed]
13. Morton, G.J.; Thatcher, B.S.; Reidelberger, R.D.; Ogimoto, K.; Wolden-Hanson, T.; Baskin, D.G.; Schwartz, M.W.; Blevins, J.E. Peripheral oxytocin suppresses food intake and causes weight loss in diet-induced obese rats. *Am. J. Physiol.-Endocrinol. Metab.* **2012**, *302*, E134–E144. [CrossRef] [PubMed]
14. Blevins, J.E.; Graham, J.L.; Morton, G.J.; Bales, K.L.; Schwartz, M.W.; Baskin, D.G.; Havel, P.J. Chronic oxytocin administration inhibits food intake, increases energy expenditure, and produces weight loss in fructose-fed obese rhesus monkeys. *Am. J. Physiol. Regul. Integr. Comp. Physiol.* **2015**, *308*, R431–R438. [CrossRef]
15. Noble, E.E.; Billington, C.J.; Kotz, C.M.; Wang, C. Oxytocin in the ventromedial hypothalamic nucleus reduces feeding and acutely increases energy expenditure. *Am. J. Physiol. Regul. Integr. Comp. Physiol.* **2014**, *307*, R737–R745. [CrossRef]
16. Zhang, G.; Bai, H.; Zhang, H.; Dean, C.; Wu, Q.; Li, J.; Guariglia, S.; Meng, Q.; Cai, D. Neuropeptide exocytosis involving synaptotagmin-4 and oxytocin in hypothalamic programming of body weight and energy balance. *Neuron* **2011**, *69*, 523–535. [CrossRef]
17. Zhang, G.; Cai, D. Circadian intervention of obesity development via resting-stage feeding manipulation or oxytocin treatment. *Am. J. Physiol. Endocrinol. Metab.* **2011**, *301*, E1004–E1012. [CrossRef] [PubMed]
18. Yi, K.J.; So, K.H.; Hata, Y.; Suzuki, Y.; Kato, D.; Watanabe, K.; Aso, H.; Kasahara, Y.; Nishimori, K.; Chen, C.; et al. The regulation of oxytocin receptor gene expression during adipogenesis. *J. Neuroendocrinol.* **2015**, *27*, 335–342. [CrossRef] [PubMed]
19. Cannon, B.; Nedergaard, J. Brown adipose tissue: Function and physiological significance. *Physiol. Rev.* **2004**, *84*, 277–359. [CrossRef]
20. Morrison, S.F.; Madden, C.J.; Tupone, D. Central neural regulation of brown adipose tissue thermogenesis and energy expenditure. *Cell Metab.* **2014**, *19*, 741–756. [CrossRef]
21. Kotz, C.M.; Perez-Leighton, C.E.; Teske, J.A.; Billington, C.J. Spontaneous Physical Activity Defends Against Obesity. *Curr. Obes. Rep.* **2017**, *6*, 362–370. [CrossRef] [PubMed]
22. Periasamy, M.; Herrera, J.L.; Reis, F.C.G. Skeletal Muscle Thermogenesis and Its Role in Whole Body Energy Metabolism. *Diabetes Metab. J.* **2017**, *41*, 327–336. [CrossRef] [PubMed]
23. Conte, E.; Romano, A.; De Bellis, M.; de Ceglia, M.; Rosaria Carratù, M.; Gaetani, S.; Maqoud, F.; Tricarico, D.; Camerino, C. Oxtr/TRPV1 expression and acclimation of skeletal muscle to cold-stress in male mice. *J. Endocrinol.* **2021**, *249*, 135–148. [CrossRef]
24. Elabd, C.; Cousin, W.; Upadhyayula, P.; Chen, R.Y.; Chooljian, M.S.; Li, J.; Kung, S.; Jiang, K.P.; Conboy, I.M. Oxytocin is an age-specific circulating hormone that is necessary for muscle maintenance and regeneration. *Nat. Commun.* **2014**, *5*, 4082. [CrossRef]
25. Espinoza, S.E.; Lee, J.L.; Wang, C.P.; Ganapathy, V.; MacCarthy, D.; Pascucci, C.; Musi, N.; Volpi, E. Intranasal Oxytocin Improves Lean Muscle Mass and Lowers LDL Cholesterol in Older Adults with Sarcopenic Obesity: A Pilot Randomized Controlled Trial. *J. Am. Med. Dir. Assoc.* **2021**, *22*, 1877–1882. [CrossRef]
26. Stanic, S.; Bardova, K.; Janovska, P.; Rossmeisl, M.; Kopecky, J.; Zouhar, P. Prolonged FGF21 treatment increases energy expenditure and induces weight loss in obese mice independently of UCP1 and adrenergic signaling. *Biochem. Pharmacol.* **2024**, *221*, 116042. [CrossRef]
27. Clemmensen, C.; Jall, S.; Kleinert, M.; Quarta, C.; Gruber, T.; Reber, J.; Sachs, S.; Fischer, K.; Feuchtinger, A.; Karlas, A.; et al. Publisher Correction: Coordinated targeting of cold and nicotinic receptors synergistically improves obesity and type 2 diabetes. *Nat. Commun.* **2018**, *9*, 4975. [CrossRef] [PubMed]
28. Haynes, W.G.; Morgan, D.A.; Walsh, S.A.; Mark, A.L.; Sivitz, W.I. Receptor-mediated regional sympathetic nerve activation by leptin. *J. Clin. Investig.* **1997**, *100*, 270–278. [CrossRef]
29. Hernandez, A.; Obregon, M.J. Triiodothyronine amplifies the adrenergic stimulation of uncoupling protein expression in rat brown adipocytes. *Am. J. Physiol. Endocrinol. Metab.* **2000**, *278*, E769–E777. [CrossRef]
30. Li, Y.; Schnabl, K.; Gabler, S.M.; Willershauser, M.; Reber, J.; Karlas, A.; Laurila, S.; Lahesmaa, M.; Din, M.U.; Bast-Habersbrunner, A.; et al. Secretin-Activated Brown Fat Mediates Prandial Thermogenesis to Induce Satiation. *Cell* **2018**, *175*, 1561–1574.e12. [CrossRef]
31. Laurila, S.; Sun, L.; Lahesmaa, M.; Schnabl, K.; Laitinen, K.; Klen, R.; Li, Y.; Balaz, M.; Wolfrum, C.; Steiger, K.; et al. Secretin activates brown fat and induces satiation. *Nat. Metab.* **2021**, *3*, 798–809. [CrossRef]
32. Cereijo, R.; Villarroya, J.; Villarroya, F. Non-sympathetic control of brown adipose tissue. *Int. J. Obes. Supplments* **2015**, *5*, S40–S44. [CrossRef]
33. Lopez, M.; Alvarez, C.V.; Nogueiras, R.; Dieguez, C. Energy balance regulation by thyroid hormones at central level. *Trends Mol. Med.* **2013**, *19*, 418–427. [CrossRef]
34. Roberts, Z.S.; Wolden-Hanson, T.H.; Matsen, M.E.; Ryu, V.; Vaughan, C.H.; Graham, J.L.; Havel, P.J.; Chukri, D.W.; Schwartz, M.W.; Morton, G.J.; et al. Chronic Hindbrain Administration of Oxytocin is Sufficient to Elicit Weight Loss in Diet-Induced Obese Rats. *Am. J. Physiol. Regul. Integr. Comp. Physiol.* **2017**, *313*, R357–R371. [CrossRef] [PubMed]

35. Edwards, M.M.; Nguyen, H.K.; Dodson, A.D.; Herbertson, A.J.; Wolden-Hanson, T.; Wietecha, T.; Honeycutt, M.K.; Slattery, J.D.; O'Brien, K.D.; Graham, J.L.; et al. Sympathetic innervation of interscapular brown adipose tissue is not a predominant mediator of oxytocin-elicited reductions of body weight and adiposity in male diet-induced obese mice. *Front. Endocrinol.* **2024**, *15*, 1440070. [CrossRef] [PubMed]
36. Ong, Z.Y.; Bongiorno, D.M.; Hernando, M.A.; Grill, H.J. Effects of Endogenous Oxytocin Receptor Signaling in Nucleus Tractus Solitarius on Satiation-Mediated Feeding and Thermogenic Control in Male Rats. *Endocrinology* **2017**, *158*, 2826–2836. [CrossRef] [PubMed]
37. Sutton, A.K.; Pei, H.; Burnett, K.H.; Myers, M.G., Jr.; Rhodes, C.J.; Olson, D.P. Control of food intake and energy expenditure by Nos1 neurons of the paraventricular hypothalamus. *J. Neurosci.* **2014**, *34*, 15306–15318. [CrossRef] [PubMed]
38. Yuan, J.; Zhang, R.; Wu, R.; Gu, Y.; Lu, Y. The effects of oxytocin to rectify metabolic dysfunction in obese mice are associated with increased thermogenesis. *Mol. Cell Endocrinol.* **2020**, *514*, 110903. [CrossRef]
39. Kasahara, Y.; Sato, K.; Takayanagi, Y.; Mizukami, H.; Ozawa, K.; Hidema, S.; So, K.H.; Kawada, T.; Inoue, N.; Ikeda, I.; et al. Oxytocin receptor in the hypothalamus is sufficient to rescue normal thermoregulatory function in male oxytocin receptor knockout mice. *Endocrinology* **2013**, *154*, 4305–4315. [CrossRef]
40. Kasahara, Y.; Tateishi, Y.; Hiraoka, Y.; Otsuka, A.; Mizukami, H.; Ozawa, K.; Sato, K.; Hidema, S.; Nishimori, K. Role of the Oxytocin Receptor Expressed in the Rostral Medullary Raphe in Thermoregulation During Cold Conditions. *Front. Endocrinol.* **2015**, *6*, 180. [CrossRef]
41. Harshaw, C.; Leffel, J.K.; Alberts, J.R. Oxytocin and the warm outer glow: Thermoregulatory deficits cause huddling abnormalities in oxytocin-deficient mouse pups. *Horm. Behav.* **2018**, *98*, 145–158. [CrossRef]
42. Xi, D.; Long, C.; Lai, M.; Casella, A.; O'Lear, L.; Kublaoui, B.; Roizen, J.D. Ablation of Oxytocin Neurons Causes a Deficit in Cold Stress Response. *J. Endocr. Soc.* **2017**, *1*, 1041–1055. [CrossRef]
43. Wu, Z.; Xu, Y.; Zhu, Y.; Sutton, A.K.; Zhao, R.; Lowell, B.B.; Olson, D.P.; Tong, Q. An obligate role of oxytocin neurons in diet induced energy expenditure. *PLoS ONE* **2012**, *7*, e45167. [CrossRef]
44. Camerino, C. Low sympathetic tone and obese phenotype in oxytocin-deficient mice. *Obesity* **2009**, *17*, 980–984. [CrossRef] [PubMed]
45. Takayanagi, Y.; Kasahara, Y.; Onaka, T.; Takahashi, N.; Kawada, T.; Nishimori, K. Oxytocin receptor-deficient mice developed late-onset obesity. *Neuroreport* **2008**, *19*, 951–955. [CrossRef] [PubMed]
46. Anekonda, V.T.; Thompson, B.W.; Ho, J.M.; Roberts, Z.S.; Edwards, M.M.; Nguyen, H.K.; Dodson, A.D.; Wolden-Hanson, T.; Chukri, D.W.; Herbertson, A.J.; et al. Hindbrain Administration of Oxytocin Reduces Food Intake, Weight Gain and Activates Catecholamine Neurons in the Hindbrain Nucleus of the Solitary Tract in Rats. *J. Clin. Med.* **2021**, *10*, 5078. [CrossRef] [PubMed]
47. Blevins, J.E.; Schwartz, M.W.; Baskin, D.G. Evidence that paraventricular nucleus oxytocin neurons link hypothalamic leptin action to caudal brain stem nuclei controlling meal size. *Am. J. Physiol. Regul. Integr. Comp. Physiol.* **2004**, *287*, R87–R96. [CrossRef] [PubMed]
48. Blevins, J.E.; Thompson, B.W.; Anekonda, V.T.; Ho, J.M.; Graham, J.L.; Roberts, Z.S.; Hwang, B.H.; Ogimoto, K.; Wolden-hanson, T.H.; Nelson, J.O.; et al. Chronic CNS oxytocin signaling preferentially induces fat loss in high fat diet-fed rats by enhancing satiety responses and increasing lipid utilization. *Am. J. Physiol. Integr. Comp. Physiol.* **2016**, *310*, R640–R658. [CrossRef]
49. Morton, G.J.; Matsen, M.E.; Bracy, D.P.; Meek, T.H.; Nguyen, H.T.; Stefanovski, D.; Bergman, R.N.; Wasserman, D.H.; Schwartz, M.W. FGF19 action in the brain induces insulin-independent glucose lowering. *J. Clin. Investig.* **2013**, *123*, 4799–4808. [CrossRef]
50. Paxinos, G.; Watson, C. (Eds.) *The Rat Brain in Stereotaxic Coordinates*; Academic Press: Burlington, MA, USA, 2007.
51. Dorfman, M.D.; Krull, J.E.; Douglass, J.D.; Fasnacht, R.; Lara-Lince, F.; Meek, T.H.; Shi, X.; Damian, V.; Nguyen, H.T.; Matsen, M.E.; et al. Sex differences in microglial CX3CR1 signalling determine obesity susceptibility in mice. *Nat. Commun.* **2017**, *8*, 14556. [CrossRef]
52. Edwards, M.M.; Nguyen, H.K.; Herbertson, A.J.; Dodson, A.D.; Wietecha, T.; Wolden-Hanson, T.; Graham, J.L.; O'Brien, K.D.; Havel, P.J.; Blevins, J.E. Chronic Hindbrain Administration of Oxytocin Elicits Weight Loss in Male Diet-Induced Obese Mice. *Am. J. Physiol. Regul. Integr. Comp. Physiol.* **2021**, *320*, R471–R487. [CrossRef]
53. Paxinos, G.; Franklin, K.B.J. (Eds.) *The Mouse Brain in Stereotaxic Coordinates*; Academic Press: San Diego, CA, USA, 2001.
54. Edwards, M.M.; Nguyen, H.K.; Dodson, A.D.; Herbertson, A.J.; Wietecha, T.A.; Wolden-Hanson, T.; Graham, J.L.; Honeycutt, M.K.; Slattery, J.D.; O'Brien, K.D.; et al. Effects of combined oxytocin and beta-3 receptor agonist (CL 316243) treatment on body weight and adiposity in male diet-induced obese rats. *Front. Physiol.* **2021**, *12*, 725912. [CrossRef]
55. Brito, M.N.; Brito, N.A.; Baro, D.J.; Song, C.K.; Bartness, T.J. Differential activation of the sympathetic innervation of adipose tissues by melanocortin receptor stimulation. *Endocrinology* **2007**, *148*, 5339–5347. [CrossRef]
56. Vaughan, C.H.; Shrestha, Y.B.; Bartness, T.J. Characterization of a novel melanocortin receptor-containing node in the SNS outflow circuitry to brown adipose tissue involved in thermogenesis. *Brain Res.* **2011**, *1411*, 17–27. [CrossRef]
57. Depocas, F.; Foster, D.O.; Zaror-Behrens, G.; Lacelle, S.; Nadeau, B. Recovery of function in sympathetic nerves of interscapular brown adipose tissue of rats treated with 6-hydroxydopamine. *Can. J. Physiol. Pharmacol.* **1984**, *62*, 1327–1332. [CrossRef]

58. Wang, L.Y.; Murphy, R.R.; Hanscom, B.; Li, G.; Millard, S.P.; Petrie, E.C.; Galasko, D.R.; Sikkema, C.; Raskind, M.A.; Wilkinson, C.W.; et al. Cerebrospinal fluid norepinephrine and cognition in subjects across the adult age span. *Neurobiol. Aging* **2013**, *34*, 2287–2292. [CrossRef]
59. Williams, D.L.; Bowers, R.R.; Bartness, T.J.; Kaplan, J.M.; Grill, H.J. Brainstem melanocortin 3/4 receptor stimulation increases uncoupling protein gene expression in brown fat. *Endocrinology* **2003**, *144*, 4692–4697. [CrossRef]
60. Product Data-DIO Series Diets. Available online: https://researchdiets.com/system/refinery/resources/W1siZiIsIjIwMTkvMTEvMjIvOWFnMmFodzBlMV9SRElfcHJvZHVjdF9kYXRhX2Rpb19zZXJpZXNfZGlldHNfd2ViNC5wZGYiXV0/RDI_product_data_dio_series%20diets_web4.pdf (accessed on 30 September 2024).
61. Suarez, J.; Rivera, P.; Arrabal, S.; Crespillo, A.; Serrano, A.; Baixeras, E.; Pavon, F.J.; Cifuentes, M.; Nogueiras, R.; Ballesteros, J.; et al. Oleoylethanolamide enhances beta-adrenergic-mediated thermogenesis and white-to-brown adipocyte phenotype in epididymal white adipose tissue in rat. *Dis. Models Mech.* **2014**, *7*, 129–141.
62. De Boer, S.F.; Van der Gugten, J. Daily Variations in Plasma Noradrenaline, Adrenaline and Corticosterone Concentrations in Rats. *Physiol. Behav.* **1987**, *40*, 323–328. [CrossRef]
63. Davidovic, V.; Petrovic, V.M. Diurnal variations in the catecholamine content in rat tissues Effects of exogenous noradrenaline. *Arch. Int. Physiol. Biochim.* **1981**, *89*, 457–460. [CrossRef]
64. Bremer, A.A.; Stanhope, K.L.; Graham, J.L.; Cummings, B.P.; Wang, W.; Saville, B.R.; Havel, P.J. Fructose-fed rhesus monkeys: A nonhuman primate model of insulin resistance, metabolic syndrome, and type 2 diabetes. *Clin. Transl. Sci.* **2011**, *4*, 243–252. [CrossRef]
65. Blevins, J.E.; Moralejo, D.H.; Wolden-Hanson, T.H.; Thatcher, B.S.; Ho, J.M.; Kaiyala, K.J.; Matsumoto, K. Alterations in activity and energy expenditure contribute to lean phenotype in Fischer 344 rats lacking the cholecystokinin-1 receptor gene. *Am. J. Physiol. Regul. Integr. Comp. Physiol.* **2012**, *303*, R1231–R1240. [CrossRef]
66. Cummings, B.P.; Digitale, E.K.; Stanhope, K.L.; Graham, J.L.; Baskin, D.G.; Reed, B.J.; Sweet, I.R.; Griffen, S.C.; Havel, P.J. Development and characterization of a novel rat model of type 2 diabetes mellitus: The UC Davis type 2 diabetes mellitus UCD-T2DM rat. *Am. J. Physiol. Regul. Integr. Comp. Physiol.* **2008**, *295*, R1782–R1793. [CrossRef]
67. Livak, K.J.; Schmittgen, T.D. Analysis of relative gene expression data using real-time quantitative PCR and the 2(-Delta Delta C(T)) Method. *Methods* **2001**, *25*, 402–408. [CrossRef]
68. Rinaman, L.; Rothe, E.E. GLP-1 receptor signaling contributes to anorexigenic effect of centrally administered oxytocin in rats. *Am. J. Physiol. Regul. Integr. Comp. Physiol.* **2002**, *283*, R99–R106. [CrossRef]
69. Tortoriello, D.V.; McMinn, J.; Chua, S.C. Dietary-induced obesity and hypothalamic infertility in female DBA/2J mice. *Endocrinology* **2004**, *145*, 1238–1247. [CrossRef]
70. Totten, M.S.; Wallace, C.W.; Pierce, D.M.; Fordahl, S.C.; Erikson, K.M. The impact of a high-fat diet on physical activity and dopamine neurochemistry in the striatum is sex and strain dependent in C57BL/6J and DBA/2J mice. *Nutr. Neurosci.* **2022**, *25*, 2601–2615. [CrossRef]
71. Susulic, V.S.; Frederich, R.C.; Lawitts, J.; Tozzo, E.; Kahn, B.B.; Harper, M.E.; Himms-Hagen, J.; Flier, J.S.; Lowell, B.B. Targeted disruption of the beta 3-adrenergic receptor gene. *J. Biol. Chem.* **1995**, *270*, 29483–29492. [CrossRef]
72. Rothwell, N.J.; Stock, M.J.; Sudera, D.K. Beta-adrenoreceptors in rat brown adipose tissue: Proportions of beta 1- and beta 2-subtypes. *Am. J. Physiol.* **1985**, *248*, E397–E402. [CrossRef]
73. Levin, B.E.; Sullivan, A.C. Beta-1 receptor is the predominant beta-adrenoreceptor on rat brown adipose tissue. *J. Pharmacol. Exp. Ther.* **1986**, *236*, 681–688.
74. Ueta, C.B.; Fernandes, G.W.; Capelo, L.P.; Fonseca, T.L.; Maculan, F.D.; Gouveia, C.H.; Brum, P.C.; Christoffolete, M.A.; Aoki, M.S.; Lancellotti, C.L.; et al. beta(1) Adrenergic receptor is key to cold- and diet-induced thermogenesis in mice. *J. Endocrinol.* **2012**, *214*, 359–365. [CrossRef]
75. Atgie, C.; D'Allaire, F.; Bukowiecki, L.J. Role of beta1- and beta3-adrenoceptors in the regulation of lipolysis and thermogenesis in rat brown adipocytes. *Am. J. Physiol.* **1997**, *273*, C1136–C1142. [CrossRef]
76. Straat, M.E.; Hoekx, C.A.; van Velden, F.H.P.; Arias-Bouda, L.M.P.; Dumont, L.; Blondin, D.P.; Boon, M.R.; Martinez-Tellez, B.; Rensen, P.C.N. Stimulation of the beta-2-adrenergic receptor with salbutamol activates human brown adipose tissue. *Cell Rep. Med.* **2023**, *4*, 100942. [CrossRef]
77. Blondin, D.P.; Nielsen, S.; Kuipers, E.N.; Severinsen, M.C.; Jensen, V.H.; Miard, S.; Jespersen, N.Z.; Kooijman, S.; Boon, M.R.; Fortin, M.; et al. Human Brown Adipocyte Thermogenesis Is Driven by beta2-AR Stimulation. *Cell Metab.* **2020**, *32*, 287–300.e7. [CrossRef]
78. Ishida, Y.; Matsushita, M.; Yoneshiro, T.; Saito, M.; Fuse, S.; Hamaoka, T.; Kuroiwa, M.; Tanaka, R.; Kurosawa, Y.; Nishimura, T.; et al. Genetic evidence for involvement of beta2-adrenergic receptor in brown adipose tissue thermogenesis in humans. *Int. J. Obes.* **2024**, *48*, 1110–1117. [CrossRef]
79. Tate, K.M.; Briend-Sutren, M.M.; Emorine, L.J.; Delavier-Klutchko, C.; Marullo, S.; Strosberg, A.D. Expression of three human beta-adrenergic-receptor subtypes in transfected Chinese hamster ovary cells. *Eur. J. Biochem./FEBS* **1991**, *196*, 357–361. [CrossRef]
80. Reed, N.; Fain, J.N. Stimulation of respiration in brown fat cells by epinephrine, dibutyryl-3′,5′-adenosine monophosphate, and m-chloro(carbonyl cyanide)phenylhydrazone. *J. Biol. Chem.* **1968**, *243*, 2843–2848. [CrossRef]
81. Strack, A.M.; Sawyer, W.B.; Platt, K.B.; Loewy, A.D. CNS cell groups regulating the sympathetic outflow to adrenal gland as revealed by transneuronal cell body labeling with pseudorabies virus. *Brain Res.* **1989**, *491*, 274–296. [CrossRef]

82. Dum, R.P.; Levinthal, D.J.; Strick, P.L. The mind-body problem: Circuits that link the cerebral cortex to the adrenal medulla. *Proc. Natl. Acad. Sci. USA* **2019**, *116*, 26321–26328. [CrossRef]
83. de Jong, J.M.; Larsson, O.; Cannon, B.; Nedergaard, J. A stringent validation of mouse adipose tissue identity markers. *Am. J. Physiol. Endocrinol. Metab.* **2015**, *308*, E1085–E1105. [CrossRef]
84. Kalinovich, A.V.; de Jong, J.M.; Cannon, B.; Nedergaard, J. UCP1 in adipose tissues: Two steps to full browning. *Biochimie* **2017**, *134*, 127–137. [CrossRef]
85. Nguyen, N.L.; Barr, C.L.; Ryu, V.; Cao, Q.; Xue, B.; Bartness, T.J. Separate and shared sympathetic outflow to white and brown fat coordinately regulate thermoregulation and beige adipocyte recruitment. *Am. J. Physiol. Regul. Integr. Comp. Physiol.* **2016**, *312*, R132–R145. [CrossRef]
86. Doslikova, B.; Tchir, D.; McKinty, A.; Zhu, X.; Marks, D.L.; Baracos, V.E.; Colmers, W.F. Convergent neuronal projections from paraventricular nucleus, parabrachial nucleus, and brainstem onto gastrocnemius muscle, white and brown adipose tissue in male rats. *J. Comp. Neurol.* **2019**, *527*, 2826–2842. [CrossRef]
87. Shi, H.; Bartness, T.J. Neurochemical phenotype of sympathetic nervous system outflow from brain to white fat. *Brain Res. Bull.* **2001**, *54*, 375–385. [CrossRef]
88. Rinaman, L. Oxytocinergic inputs to the nucleus of the solitary tract and dorsal motor nucleus of the vagus in neonatal rats. *J. Comp. Neurol.* **1998**, *399*, 101–109. [CrossRef]
89. Sawchenko, P.E.; Swanson, L.W. Immunohistochemical identification of neurons in the paraventricular nucleus of the hypothalamus that project to the medulla or to the spinal cord in the rat. *J. Comp. Neurol.* **1982**, *205*, 260–272. [CrossRef]
90. Bi, S. Dorsomedial hypothalamic NPY modulation of adiposity and thermogenesis. *Physiol. Behav.* **2013**, *121*, 56–60. [CrossRef]
91. Sakamoto, T.; Sugimoto, S.; Uekita, T. Effects of intraperitoneal and intracerebroventricular injections of oxytocin on social and emotional behaviors in pubertal male mice. *Physiol. Behav.* **2019**, *212*, 112701. [CrossRef]
92. Fischer, A.W.; Schlein, C.; Cannon, B.; Heeren, J.; Nedergaard, J. Intact innervation is essential for diet-induced recruitment of brown adipose tissue. *Am. J. Physiol. Endocrinol. Metab.* **2019**, *316*, E487–E503. [CrossRef]
93. Bal, N.C.; Maurya, S.K.; Singh, S.; Wehrens, X.H.; Periasamy, M. Increased Reliance on Muscle-based Thermogenesis upon Acute Minimization of Brown Adipose Tissue Function. *J. Biol. Chem.* **2016**, *291*, 17247–17257. [CrossRef]
94. Bartness, T.J.; Vaughan, C.H.; Song, C.K. Sympathetic and sensory innervation of brown adipose tissue. *Int. J. Obes.* **2010**, *34* (Suppl. S1), S36–S42. [CrossRef]
95. Hicks, C.; Ramos, L.; Reekie, T.; Misagh, G.H.; Narlawar, R.; Kassiou, M.; McGregor, I.S. Body temperature and cardiac changes induced by peripherally administered oxytocin, vasopressin and the non-peptide oxytocin receptor agonist WAY 267,464: A biotelemetry study in rats. *Br. J. Pharmacol.* **2014**, *171*, 2868–2887. [CrossRef]
96. Kohli, S.; King, M.V.; Williams, S.; Edwards, A.; Ballard, T.M.; Steward, L.J.; Alberati, D.; Fone, K.C.F. Oxytocin attenuates phencyclidine hyperactivity and increases social interaction and nucleus accumben dopamine release in rats. *Neuropsychopharmacol. Off. Publ. Am. Coll. Neuropsychopharmacol.* **2019**, *44*, 295–305. [CrossRef]
97. Maejima, Y.; Aoyama, M.; Sakamoto, K.; Jojima, T.; Aso, Y.; Takasu, K.; Takenosihita, S.; Shimomura, K. Impact of sex, fat distribution and initial body weight on oxytocin's body weight regulation. *Sci. Rep.* **2017**, *7*, 8599. [CrossRef] [PubMed]

current issues in
molecular biology

MDPI

Article

Oxytocin Exhibits Neuroprotective Effects on Hippocampal Cultures under Severe Oxygen–Glucose Deprivation Conditions

Mara Ioana Ionescu [1,†], Ioana-Florentina Grigoras [1,2,†], Rosana-Bristena Ionescu [1,3,4], Diana Maria Chitimus [1], Robert Mihai Haret [1,5], Bogdan Ianosi [1,6], Mihai Ceanga [1,7,*] and Ana-Maria Zagrean [1,*]

1 Department of Functional Sciences, Division of Physiology II-Neuroscience, Carol Davila University of Medicine and Pharmacy, 050474 Bucharest, Romania; mara-ioana.iesanu@umfcd.ro (M.I.I.); ioana.grigoras@ndcn.ox.ac.uk (I.-F.G.); rbi24@cam.ac.uk (R.-B.I.); diana.chitimus@rez.umfcd.ro (D.M.C.); robertharet@gmail.com (R.M.H.); ianosi.bogdan@gmail.com (B.I.)
2 Wellcome Centre for Integrative Neuroimaging, Functional MRI of the Brain, Nuffield Department of Clinical Neurosciences, University of Oxford, Oxford OX3 9DU, UK
3 Department of Clinical Neurosciences, University of Cambridge, Cambridge CB2 1TN, UK
4 NIHR Biomedical Research Centre, University of Cambridge, Cambridge CB2 0QQ, UK
5 Department of Ophthalmology, University Medical Center Gottingen, 37075 Gottingen, Germany
6 Department of Neurology, Stroke Unit, Neuromed Campus, Kepler University Hospital, 4020 Linz, Austria
7 Section of Translational Neuroimmunology, Department of Neurology, Jena University Hospital, 07747 Jena, Germany
* Correspondence: mihai.ceanga@med.uni-jena.de (M.C.); ana-maria.zagrean@umfcd.ro (A.-M.Z.)
† These authors contributed equally to this work.

Abstract: Perinatal asphyxia (PA) and hypoxic-ischemic encephalopathy can result in severe, long-lasting neurological deficits. In vitro models, such as oxygen–glucose deprivation (OGD), are used experimentally to investigate neuronal response to metabolic stress. However, multiple variables can affect the severity level of OGD/PA and may confound any measured treatment effect. Oxytocin (OXT) has emerged as a potential neuroprotective agent against the deleterious effects of PA. Previous studies have demonstrated OXT's potential to enhance neuronal survival in immature hippocampal cultures exposed to OGD, possibly by modulating gamma-aminobutyric acid-A receptor activity. Moreover, OXT's precise impact on developing hippocampal neurons under different severities of OGD/PA remains uncertain. In this study, we investigated the effects of OXT (0.1 μM and 1 μM) on 7-day-old primary rat hippocampal cultures subjected to 2 h OGD/sham normoxic conditions. Cell culture viability was determined using the resazurin assay. Our results indicate that the efficacy of 1 μM OXT treatment varied according to the severity of the OGD-induced lesion, exhibiting a protective effect ($p = 0.022$) only when cellular viability dropped below 49.41% in non-treated OGD cultures compared to normoxic ones. Furthermore, administration of 0.1 μM OXT did not yield significant effects, irrespective of lesion severity ($p > 0.05$). These findings suggest that 1 μM OXT treatment during OGD confers neuroprotection exclusively in severe lesions in hippocampal neurons after 7 days in vitro. Further research is warranted to elucidate the mechanisms involved in OXT-mediated neuroprotection.

Keywords: perinatal asphyxia; hypoxic-ischemic encephalopathy; oxytocin; oxygen–glucose deprivation; hippocampal cell cultures; GABA

1. Introduction

During the peripartum phase, the fetus is at risk of perinatal asphyxia (PA), which is associated with elevated mortality and morbidity rates in infants. PA can culminate in severe systemic and neurological manifestations, with neonatal hypoxic-ischemic encephalopathy (HIE) notably impacting the vulnerable and hypoxia-prone developing brain [1]. Hence, there is a pressing necessity to find new therapeutic modalities for addressing this pathology.

Oxytocin (OXT) is a hypothalamic neurohormone that is secreted by the posterior pituitary gland and has an important role in promoting birth, lactation, and mother-infant bonding [2–4]. During pregnancy and labor, OXT is secreted in pulses, which increase in frequency, duration, and amplitude throughout labor, leading to a 3–4 times increase in the OXT plasma levels at birth [5]. Importantly, OXT crosses the placenta and the neonatal blood-brain barrier [6], reaching effective concentrations in the neonatal central nervous system during labor [7]. Intrapartum OXT exposure deficiency has been correlated with the extent of neonatal brain injury, suggesting that OXT is a promising therapeutic avenue in HIE [8]. Increasing evidence underscores the central role of the maternal neurohormone OXT in facilitating neuroprotection against brain damage induced by PA. OXT may protect fetal neurons through the regulation of microglial activation [9,10] and modulation of brain excitability, facilitating an excitatory-to-inhibitory gamma-aminobutyric acid (GABA) switch [6,11–13] during childbirth.

GABA exerts a pivotal role in the central nervous system, being the major inhibitory neurotransmitter in the adult mammalian brain [14]. $GABA_A$ receptors ($GABA_ARs$) are chloride (Cl^-) channels that hyperpolarize mature neurons, counterbalancing excitatory input [15]. However, during the fetal period, GABA elicits a depolarizing response as a consequence of the high intracellular Cl^- concentration expressed in immature neurons [16,17], transiently providing the major excitatory input [18,19].

The modulation of GABA-mediated inhibition is crucial for neuronal development and function, with intracellular Cl^- concentration playing an essential role [20]. During early development, the Na-K-Cl cotransporter 1 (NKCC1) predominates, leading to Cl^- accumulation and excitatory GABAergic signaling [21]. However, as the K-Cl cotransporter 2 (KCC2) expression increases before birth, there is a reduction in intracellular Cl^- levels, facilitating an inhibitory GABAergic switch in mature neurons [6,22]. OXT influences Cl^- homeostasis by stimulating KCC2 phosphorylation, enhancing its activity, and promoting Cl^- extrusion [23,24]. This modulation, mediated by OXT receptors (OXTRs), contributes to the GABA switch during neuronal development [6,23]. Nonetheless, OXT's effects are temporally restricted, showing efficacy only during early postnatal stages [23]. Additionally, disruption of OXTR expression leads to disturbances in KCC2 upregulation and delays in the GABA switch, underscoring the significance of OXT in neuronal maturation and function [23].

In vitro experiments have extensively utilized oxygen–glucose deprivation (OGD) to study ischemic neuronal damage in both cell cultures and acute tissue slices [25]. OGD studies usually assess the cellular viability of the neurons immediately after the exposure or after a post-OGD reperfusion-like period to report the injury severity [26,27]. The post-OGD cellular viability results are difficult to compare between studies, given the variations in the exposure's actual duration and hypoxia level and the use of different cell viability assays. Thus, it is important to carefully evaluate the severity of the lesion when interpreting findings from OGD studies.

We have previously reported that modulation of Cl^- transport by OXT has a neuroprotective effect when administered to early-stage hippocampal cultures at 2 days in vitro (DIV) during OGD [11], as well as perinatally in vivo [28]. However, OXT's potential neuroprotective effect on neuronal cultures at later stages of maturation remains uncertain. As described above, immature and maturing neurons differ significantly in their intracellular Cl^- concentration and subsequent reaction to fast GABAergic neurotransmission, with OXT exerting its neuroprotective effect during the first 5 DIV [23]. Therefore, it stands to reason that OXT could have a different effect on cellular viability after OGD at DIV7 when compared to DIV2 neurons. This could be of translational significance since rodents at postnatal days 7–10 represent an equivalent developmental stage to term infants [29].

Thus, in the present study, we aim to investigate the effect of OXT at different concentrations on the viability of DIV7 rat hippocampal cell cultures subjected to OGD and reoxygenation and how this effect changes with lesion severity.

2. Materials and Methods

2.1. Primary Cultures of Rat Hippocampal Neurons

All animal procedures were carried out on Wistar rats, with the approval of the local ethics committee for animal research following the European Communities Council Directive 2010/63/EU on the protection of animals used for scientific purposes. Primary hippocampal cell cultures were prepared from postnatal day zero (P0) Wistar rat pups using a previously described protocol [11,30,31].

The hippocampi were isolated from the meninges and vascular plexuses under a dissecting microscope. Next, the dissociated cells were obtained by mechanical trituration and a 20 min incubation in papain (Worthington, 3.8 U/mL, Lakewood, CA, USA) for further enzymatic dissociation. The resulting cell suspension was kept in a culture medium (CM) containing Neurobasal-A (Gibco, Thermo Fischer Scientific, Waltham, MA, USA) supplemented with B-27 (Invitrogen, Cat. No. 17504044, Carlsbad, CA, USA), l-glutamine (HyClone, 0.5 mM, Washington, DC, USA), and antibiotic-antimycotic (Invitrogen, Cat. No. 15240062). The cells were then plated on poly-D-lysine-coated (Sigma-Aldrich, 70,000–150,000 kDa, 100 μg/mL, Burlington, MA, USA) 24-well plates at a cell density of 150,000 cells/well. Cultures were maintained at 37 °C in a 5% CO_2, humidified atmosphere. Following a 5-day incubation period, half of the CM volume in each well (250 μL) was removed and replaced with 400 μL freshly prepared CM.

This protocol enables the cultivation of relatively pure hippocampal cultures characterized by a glial cell population of less than 5%, as documented in previous studies [32,33].

2.2. Exposure to Oxygen–Glucose Deprivation

OGD was performed at DIV7 as previously described [31,34]. We employed hippocampal cultures at DIV7 as this age corresponds to the developmental stage typically studied in rodent models of HIE [35,36]. These models traditionally use rodents aged between postnatal days 7 and 10, which approximate the developmental stage of term human infants, historically established through postmortem brain weight measurements [37].

CM was removed from the wells and replaced with a deoxygenated experimental medium without glucose (EM-G), consisting of Neurobasal-A lacking both glucose and sodium pyruvate (Invitrogen), supplemented with HEPES (Sigma-Aldrich, 10 mM) and l-glutamine (HyClone, 0.5 mM). The plates were then promptly moved to a modular chamber (Billups-Rothenberg, San Diego, CA, USA), which was perfused with 100% N_2 for 10 min, then hermetically sealed and kept in an incubator at 37 °C for 2 h. Control cultures were kept in an experimental medium consisting of Neurobasal-A with 25 mM glucose (EM + G) and sham normoxic conditions for 2 h. The experimental design is illustrated in Figure 1.

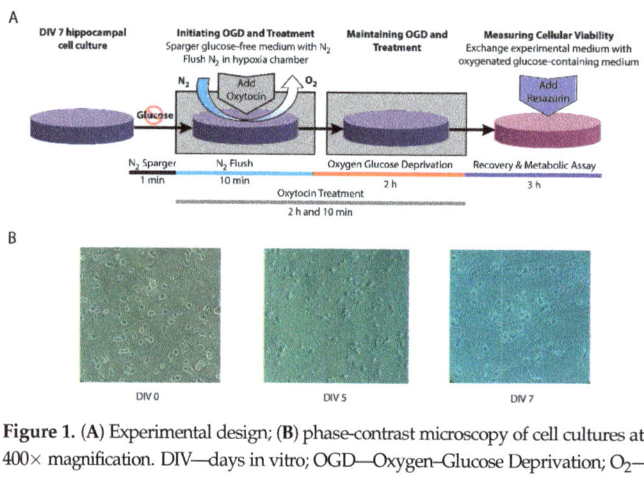

Figure 1. (**A**) Experimental design; (**B**) phase-contrast microscopy of cell cultures at DIV0, DIV5, and DIV7, 400× magnification. DIV—days in vitro; OGD—Oxygen–Glucose Deprivation; O_2—oxygen; N_2—nitrogen.

2.3. Treatment with Oxytocin

OXT treatments were applied during the 2 h exposure to OGD/sham normoxic conditions. Consistent with literature precedents, the DIV7 hippocampal cultures were treated with two different OXT concentrations previously employed in studies: 0.1 μM [23, 38] and 1 μM [39,40]. Prior investigations have demonstrated a dose-dependent effect of OXT, with efficacy observed starting at 0.1 μM and reaching its peak at 1 μM [36].

2.4. Assessment of Cellular Metabolism and Viability

Post-exposure, the reoxygenated cell cultures were incubated at 37 °C and 5% CO_2 for a 3-h time frame in a resazurin-containing medium to investigate their metabolic viability. Resazurin (Sigma-Aldrich) is a cell-permeable redox indicator that is irreversibly reduced in the mitochondria of viable cells to resorufin [11,31]. Thus, for the resazurin assay, EM-/+G was replaced with fresh assessment medium containing Neurobasal-A without phenol red (Gibco, Thermo Fischer Scientific), supplemented with HEPES (10 mM), l-glutamine (HyClone, 0.5 mM), and resazurin (Sigma-Aldrich, 100 μM). Resorufin fluorescence intensity was read at baseline and after a 3-h reoxygenation period at 535 nm excitation and 595 nm emission using a multimode detector (DTX880, Beckman Coulter, Brea, CA, USA).

2.5. Data Analysis

Data analysis and figures were rendered using GraphPad Prism 7.00 (GraphPad Software Inc. 9.3.0). The cellular viability for each condition was calculated as percentages compared to control values (normoxia + no treatment condition), considered 100%. The statistical unit was the number of culture plates tested per condition (n). All variables were tested for normality of distribution using the Shapiro–Wilk test, and central tendencies are reported as mean ± standard deviation (SD) unless otherwise stated. For normoxic data, statistical significance was evaluated using ordinary one-way ANOVA with a main factor of treatment (no treatment, 0.1 μM OXT, 1 μM OXT). For OGD data, statistical significance was evaluated using linear regression and paired t-tests. The significance threshold was set at $\alpha = 0.05$.

3. Results

3.1. Oxytocin Did Not Affect the Viability of DIV7 Hippocampal Cell Cultures under Normoxic Conditions

The effect of OXT treatment on DIV7 hippocampal cell cultures under normoxic conditions was measured using ordinary one-way ANOVA with a main factor of treatment (no treatment, 0.1 μM OXT, 1 μM OXT). The effect of OXT treatment on the DIV7 cell cultures was not statistically significant ($F(2, 23) = 1.289$, $p = 0.294$; Table 1).

Table 1. Mean cellular viability of DIV7 hippocampal neurons after OXT treatment under normoxic conditions.

	Mean Viability	SD	Number of Plates	Total Number of Wells	*p* Value Paired *t*-Test
Non-treated normoxia	100%	0	17	30	
0.1 μM OXT	101.0%	5.532%	7	28	0.605
1 μM OXT	104.3%	0.870%	2	7	0.341

DIV—days in vitro; SD—standard deviation; OXT—oxytocin.

3.2. Cellular Viability after 2 h of OGD Followed a Normal Distribution

The Shapiro–Wilk test was used to check for normal distribution of cellular viability following 2 h of OGD exposure with no treatment. The viabilities were normally distributed ($p = 0.823$, mean = 49.98%, SD = 12.72%, $n = 21$; Figure 2).

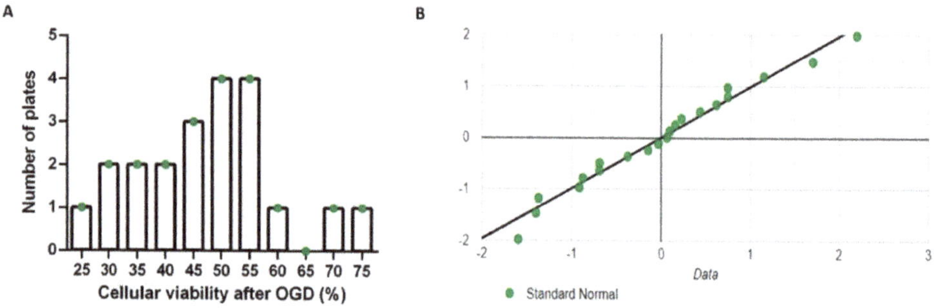

Figure 2. Cellular viability after 2 h of oxygen–glucose deprivation (OGD) follows a normal distribution. (**A**) Histogram of cellular viability (% of normoxic control condition) after 2 h of OGD exposure; (**B**) quantile–quantile plot.

3.3. Oxytocin's Effect on Cellular Viability Changed with OGD Severity

For each plate exposed to OGD, we calculated the mean viability of non-treated OGD wells (NT-OGD) and the mean viability of the corresponding OXT-treated wells. We then calculated the percentage change in viability in OXT-treated wells compared to the non-treated ones as (OXT-NT)/NT. We can thus report plate-wise the effect of OXT treatment with respect to the OGD severity of the NT-wells. When cellular viability is higher in OXT-treated wells than in non-treated wells, this percentage change has a positive value, which would indicate neuroprotection; when cellular viability is lower in OXT-treated wells than in non-treated wells, this percentage change would have a negative value (Figure 3A).

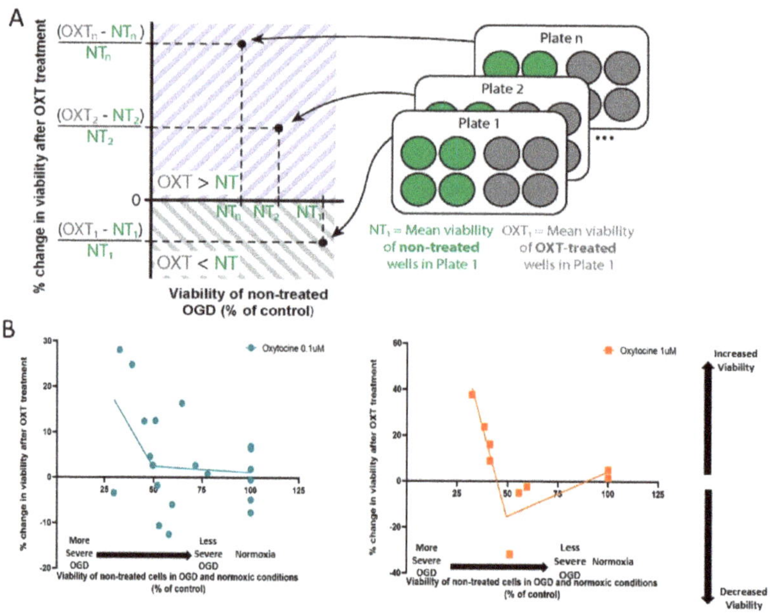

Figure 3. Oxytocin's effect on cellular viability in relation to non-treated OGD viability, as measured by means of resazurin assay. (**A**) Schematic of data analysis and graphical representation of results. (**B**) Segmented regression of change in viability due to OXT treatment in non-treated cells in OGD and normoxic conditions: left—0.1 μM OXT, right—1 μM OXT. OGD—Oxygen–Glucose Deprivation; OXT—oxytocin.

We analyzed the relationship between the percentage of change in viability due to OXT treatment and OGD severity using linear regression and correlation analysis. The change in viability after 1 µM OXT treatment was negatively correlated with OGD severity (p_{slope} = 0.0393, R^2 = 0.606, *n* = 7 culture plates with 38 wells). The change in viability after 0.1 µM OXT treatment was not correlated with OGD severity (p_{slope} = 0.206, R^2 = 0.128, *n* = 14 culture plates with 41 wells).

The linear regression line intersected the abscissa at an OGD viability of 49.41%. Based on this breakpoint of 49.41% viability, segmented regression was further performed for both OXT concentrations in OGD and normoxic conditions (Figure 3B). For OGD viabilities <49.41%, 1 µM OXT was associated with increases in the viability of cells exposed to OGD, showing a neuroprotective effect. For OGD viabilities >49.41%, 1 µM OXT was not associated with a significant effect (Figure 3B). For OGD viabilities <49.41%, 0.1 µM OXT tends to increase the viability of cells, with no further impact for OGD viabilities >49.41% (Figure 3B).

3.4. Administration of 1 µM Oxytocin Increased the Viability of DIV7 Hippocampal Neurons Exposed to Severe OGD but Not of Those Exposed to Moderate OGD

As described above, we used linear regression to characterize the relationship between OXT-induced changes in viability and OGD severity. The linear regression line intersected the abscissa at an NT-OGD viability of 49.41%.

This cutoff was used to separate culture plates into two categories of OGD strength. Plates with mean viability of NT-OGD wells <49.41% were considered to have suffered a severe post-OGD lesion, while plates with mean viability of NT-OGD wells >49.41% were considered to have suffered a moderate post-OGD lesion.

For each plate exposed to OGD, we calculated the mean viability of NT-OGD wells and the mean viability of 1 µM OXT-treated wells (Figure 4A). The effect of 1 µM OXT treatment on DIV7 hippocampal cell cultures was compared to the viability of the corresponding NT-OGD using separate paired t-tests for moderate and severe OGD.

After a severe lesion, 1 µM OXT treatment significantly increased the viability of DIV7 hippocampal cell cultures (mean difference = 7.88%, 95% CI = [2.19; 13.57], *p* = 0.021). However, it had no effect on the viability of cultures exposed to moderate OGD (mean difference = 5.93%, 95% CI = [−8.99; 20.84], *p* = 0.286) (Figure 4B, Table 2).

Table 2. Mean cellular viability of DIV7 hippocampal neurons exposed to OGD with or without 1 µM OXT.

		Mean Cellular Viability	SD	Number of Plates	Total Number of Wells	*p* Value Paired *t*-Test
Moderate OGD	No treatment	55.39%	4.2%	3	19	
	1 µM OXT	48.58%	12.15%	3	15	0.286
Severe OGD	No treatment	38.17%	4.25%	4	21	
	1 µM OXT	46.05%	1.78%	4	23	0.021

DIV—days in vitro; SD—standard deviation; OXT—oxytocin; OGD—oxygen–glucose deprivation.

A

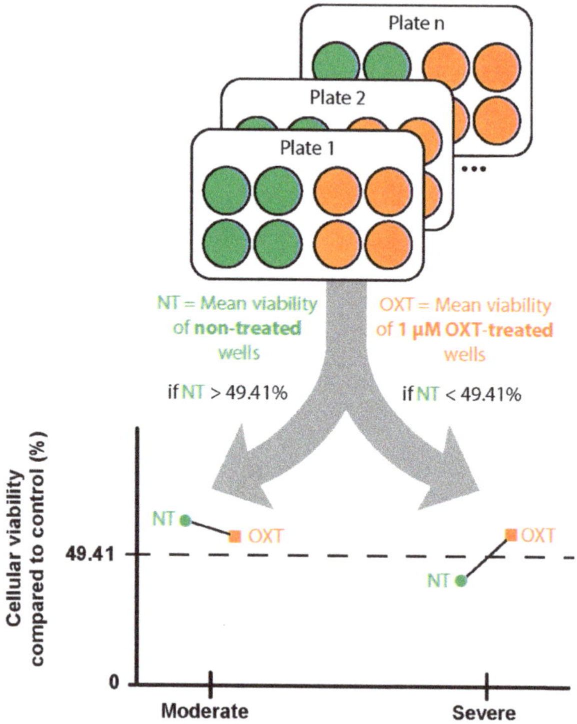

B

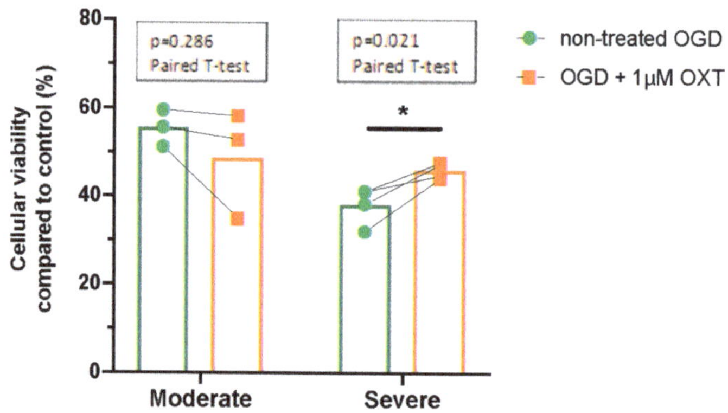

Figure 4. One µM of oxytocin increases the viability of cultures exposed to severe OGD but not of those exposed to moderate OGD. Schematic of data analysis and graphical representation of results (**A**). Viability of plates exposed to moderate and severe OGD, with or without OXT treatment (**B**). Individual data points are the mean cellular viability per culture plate. * $p < 0.05$. OGD—Oxygen–Glucose Deprivation; OXT—oxytocin.

3.5. *Administration of 0.1 µM Oxytocin Had No Significant Effect on the Viability of Cultures Exposed to OGD, Regardless of OGD Severity*

For each plate exposed to OGD, we calculated the mean viability of NT-OGD wells and the mean viability of 0.1 µM OXT-treated wells (Figure 5A). The effect of 0.1 µM

OXT treatment on DIV7 hippocampal cell cultures was compared to the viability of the corresponding NT-OGD using separate paired t-tests for moderate and severe OGD.

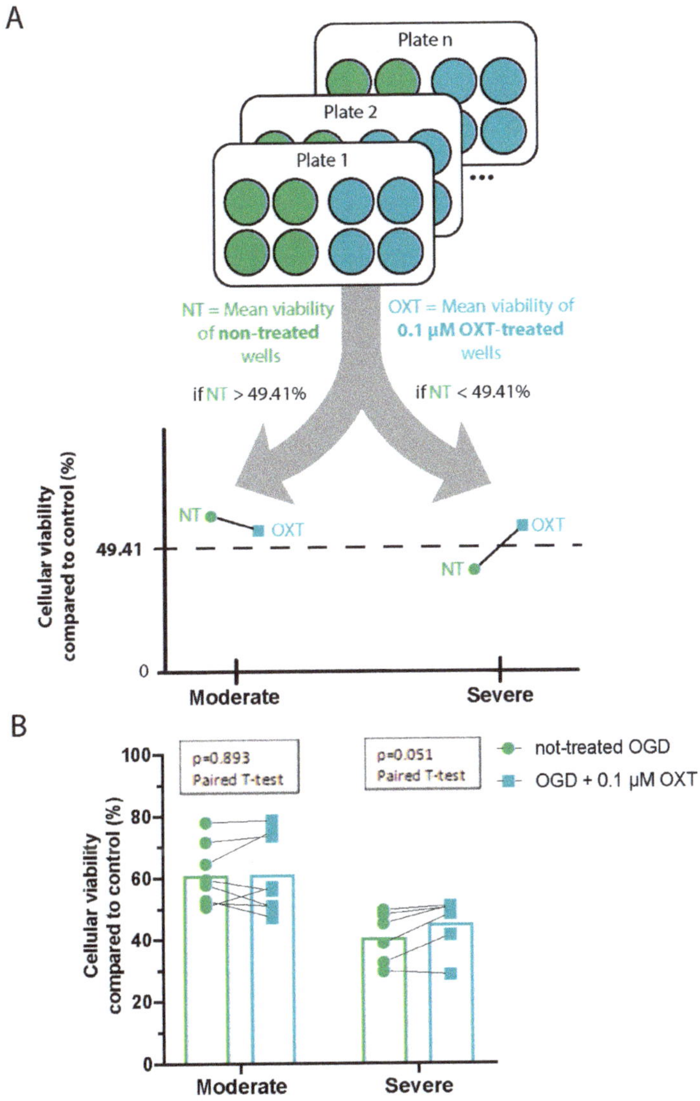

Figure 5. Oxytocin (0.1 µM) has no effect on the viability of cultures exposed to OGD, regardless of OGD severity. Schematic of data analysis and graphical representation of results (**A**). Viability of plates exposed to moderate and severe OGD, with or without OXT treatment (**B**). Individual data points are the mean cellular viability per culture plate. OXT—oxytocin; OGD—Oxygen-Glucose Deprivation; OXT—oxytocin.

At 0.1 µM, OXT treatment did not significantly change the viability of DIV7 hippocampal cell cultures exposed to moderate OGD (mean difference = 0.30%, 95% CI = [−4.732; 5.324], $p = 0.893$). For severe OGD lesions, there was a trend towards neuroprotection, which failed to reach statistical significance (mean difference = 4.50%, 95% CI = [−0.054; 9.061], $p = 0.051$; Figure 5B, Table 3).

Table 3. Mean cellular viability of DIV7 hippocampal neurons exposed to OGD with or without 0.1 μM OXT.

		Mean Cellular Viability	SD	Number of Plates	Total Number of Wells	*p* Value Paired *t*-Test
Moderate OGD	No treatment	60.89%	9.85%	8	27	
	0.1 μM OXT	61.18%	12.57%	8	22	0.893
Severe OGD	No treatment	40.64%	8.31%	6	22	
	0.1 μM OXT	45.13%	8.8%	6	26	0.051

DIV—days in vitro; SD—standard deviation; OXT—oxytocin; OGD—oxygen-glucose deprivation.

4. Discussion

Here, we investigated the effect of OXT on DIV7 hippocampal neurons following OGD and how this effect depends on OGD severity, considering that $GABA_A$Rs have shifted to inhibitory effects in most neurons at this stage. The pathophysiology of brain ischemia is incompletely understood, aggravated in the developing brain by the high rate of simultaneous developmental changes. The energy deficit during ischemia influences the activity of ion pumps and results in electrolyte imbalances, which can, in turn, affect the activity of receptor channels, such as $GABA_A$ channels [41].

In this study, we focused on hippocampal cultures due to the hippocampus' vulnerability to hypoxia and its critical role in memory and learning, which are often impaired in individuals experiencing PA [42]. DIV7 hippocampal cultures, associated with a physiological excitatory-to-inhibitory shift in GABAergic signaling [43] and synapse formation [44,45], represent a specific developmental window [46]. Previous studies show that neurogenesis and synaptic integration in dissociated hippocampal cultures continue within 10 days of cells in vitro [46]. This underscores the relevance of our model, as the DIV7 cultures capture a critical period of early network formation and synaptic activity, aligning with the developmental stage of term infants [36]. Additionally, while DIV7 cultures exhibit single spikes and lower synaptic density compared to older stages, they still represent a fundamental phase of network maturation [47]. Moreover, there is evidence of a hyperpolarizing effect of $GABA_A$R from both DIV6 primary rat neuronal cells and hippocampal slices from post-natal day 7 rats [36,39]. The preservation of this GABA switch across various brain structures and species, spanning from frogs, turtles, and rats to birds and likely extending to humans, indicates its evolutionary conservation and critical role in brain development [29,48].

OXT plays an important role in orchestrating the developmental switch from excitatory to inhibitory GABA signaling in the mammalian brain, but only within a limited temporal window during the peripartum period [6]. While prior research has highlighted the neuroprotective potential of OXT in immature hippocampal cell cultures [11], probably attributed to its facilitation of the physiological GABA switch, its impact on neurons exposed to OGD after this period remains uncertain. Notably, significant changes to Cl^- homeostasis have been reported to take place during the first week in vitro: KCC2 levels are reported to increase 15 times and NKCC1 to decrease four times at DIV6 compared to DIV1 [23], which is likely to impact the effect of OXT. Thus, this specific phase is crucial for studying OXT's effects on neuronal connectivity and function. By focusing on this developmental stage, we provide valuable insights into OXT's effects, which can be further validated in more complex in vivo models.

Our results showed that OXT treatment did not significantly impact the viability of DIV7 hippocampal cell cultures in control normoxic cultures. This is consistent with previous literature: when administered after DIV5 in normoxic conditions, OXT treatment did not lead to significant changes in chloride homeostasis in hippocampal neurons [23].

After 2-h OGD, we observed a wide range of values for mean cellular viability (between 25% and 75% of control viability), highlighting the variability in OGD vulnerability between cultures. This may be due to phenotypical differences in the pups or the method itself, including slight differences in hypoxia levels. Similar variability is seen in the wider OGD literature, perhaps due to non-standardized OGD and/or reperfusion durations, as well as different viability assays [25]. For example, cellular viability can range from 50% after a 30 min OGD [49] to approximately 60% after a 2 h OGD [50], with other studies reporting 52% viability after a 90 min OGD [51] and as many as 70% pyknotic nuclei reflecting apoptosis markers after a 4 h OGD exposure [52]. This variability may also reflect the real-life complexity of ischemic conditions where unpredictability is also an issue, emphasizing the need for careful consideration of sample size and standardized reporting of cellular viability after OGD exposure. Nevertheless, these simplistic models inadequately replicate the complex conditions observed in human scenarios.

Our study's primary finding is that 1 μM OXT treatment during OGD had a significant neuroprotective effect only in hippocampal cultures severely affected by OGD, where neuron viability was less than 49.41%. In cultures with moderate lesions, where viability was above 49.41%, OXT did not exhibit a significant protective effect.

These findings may offer insights regarding previous conflicting reports using animals/neuronal cultures at different maturation stages and different severity lesions. Kaneko et al. reported that administering 1 μM OXT in DIV3 rat hippocampal cell cultures for 3 days before OGD led to a neuroprotective effect, but this effect was not observed if OXT was administered during OGD at DIV6 [39]. They reported approximately 70% cellular viability after OGD, a severity level categorized as moderate in our study, where we did not find significant neuroprotection. Furthermore, Kaneko et al. used cultures with 60% glial cells compared to our method's <5% [53], suggesting that lesion severity may depend on the specific type of tissue or cells utilized. The neuroprotective effect of 1 μM OXT was also reported in an ischemia model consisting of exposure to OGD of hippocampal slices from 7 to 10-day-old rat pups, resembling hypoxic-ischemic lesions [36]. In this study, OXT mitigated neuronal death in a dose-dependent manner and delayed the anoxic depolarization in hippocampal CA1 neurons, likely by increasing GABA release through OXTR activation, enhancing the local inhibitory effect [36]. However, this study did not specify the precise viability of neurons.

Our findings are consistent with previous work showing that intranasal 1 μM OXT has a neuroprotective effect when administered to postnatal day-6 rat pups following an in vivo PA model [28]. Moreover, a clinical study linked the absence of intrapartum OXT exposure with the severity of neonatal brain injury, underscoring the OXT's crucial role in severe PA outcomes [8].

Our current results align with reports suggesting that after OGD-induced ischemic lesions, neurons could return to a functionally immature phenotype, characterized by higher intracellular Cl^- concentrations and a subsequent decrease in GABAergic inhibition [54–57]. In both ischemic and traumatic injuries in the adult brain, GABA can return to a functionally immature phenotype [58,59]. This shift is driven by changes in the expression and activity of cation/chloride co-transporters. Specifically, KCC2 mRNA and protein levels are down-regulated within the first hours following ischemia [55,56,60], while NKCC1 activity increases through phosphorylation during 2 h reoxygenation [61]. This subsequently leads to an increased intracellular Cl^- concentration [57], which may result in reduced Cl^- influx through $GABA_AR$ and decreased GABAergic inhibition [55,62]. As the severity of the injury increases, KCC2 is more strongly downregulated [56], and GABAergic inhibition appears to decrease, being eventually suppressed in anoxic conditions [63]. The severity of the ischemic injury seems, therefore, to play an important role in the magnitude of GABAergic dysregulation. Moreover, given that these alterations occur within minutes to hours following the ischemic event, it is plausible that similar changes may also manifest in our model, which assessed 2 h OGD and 3 h reoxygenation.

OXT's role in Cl^- homeostasis in immature neurons includes stimulating KCC2 phosphorylation and activity [39], and it might regain this effect in post-OGD functionally immature neuronal phenotypes. By enhancing KCC2 activity in these neurons, OXT would contribute to correcting the post-ischemic chloride imbalance and, therefore, exert a neuroprotective effect.

We report that 0.1 μM OXT did not have a statistically significant effect on DIV7 hippocampal cell cultures exposed to OGD, regardless of the severity of the lesion, though we could observe a trend towards neuroprotection in the severe lesions. Previous studies on immature neurons showed significant neuroprotective effects of 0.1 μM OXT on cell cultures after OGD exposure, as well as important effects on neuronal development and function [23,38]. This difference may arise from the type of neurons used in these studies. It is likely that, even though 0.1 μM OXT has significant effects on immature neurons, the concentration might be too low to elicit similar effects in neurons at this developmental stage since we could notice a trend towards neuroprotection.

These study's findings are important for clinical practice as they suggest that a treatment effect may depend on the ischemic lesion's severity. Therapeutic options in PA and subsequent HIE are currently limited, and translating the results from animal studies to clinical practice has proven extremely difficult. Despite advances in medical care, only hypothermia has shown neuroprotective benefits for term neonates with HIE after PA [64]. However, there is little evidence on the efficacy of hypothermia in reducing mortality and mitigating neurodevelopmental disability among infants with PA [65], creating a real need to develop treatment solutions for neuroprotection in PA. Furthermore, based on our present finding, it is advisable to categorize patients with PA according to the severity of their lesions, as some treatments may demonstrate efficacy specifically in cases of severe injury.

Study Limitations and Future Perspectives

This study utilized in vitro hippocampal cell cultures, which do not fully recapitulate the complexity of the in vivo brain environment. Recent evidence indicates significant transcriptional disparities between primary hippocampal cultures and their in vivo tissue counterparts. Additionally, focusing solely on the hippocampus does not account for other brain areas affected by PA, where OXT is also known to influence neuronal activity.

The 2-h OGD model used here, while offering controlled conditions, does not fully mimic the multifaceted pathophysiology of PA observed clinically. The resazurin reduction method employed to assess cell metabolism has limitations, such as potential resazurin pool depletion with high numbers of viable cells and the generation of non-fluorescent hydroresorufin, which may lead to underestimating cell viability. We acknowledge the limitation of using a single viability assessment method. Despite inherent variability between cell cultures and OGD exposures, which could introduce confounding factors, our findings significantly advance the understanding of neuroprotection in OGD/PA-associated lesions.

Future research should include neurons from other brain regions to provide a comprehensive view of OXT's neuroprotective effects and uncover additional protective mechanisms across various neural substrates. Incorporating cortical neurons, which are sensitive to hypoxia and modulated by OXT, could enhance the broader applicability of our findings. In vivo models are crucial to assess OXT's effectiveness in an intact nervous system, addressing the limitations of our in vitro approach and exploring OXT delivery methods that minimize maternal impact. This research will bridge the gap between basic neuroscience and clinical applications, contributing to therapeutic strategies for neonatal care following PA.

5. Conclusions

In conclusion, this study provides insights into the neuroprotective role of OXT in the context of OGD as a model of HIE. We find that 1 μM OXT treatment during 2 h OGD exerts a neuroprotective effect on DIV7 hippocampal cell cultures only in case of a severe OGD-induced lesion, a condition which can be accompanied by a change in the neuronal phenotype into a functionally immature one. OXT treatment had no significant effect when administered during moderate OGD exposure or in normoxic conditions. Our results suggest that OXT might activate neuroprotective mechanisms in the developing neurons at DIV7 impacted by severe ischemia.

Our findings underscore the importance of carefully considering the developmental stage of neurons and the severity of the ischemic insult in assessing the efficacy of OXT treatment. Furthermore, our results from in vitro models may inform further in vivo research investigating the underlying mechanisms of OXT-mediated neuroprotection and clinically relevant studies optimizing therapeutic strategies for managing PA and subsequent HIE. Ultimately, the identification of OXT as a potential therapeutic target from in vitro studies holds promise for improving outcomes in neonates at risk of neurological injury due to PA, offering hope for novel interventions in clinical practice.

Author Contributions: Conceptualization, A.-M.Z., I.-F.G., B.I. and M.C.; data curation, I.-F.G., M.I.I., R.-B.I., D.M.C. and R.M.H.; formal analysis, I.-F.G., M.I.I., R.-B.I., D.M.C., B.I. and M.C.; funding acquisition, A.-M.Z.; investigation, A.-M.Z., I.-F.G., M.I.I., R.-B.I., D.M.C. and R.M.H.; methodology, A.-M.Z., I.-F.G., B.I. and M.C.; project administration, A.-M.Z., I.-F.G. and M.I.I.; resources, A.-M.Z.; supervision, A.-M.Z.; visualization, A.-M.Z., I.-F.G. and R.M.H.; writing—original draft, A.-M.Z., I.-F.G., M.I.I., R.-B.I., R.M.H. and M.C.; Writing—review and editing, A.-M.Z., M.I.I., I.-F.G., R.-B.I., B.I. and M.C. All authors have read and agreed to the published version of the manuscript.

Funding: This research was supported by Carol Davila University of Medicine and Pharmacy, Bucharest, Romania, the Interdisziplinäres Zentrum für Klinische Forschung (IZKF) Jena, and the Foundation "Else Kröner-Fresenius-Stiftung" within the Else Kröner Research School for Physicians "AntiAge" (to M.C.).

Institutional Review Board Statement: The study protocol was approved by the Ethics Committee of Carol Davila University of Medicine and Pharmacy, Bucharest, Romania (21994/2 August 2022).

Informed Consent Statement: Not applicable.

Data Availability Statement: Data is contained within the article.

Conflicts of Interest: The authors declare no conflicts of interest.

References

1. Gillam-Krakauer, M.; Gowen, C.W., Jr. Birth Asphyxia. In *StatPearls*; StatPearls Publishing: Treasure Island, FL, USA, 2024.
2. Zagrean, A.-M.; Georgescu, I.-A.; Iesanu, M.I.; Ionescu, R.-B.; Haret, R.M.; Panaitescu, A.M.; Zagrean, L. Chapter Three—Oxytocin and Vasopressin in the Hippocampus. In *Hormones, Regulators and Hippocampus*; Vitamins and Hormones; Litwack, G., Ed.; Academic Press: Cambridge, MA, USA, 2022; Volume 118, pp. 83–127.
3. Russell, J.A.; Leng, G.; Douglas, A.J. The Magnocellular Oxytocin System, the Fount of Maternity: Adaptations in Pregnancy. *Front. Neuroendocrinol.* **2003**, *24*, 27–61. [CrossRef]
4. Krol, K.M.; Moulder, R.G.; Lillard, T.S.; Grossmann, T.; Connelly, J.J. Epigenetic Dynamics in Infancy and the Impact of Maternal Engagement. *Sci. Adv.* **2019**, *5*, eaay0680. [CrossRef]
5. Besser, G.M.; Mortimer, C.H. Hypothalamic Regulatory Hormones: A Review. *J. Clin. Pathol.* **1974**, *27*, 173–184. [CrossRef]
6. Tyzio, R.; Cossart, R.; Khalilov, I.; Minlebaev, M.; Hübner, C.A.; Represa, A.; Ben-Ari, Y.; Khazipov, R. Maternal Oxytocin Triggers a Transient Inhibitory Switch in GABA Signaling in the Fetal Brain during Delivery. *Science* **2006**, *314*, 1788–1792. [CrossRef]
7. Walter, M.H.; Abele, H.; Plappert, C.F. The Role of Oxytocin and the Effect of Stress During Childbirth: Neurobiological Basics and Implications for Mother and Child. *Front. Endocrinol.* **2021**, *12*, 742236. [CrossRef]
8. Wayock, C.P.; Meserole, R.L.; Saria, S.; Jennings, J.M.; Huisman, T.A.G.M.; Northington, F.J.; Graham, E.M. Perinatal Risk Factors for Severe Injury in Neonates Treated with Whole-Body Hypothermia for Encephalopathy. *Am. J. Obstet. Gynecol.* **2014**, *211*, 41.e1–41.e8. [CrossRef]
9. Mairesse, J.; Zinni, M.; Pansiot, J.; Hassan-Abdi, R.; Demene, C.; Colella, M.; Charriaut-Marlangue, C.; Rideau Batista Novais, A.; Tanter, M.; Maccari, S.; et al. Oxytocin Receptor Agonist Reduces Perinatal Brain Damage by Targeting Microglia. *Glia* **2019**, *67*, 345–359. [CrossRef]
10. Knoop, M.; Possovre, M.-L.; Jacquens, A.; Charlet, A.; Baud, O.; Darbon, P. The Role of Oxytocin in Abnormal Brain Development: Effect on Glial Cells and Neuroinflammation. *Cells* **2022**, *11*, 3899. [CrossRef]
11. Ceanga, M.; Spataru, A.; Zagrean, A.-M. Oxytocin Is Neuroprotective against Oxygen-Glucose Deprivation and Reoxygenation in Immature Hippocampal Cultures. *Neurosci. Lett.* **2010**, *477*, 15–18. [CrossRef]
12. Cao, W.; Pavlinec, C.; Gravenstein, N.; Seubert, C.N.; Martynyuk, A.E. Roles of Aldosterone and Oxytocin in Abnormalities Caused by Sevoflurane Anesthesia in Neonatal Rats. *Anesthesiology* **2012**, *117*, 791–800. [CrossRef]

13. Ben-Ari, Y. Oxytocin and Vasopressin, and the GABA Developmental Shift During Labor and Birth: Friends or Foes? *Front. Cell. Neurosci.* **2018**, *12*, 254. [CrossRef]
14. Farrant, M.; Kaila, K. The Cellular, Molecular and Ionic Basis of GABA(A) Receptor Signalling. *Prog. Brain Res.* **2007**, *160*, 59–87. [CrossRef]
15. Olsen, R.W.; Sieghart, W. GABA A Receptors: Subtypes Provide Diversity of Function and Pharmacology. *Neuropharmacology* **2009**, *56*, 141–148. [CrossRef]
16. Watanabe, M.; Fukuda, A. Development and Regulation of Chloride Homeostasis in the Central Nervous System. *Front. Cell Neurosci.* **2015**, *9*, 371. [CrossRef]
17. Ben-Ari, Y.; Gaiarsa, J.-L.; Tyzio, R.; Khazipov, R. GABA: A Pioneer Transmitter That Excites Immature Neurons and Generates Primitive Oscillations. *Physiol. Rev.* **2007**, *87*, 1215–1284. [CrossRef]
18. Tyzio, R.; Represa, A.; Jorquera, I.; Ben-Ari, Y.; Gozlan, H.; Aniksztejn, L. The Establishment of GABAergic and Glutamatergic Synapses on CA1 Pyramidal Neurons is Sequential and Correlates with the Development of the Apical Dendrite. *J. Neurosci.* **1999**, *19*, 10372–10382. [CrossRef]
19. Ben-Ari, Y. Excitatory Actions of Gaba during Development: The Nature of the Nurture. *Nat. Rev. Neurosci.* **2002**, *3*, 728–739. [CrossRef]
20. Kahle, K.T.; Staley, K.J. The Bumetanide-Sensitive Na-K-2Cl Cotransporter NKCC1 as a Potential Target of a Novel Mechanism-Based Treatment Strategy for Neonatal Seizures. *Neurosurg. Focus.* **2008**, *25*, E22. [CrossRef]
21. Dzhala, V.I.; Talos, D.M.; Sdrulla, D.A.; Brumback, A.C.; Mathews, G.C.; Benke, T.A.; Delpire, E.; Jensen, F.E.; Staley, K.J. NKCC1 Transporter Facilitates Seizures in the Developing Brain. *Nat. Med.* **2005**, *11*, 1205–1213. [CrossRef]
22. Yamada, J.; Okabe, A.; Toyoda, H.; Kilb, W.; Luhmann, H.J.; Fukuda, A. Cl^- Uptake Promoting Depolarizing GABA Actions in Immature Rat Neocortical Neurones Is Mediated by NKCC1. *J. Physiol.* **2004**, *557*, 829–841. [CrossRef]
23. Leonzino, M.; Busnelli, M.; Antonucci, F.; Verderio, C.; Mazzanti, M.; Chini, B. The Timing of the Excitatory-to-Inhibitory GABA Switch Is Regulated by the Oxytocin Receptor via KCC2. *Cell Rep.* **2016**, *15*, 96–103. [CrossRef]
24. Bos, R.; Sadlaoud, K.; Boulenguez, P.; Buttigieg, D.; Liabeuf, S.; Brocard, C.; Haase, G.; Bras, H.; Vinay, L. Activation of 5-HT2A Receptors Upregulates the Function of the Neuronal K-Cl Cotransporter KCC2. *Proc. Natl. Acad. Sci. USA* **2013**, *110*, 348–353. [CrossRef]
25. Tasca, C.I.; Dal-Cim, T.; Cimarosti, H. In Vitro Oxygen-Glucose Deprivation to Study Ischemic Cell Death. *Methods Mol. Biol.* **2015**, *1254*, 197–210. [CrossRef]
26. Zăgrean, A.M.; Spataru, A.; Ceangă, M.; Zăgrean, L. The Single versus Combinatorial Effects of MK-801, CNQX, Nifedipine and AP-3 on Primary Cultures of Cerebellar Granule Cells in an Oxygen-Glucose Deprivation Model. *Rom. J. Morphol. Embryol.* **2014**, *55*, 811–816.
27. Spataru, A.; Le Duc, D.; Zagrean, L.; Zagrean, A.-M. Ethanol Exposed Maturing Rat Cerebellar Granule Cells Show Impaired Energy Metabolism and Increased Cell Death after Oxygen-Glucose Deprivation. *Neural Regen. Res.* **2019**, *14*, 485–490. [CrossRef]
28. Panaitescu, A.M.; Isac, S.; Pavel, B.; Ilie, A.S.; Ceanga, M.; Totan, A.; Zagrean, L.; Peltecu, G.; Zagrean, A.M. Oxytocin Reduces Seizure Burden and Hippocampal Injury in a Rat Model of Perinatal Asphyxia. *Acta Endocrinol.* **2018**, *14*, 315–319. [CrossRef]
29. Peerboom, C.; Wierenga, C.J. The Postnatal GABA Shift: A Developmental Perspective. *Neurosci. Biobehav. Rev.* **2021**, *124*, 179–192. [CrossRef]
30. Brewer, G.J.; Torricelli, J.R. Isolation and Culture of Adult Neurons and Neurospheres. *Nat. Protoc.* **2007**, *2*, 1490–1498. [CrossRef]
31. Zagrean, A.-M.; Grigoras, I.-F.; Iesanu, M.I.; Ionescu, R.-B.; Chitimus, D.M.; Haret, R.M.; Ianosi, B.; Ceanga, M.; Zagrean, L. Neuronal Transmembrane Chloride Transport Has a Time-Dependent Influence on Survival of Hippocampal Cultures to Oxygen-Glucose Deprivation. *Brain Sci.* **2019**, *9*, 360. [CrossRef]
32. Brewer, G.J.; Torricelli, J.R.; Evege, E.K.; Price, P.J. Optimized Survival of Hippocampal Neurons in B27-Supplemented Neurobasal^{TM}, a New Serum-Free Medium Combination. *J. Neurosci. Res.* **1993**, *35*, 567–576. [CrossRef]
33. Sahu, M.P.; Nikkilä, O.; Lågas, S.; Kolehmainen, S.; Castrén, E. Culturing Primary Neurons from Rat Hippocampus and Cortex. *Neuronal Signal.* **2019**, *3*, NS20180207. [CrossRef]
34. Le Duc, D.; Spataru, A.; Ceanga, M.; Zagrean, L.; Schöneberg, T.; Toescu, E.C.; Zagrean, A.-M. Developmental Exposure to Ethanol Increases the Neuronal Vulnerability to Oxygen-Glucose Deprivation in Cerebellar Granule Cell Cultures. *Brain Res.* **2015**, *1614*, 1–13. [CrossRef]
35. Lyu, H.; Sun, D.M.; Ng, C.P.; Chen, J.F.; He, Y.Z.; Lam, S.Y.; Zheng, Z.Y.; Askarifirouzjaei, H.; Wang, C.C.; Young, W.; et al. A New Hypoxic Ischemic Encephalopathy Model in Neonatal Rats. *Heliyon* **2021**, *7*, e08646. [CrossRef]
36. Wu, Z.; Xie, C.; Kuang, H.; Wu, J.; Chen, X.; Liu, H.; Liu, T. Oxytocin Mediates Neuroprotection against Hypoxic-Ischemic Injury in Hippocampal CA1 Neuron of Neonatal Rats. *Neuropharmacology* **2021**, *187*, 108488. [CrossRef]
37. Semple, B.D.; Blomgren, K.; Gimlin, K.; Ferriero, D.M.; Noble-Haeusslein, L.J. Brain Development in Rodents and Humans: Identifying Benchmarks of Maturation and Vulnerability to Injury across Species. *Prog. Neurobiol.* **2013**, *106–107*, 1–16. [CrossRef]
38. Ripamonti, S.; Ambrozkiewicz, M.C.; Guzzi, F.; Gravati, M.; Biella, G.; Bormuth, I.; Hammer, M.; Tuffy, L.P.; Sigler, A.; Kawabe, H.; et al. Transient Oxytocin Signaling Primes the Development and Function of Excitatory Hippocampal Neurons. *Elife* **2017**, *6*, e22466. [CrossRef]
39. Kaneko, Y.; Pappas, C.; Tajiri, N.; Borlongan, C.V. Oxytocin Modulates GABAAR Subunits to Confer Neuroprotection in Stroke In Vitro. *Sci. Rep.* **2016**, *6*, 35659. [CrossRef]
40. Lin, Y.-T.; Chen, C.-C.; Huang, C.-C.; Nishimori, K.; Hsu, K.-S. Oxytocin Stimulates Hippocampal Neurogenesis via Oxytocin Receptor Expressed in CA3 Pyramidal Neurons. *Nat. Commun.* **2017**, *8*, 537. [CrossRef]

41. Haugstad, T.S.; Karlsen, H.E.; Krajtči, P.; Due-Tønnessen, B.; Larsen, M.; Sandberg, C.; Sand, O.; Brandtzaeg, P.; Langmoen, I.A. Efflux of Gamma-Aminobutyric Acid Caused by Changes in Ion Concentrations and Cell Swelling Simulating the Effect of Cerebral Ischaemia. *Acta Neurochir.* **1997**, *139*, 453–463. [CrossRef]
42. Annink, K.V.; de Vries, L.S.; Groenendaal, F.; van den Heuvel, M.P.; van Haren, N.E.M.; Swaab, H.; van Handel, M.; Jongmans, M.J.; Benders, M.J.; van der Aa, N.E. The Long-Term Effect of Perinatal Asphyxia on Hippocampal Volumes. *Pediatr. Res.* **2019**, *85*, 43–49. [CrossRef]
43. Valeeva, G.; Valiullina, F.; Khazipov, R. Excitatory Actions of GABA in the Intact Neonatal Rodent Hippocampus In Vitro. *Front. Cell. Neurosci.* **2013**, *7*, 20. [CrossRef]
44. Moutin, E.; Hemonnot, A.-L.; Seube, V.; Linck, N.; Rassendren, F.; Perroy, J.; Compan, V. Procedures for Culturing and Genetically Manipulating Murine Hippocampal Postnatal Neurons. *Front. Synaptic Neurosci.* **2020**, *12*, 19. [CrossRef]
45. Grabrucker, A.; Vaida, B.; Bockmann, J.; Boeckers, T.M. Synaptogenesis of Hippocampal Neurons in Primary Cell Culture. *Cell Tissue Res.* **2009**, *338*, 333–341. [CrossRef]
46. Cheyne, J.E.; Grant, L.; Butler-Munro, C.; Foote, J.W.; Connor, B.; Montgomery, J.M. Synaptic Integration of Newly Generated Neurons in Rat Dissociated Hippocampal Cultures. *Mol. Cell Neurosci.* **2011**, *47*, 203–214. [CrossRef]
47. Biffi, E.; Regalia, G.; Menegon, A.; Ferrigno, G.; Pedrocchi, A. The Influence of Neuronal Density and Maturation on Network Activity of Hippocampal Cell Cultures: A Methodological Study. *PLoS ONE* **2013**, *8*, e83899. [CrossRef]
48. Myers, M.M.; Grieve, P.G.; Izraelit, A.; Fifer, W.P.; Isler, J.R.; Darnall, R.A.; Stark, R.I. Developmental Profiles of Infant EEG: Overlap with Transient Cortical Circuits. *Clin. Neurophysiol.* **2012**, *123*, 1502–1511. [CrossRef]
49. Chung, H.; Kim, E.; Lee, D.H.; Seo, S.; Ju, S.; Lee, D.; Kim, H.; Park, S. Ghrelin Inhibits Apoptosis in Hypothalamic Neuronal Cells during Oxygen-Glucose Deprivation. *Endocrinology* **2007**, *148*, 148–159. [CrossRef]
50. Ye, R.; Li, N.; Han, J.; Kong, X.; Cao, R.; Rao, Z.; Zhao, G. Neuroprotective Effects of Ginsenoside Rd against Oxygen-Glucose Deprivation in Cultured Hippocampal Neurons. *Neurosci. Res.* **2009**, *64*, 306–310. [CrossRef]
51. Iijima, T.; Mishima, T.; Akagawa, K.; Iwao, Y. Mitochondrial Hyperpolarization after Transient Oxygen-Glucose Deprivation and Subsequent Apoptosis in Cultured Rat Hippocampal Neurons. *Brain Res.* **2003**, *993*, 140–145. [CrossRef]
52. Culmsee, C.; Zhu, C.; Landshamer, S.; Becattini, B.; Wagner, E.; Pellecchia, M.; Blomgren, K.; Plesnila, N. Apoptosis-Inducing Factor Triggered by Poly(ADP-Ribose) Polymerase and Bid Mediates Neuronal Cell Death after Oxygen-Glucose Deprivation and Focal Cerebral Ischemia. *J. Neurosci.* **2005**, *25*, 10262–10272. [CrossRef]
53. Limatola, C.; Lauro, C.; Catalano, M.; Ciotti, M.T.; Bertollini, C.; Di Angelantonio, S.; Ragozzino, D.; Eusebi, F. Chemokine CX3CL1 Protects rat Hippocampal Neurons against Glutamate-Mediated Excitotoxicity. *J. Neuroimmunol.* **2005**, *166*, 19–28. [CrossRef]
54. Wu, H.; Che, X.; Tang, J.; Ma, F.; Pan, K.; Zhao, M.; Shao, A.; Wu, Q.; Zhang, J.; Hong, Y. The K(+)-Cl(−) Cotransporter KCC2 and Chloride Homeostasis: Potential Therapeutic Target in Acute Central Nervous System Injury. *Mol. Neurobiol.* **2016**, *53*, 2141–2151. [CrossRef]
55. Galeffi, F.; Sah, R.; Pond, B.B.; George, A.; Schwartz-Bloom, R.D. Changes in Intracellular Chloride after Oxygen–Glucose Deprivation of the Adult Hippocampal Slice: Effect of Diazepam. *J. Neurosci.* **2004**, *24*, 4478–4488. [CrossRef]
56. Jaenisch, N.; Witte, O.W.; Frahm, C. Downregulation of Potassium Chloride Cotransporter KCC2 after Transient Focal Cerebral Ischemia. *Stroke* **2010**, *41*, e151–e159. [CrossRef]
57. Inglefield, J.R.; Schwartz-Bloom, R.D. Optical Imaging of Hippocampal Neurons with a Chloride-Sensitive Dye: Early Effects of In Vitro Ischemia. *J. Neurochem.* **1998**, *70*, 2500–2509. [CrossRef]
58. van den Pol, A.N.; Obrietan, K.; Chen, G. Excitatory Actions of GABA after Neuronal Trauma. *J. Neurosci.* **1996**, *16*, 4283–4292. [CrossRef]
59. Schwartz-Bloom, R.D.; Sah, R. Gamma-Aminobutyric Acid(A) Neurotransmission and Cerebral Ischemia. *J. Neurochem.* **2001**, *77*, 353–371. [CrossRef]
60. Toda, T.; Ishida, K.; Kiyama, H.; Yamashita, T.; Lee, S. Down-Regulation of KCC2 Expression and Phosphorylation in Motoneurons, and Increases the Number of in Primary Afferent Projections to Motoneurons in Mice with Post-Stroke Spasticity. *PLoS ONE* **2014**, *9*, e114328. [CrossRef]
61. Pond, B.B.; Berglund, K.; Kuner, T.; Feng, G.; Augustine, G.J.; Schwartz-Bloom, R.D. The Chloride Transporter Na+-K+-Cl− Cotransporter Isoform-1 Contributes to Intracellular Chloride Increases after In Vitro Ischemia. *J. Neurosci.* **2006**, *26*, 1396–1406. [CrossRef]
62. Mittmann, T.; Qü, M.; Zilles, K.; Luhmann, H.J. Long-Term Cellular Dysfunction after Focal Cerebral Ischemia: In Vitro Analyses. *Neuroscience* **1998**, *85*, 15–27. [CrossRef]
63. Katchman, A.N.; Vicini, S.; Hershkowitz, N. Mechanism of Early Anoxia-Induced Suppression of the GABAA-Mediated Inhibitory Postsynaptic Current. *J. Neurophysiol.* **1994**, *71*, 1128–1138. [CrossRef]
64. Antonucci, R.; Porcella, A.; Pilloni, M.D. Perinatal Asphyxia in the Term Newborn. *J. Pediatr. Neonatal Individ. Med. (JPNIM)* **2014**, *3*, e030269. [CrossRef]
65. McGuire, W. Perinatal Asphyxia. *BMJ Clin. Evid.* **2007**, *2007*, 0320.

current issues in molecular biology

MDPI

Review

Oxytocin, Vasopressin and Stress: A Hormetic Perspective

Hans P. Nazarloo [1,*], Marcy A. Kingsbury [2], Hannah Lamont [2], Caitlin V. Dale [3], Parmida Nazarloo [4], John M. Davis [5], Eric C. Porges [6,7], Steven P. Cuffe [1] and C. Sue Carter [1,4]

1 Department of Psychiatry, College of Medicine-Jacksonville, University of Florida, 580 West 8th St., Tower II, 6th Floor, Jacksonville, FL 32209, USA; steven.cuffe@jax.ufl.edu (S.P.C.); suecarterporges@gmail.com (C.S.C.)
2 Lurie Center for Autism, Mass General Research Institute, Harvard Medical School, Charlestown, MA 02129, USA; makingsbury@mgh.harvard.edu (M.A.K.); hhl23@gsbs.rutgers.edu (H.L.)
3 Dedman College of Humanities and Sciences, Southern Methodist University, Dallas, TX 75205, USA; cate@smu.edu
4 Traumatic Stress Research Consortium, Kinsey Institute, Indiana University, Bloomington, IN 47405, USA; pnazarl@iu.edu
5 Department of Psychiatry, University of Illinois at Chicago, Chicago, IL 60607, USA; davisjm@uic.edu
6 Department of Clinical and Health Psychology, College of Public Health and Health Professions, University of Florida, Gainesville, FL 32611, USA; eporges@ufl.edu
7 Center for Cognitive Aging and Memory, Evelyn F. and William L. McKnight Brain Institute, University of Florida, Gainesville, FL 32611, USA
* Correspondence: hans.nazarloo@jax.ufl.edu; Tel.: +1-(773)-886-3560; Fax: +1-(904)-244-4486

Abstract

The purpose of this article is to examine a previously unrecognized role for the vasopressin–oxytocin (VP-OT) system in mammalian "stress-response hormesis." The current review adds hormesis to the long list of beneficial effects of OT. Hormesis, a biphasic adaptive response to low-level stressors, is introduced here to contextualize the dynamic roles of oxytocin and vasopressin. As with hormesis, the properties of the VP-OT system are context-, time-, and dose-sensitive. Here we suggest that one key to understanding hormesis is the fact that VP and OT and their receptors function as an integrated system. The VP-OT system is capable of changing and adapting to challenges over time, including challenges necessary for survival, reproduction and sociality. Prior research suggests that many beneficial effects of OT are most apparent only following stressful experiences, possibly reflecting interactions with VP, its receptors and other components of the hypothalamic–pituitary–adrenal axis. The release of OT is documented following various kinds of hormetic experiences such as birth, vigorous exercise, ischemic events and the ingestion of emetics, including psychedelics. The phasic or cyclic modulation of VP and related "stress" hormones, accompanied or followed by the release of OT, creates conditions that conform to the core principles of hormesis. This concept is reviewed here in the context of other hormones including corticotropin releasing hormone (CRH) and urocortin, as well as cytokines. In general, VP and classic "stress hormones" support an active response, helping to quickly mobilize body systems. OT interacts with all of these, and may subsequently re-establish homeostasis and precondition the organism to deal with future stressors. However, the individual history of an organism, including epigenetic modifications of classical stress hormones such as VP, can moderate the effects of OT. Oxytocin's effects also help to explain the important role of sociality in mammalian resilience and longevity. A hormetic perspective, focusing on a dynamic VP-OT system, offers new insights into emotional and physical disorders, especially those associated with the management of chronic stress, and helps us to understand the healing power of social behavior and perceived safety.

Keywords: oxytocin; vasopressin; stress; sociality; hormesis

1. Purpose and Historical Background

In this narrative review, we hypothesize that a system involving interactions between oxytocin (OT) and vasopressin (VP) plays a critical role in understanding the physiological process termed "stress-response hormesis" [1]. VP and OT jointly encode environmental experience in cellular memory and create mechanisms that can be leveraged to optimize adaptation across many domains of human health. Although not previously recognized, many of the functions attributed to VP-OT interactions are similar to those described in hormesis.

1.1. Oxytocin and Vasopressin

Clinical research in the early 20th century recognized the capacity of extracts of the pituitary gland to speed up labor or facilitate milk ejection. OT was chemically identified as a polypeptide in the 1950s, an achievement recognized with a Nobel Prize [2]. However, as recently as the 1980s, research papers continued to describe OT as a "female" reproductive hormone with "no known function in males" [3]. The narrow characterization of OT as of importance only in females and only relevant to reproduction was incomplete and almost certainly slowed research on this remarkable molecule [4,5]. The pleiotropic functions of OT are only now being recognized [6,7].

OT extends beyond social and reproductive functions, including the regulation of food intake (reducing appetite via hypothalamic pathways), inducing muscle contraction (notably during parturition and lactation), and influencing thermoregulation by activating brown adipose tissue and mediating cold stress responses [8] OT has been more recently identified as an "anti-stress" hormone [9,10] with major consequences for health, resilience [8] and longevity [11]. OT has anti-inflammatory properties and is involved in regulating oxidative stress [12,13]. Furthermore, pathways dependent on OT facilitate prosocial behaviors, creating a "social solution to the stress of life" [14]. However, OT acts against the background of many other molecules, including VP, corticotropin-releasing hormone (CRH), urocortin (UCN), and molecules of the immune system. The temporal effects of OT vary in this context. Although both oxytocin-like and vasopressin-like peptides emerged from a shared ancestral gene early in vertebrate evolution, our current understanding of their functional specialization is based largely on mammalian and rodent models. While we use evolutionary terminology to highlight adaptive roles, we acknowledge that direct functional evidence from lower vertebrates and invertebrates remains limited.

In this review, we focus on the relationship of OT with a more ancient neuropeptide, vasopressin. VP and OT arose from a common ancestor [15]. These structurally similar peptides exhibit dynamic cross-reactivity with each other's receptors [16]. Together VP and OT form a system with the capacity for adaptive and sequential changes, and with varying consequences across time. OT may be particularly relevant in the context of chronic stress, where it helps protect against endocrine dysregulation in conditions such as posttraumatic stress disorder (PTSD). This may in part be due to OT's capacity to sequentially counter the actions of VP [4,17,18] and temper physiological overreactions to threat.

Under chronic versus acute conditions, the phenotype of these responses is different: VP is associated with quick responses including mobilization and anxiety, while OT responds more slowly, and may be associated with calming, immobilization and restoration. For example, the VP-OT system is especially important in social behavior and attachment [19]. In social mammals, the presence of safe others is of particular relevance during stress-management and "sociostasis" [20]. We suggest here that when better understood, VP- and OT-mediated interactions will offer novel strategies to harness the therapeutic capacity of these neuropeptides.

1.2. Time Matters

In mammals under chronic stress, VP and OT molecules have divergent, interactive and often sequential functions. In addition, the receptors for VP and OT, along with their subcellular signaling mechanisms, are critical for enabling dynamic functional interactions between these systems [21].

Across the lifespan, time also matters. There is abundant evidence that the VP-OT system is adaptively calibrated by early life experiences [22,23]. Context also is critical in regulating the VP-OT system [24]. The hormetic perspective described below also emphasizes the importance of time, development, feedback systems and context—all processes within which VP and OT have been shown to have dynamically adaptive functions.

2. Hormesis

Hormesis as a process has been used to describe biphasic or cyclic responses to challenge across species, ranging from single-cell organisms to vertebrates, including humans [25,26]. For convenience, we are focusing our scope here on mammals. Although hormetic functions are observed in a broad array of living organisms, we recognize the specific endocrine systems described here may not generalize beyond mammals. However, as initially described [1], cellular and subcellular processes that are shared across phyla may offer unifying mechanisms that explain hormesis across levels of biological complexity.

2.1. The Role of Peptides in the Hormetic Hypothesis

Both VP and OT are released by various challenging or stressful experiences. This pattern of simultaneous or sequential release could help to explain the poorly understood, biphasic, and sometimes beneficial consequences of "stress," especially if the challenge is followed by a period of restoration. This process, sometimes known as "stress-response hormesis", has well-documented consequences for health, restoration and longevity [1,27].

VP and OT have independently been shown to play major roles in the regulation of stress [6,28]. The current review focuses on the novel hypothesis that VP and OT also are critical to "stress-response hormesis". Here we suggest that, at least in mammals, the benefits of hormesis may involve the sequential effects of VP followed by a compensatory release of OT. To our knowledge, this hypothesis has not been experimentally tested in the context of hormesis. We provide selected examples of hormesis in which interactions among OT and VP may help to explain beneficial adaptations following intense challenges, such as those observed in birth or exercise (Tables 1 and 2). We further suggest that the investigation of the hormetic role of these peptides may identify previously unidentified adaptive functions for both VP and OT, and deepen our understanding of their coordinated activity as an integrated system.

2.2. Classical Definitions of Hormesis

Hormesis is a highly conserved, biphasic biological process characterized by stimulatory or beneficial effects following a challenge, and inhibitory or deleterious effects at high doses or during prolonged exposure. Initially conceptualized within toxicology [29,30], hormesis has since evolved into a unifying paradigm that intersects diverse fields such as aging biology, neuroscience, immunology, metabolism, and exercise physiology [31] (Table 2).

Table 1. This table outlines key conceptual domains and unresolved research questions regarding the roles of oxytocin (OT) and vasopressin (VP) in stress regulation. It organizes future directions across four tiers—mechanistic foundations, neuroendocrine integration, adaptive stress responses (hormesis and allostasis), and translational/clinical potential—highlighting essential gaps in knowledge from molecular mechanisms to therapeutic applications.

Conceptual Domains	Unresolved Research Questions
Mechanistic Foundations	• How do OT and VP exhibit biphasic, dose-dependent responses? • What intracellular pathways are engaged under low vs. high stress?
Neuroendocrine Integration	• How do CRH, UCNs, VP, OT orchestrate the catabolic-anabolic shift? • What is the temporal role of each peptide in initiating or resolving stress?
Hormesis and Allostasis	• How do VP and OT affect allostatic load vs. resilience? • Can they promote parasympathetic re-engagement after acute stress?
Translational and Clinical Potential	• What is the optimal timing/dose for clinical use in PTSD, depression, aging? • Are there risks of desensitization or dysregulation with chronic exposure?

Table 2. Translational applications of CACH in human healthspan and disease modulation. This table illustrates the translational relevance of CACH across key human domains. It outlines how intermittent, tolerable stressors engage catabolic signaling, followed by anabolic restoration and remodeling. The net effect is enhanced systemic resilience, longevity, and performance, with implications for preventive and therapeutic strategies in aging, cardiometabolic health, and psychiatry.

Domain	Example Stressors/Interventions	Mechanisms Catabolic–Anabolic Cycling	Key Outcomes	Ref.
Aging & Geroscience	• Fasting • Heat/cold exposure • Exercise (e.g., resistance or endurance)	• Catabolic: Autophagy, mitophagy, SIRT1/FOXO • Anabolic: Mitochondrial biogenesis, mTORC1, IGF-1 • Cellular repair, proteostasis	• Extended healthspan • Reduced frailty • Improved mitochondrial function	[32,33]
Cardiovascular Resilience	• High-Intensity Interval Training (HIIT) • Remote ischemic preconditioning • Mild oxidative stress (e.g., ozone, hypoxia exposure)	• Catabolic: ROS → Nrf2/HSPs, shear stress signaling • Anabolic: Angiogenesis, endothelial nitric oxide production, anti-inflammatory cytokines	• Improved vascular function • Reduced endothelial senescence • Autonomic balance	[32,34]
Neuropsychiatric Health	• Controlled emotional challenge • Novelty and cognitive engagement • Therapeutic stress (e.g., exposure therapy)	• Catabolic: Acute HPA-axis activation, monoaminergic shifts • Anabolic: Synaptic remodeling, BDNF, OT/VP modulation, epigenetic plasticity	• Enhanced stress resilience • Reduced depression/anxiety • Cognitive flexibility	[33,35,36]

Hormesis is characterized by adaptive benefits arising from exposure to transient stressors. This is a biphasic phenomenon, wherein stress induces cellular and systemic mechanisms that feed forward to enhance resilience, repair, and longevity. Hormesis has garnered increasing recognition as a unifying paradigm across multiple biological disciplines [29,37,38].

In contrast to traditional views that regard stress as inherently deleterious, the hormetic model proposes that organisms can be preconditioned by moderate stress exposures to mount more effective responses to future challenges, thereby decreasing disease vulnerability and promoting systemic robustness [39,40].

2.3. *Ancient Peptides and Hormesis*

We suggest here that at the core of the neuroendocrine orchestration of stress-related hormetic adaptation are VP and OT, and their primary receptors. VP is the more primitive neuropeptide, believed to have evolved over 400 million years ago from vasotocin. OT-like molecules existed for millennia [15]. However, in its current form, OT appeared in conjunction with the evolution of mammals over 200 million years ago [15,29]. In mammals, both VP and OT are primarily synthesized in the paraventricular and supraoptic nuclei of the hypothalamus, although VP and OT usually appear in different cells [15,41]. These peptides exert widespread effects through both central and peripheral pathways, and are deeply implicated in modulating stress responsivity, social behavior, emotional processing, immune regulation, and metabolic balance.

VP is classically associated with the initiation of acute stress responses, operating primarily through the activation of the HPA axis. VP facilitates a suite of immediate survival-oriented adaptations, including elevated arousal, enhanced vigilance, cardiovascular activation, fluid retention, and mnemonic reinforcement [42,43]. While these responses are adaptive under conditions of short-term stress, sustained or dysregulated VP signaling has been implicated in pathological outcomes such as generalized anxiety, major depression, metabolic syndrome, and social dysfunction [18,35,36]. This dualistic capacity positions VP as a dose- and context-sensitive regulator. The controlled activation of VP supports the hormetic enhancement of stress tolerance, but chronic overactivation can promote allostatic overload.

OT also may be released by challenges, but OT seems to work more slowly than VP [20]. The functions of OT include the facilitation of the recovery and restorative phases of reactions to stressful experiences [10]. OT can downregulate HPA axis activity, mitigate inflammation, enhance immune surveillance, and promote parasympathetic reactivation. In mammals, these mechanisms are essential for the resolution of acute stress and the establishment of physiological homeostasis. Beyond its central neuromodulatory effects, OT has demonstrated systemic benefits, including the maintenance of gut barrier integrity, the modulation of gut microbiota, and the regulation of glucose and lipid metabolism, positioning it as a key player in metabolic resilience and healthy aging [44–46]. Furthermore, OT contributes to cognitive flexibility, emotional regulation, and the formation of prosocial bonds. OT, in conjunction with VP, maximizes the benefits of positive social experiences [20]. Collectively, OT interactions with VP may serve as buffers against chronic stress and even psychopathology [47,48].

Emerging research highlights a reciprocal and dynamic interaction between VP and OT, whereby these neuropeptides mediate the shift from acute stress activation to adaptive recovery. VP is temporally aligned with immediate threat detection and mobilization, whereas OT operates during the subsequent recalibration phase, fostering synaptic plasticity, tissue repair, and behavioral regulation. This biphasic neuropeptidergic modulation exemplifies the core principles of hormesis, enabling organisms to respond with both rapid adaptation and long-term resilience [46,49].

2.4. *Hormesis: Conceptual Foundations*

2.4.1. Oxidative Stress

At the cellular level, one of the most well-characterized manifestations of hormesis is the oxidative stress response. Subtoxic exposure to reactive oxygen species (ROS) can initiate adaptive subcellular signaling cascades that upregulate antioxidant defenses [e.g., nuclear factor erythroid 2-related factor 2 (Nrf2) activation], enhance DNA repair, and promote autophagy and proteostasis, thereby increasing cellular longevity and functional resilience [32,34,50]. This concept extends into metabolic hormesis, as observed in caloric restriction, intermittent fasting, and other forms of nutrient stress, which activate energy-sensing pathways such as AMP-activated protein kinase (AMPK), Sirtuin 1 (SIRT1), and Peroxisome proliferator-activated receptor gamma coactivator 1-alpha (PGC-1α). These in turn induce mitochondrial biogenesis, improve oxidative phosphorylation efficiency, and reduce systemic inflammation, which are hallmarks of delayed aging and metabolic flexibility [33,51–54] (Table 2).

2.4.2. HPA Axis Hormones and Hormesis

An essential upstream regulator of stress-induced hormesis is the CRH system, which coordinates central and peripheral responses to homeostatic perturbation. CRH, a hypothalamic neuropeptide released from parvocellular neurons in the paraventricular nucleus (PVN), plays a critical role in orchestrating the initial activation of the HPA axis, thereby facilitating energy mobilization, vigilance, and behavioral arousal in response to stressors. This activation represents the early catabolic phase of the hormetic cycle, promoting systemic readiness and cellular defense through glucocorticoid secretion and sympathetic stimulation [55]. Recent advances expanded the CRH paradigm to include its peripheral and centrally acting paralogs, the urocortins (UCN1, UCN2, and UCN3), which engage both CRH receptor type 1 (CRHR1) and type 2 (CRHR2) with divergent physiological outcomes.

CRHR1 activation by CRH and UCN1 is typically associated with acute stress mobilization and anxiety-like behavior, mirroring VP-mediated HPA axis activation and catabolic mobilization [56,57]. Conversely, CRHR2, primarily activated by UCN2 and UCN3, has been shown to promote stress recovery, cardiovascular stability, and anxiolytic effects by functioning as a molecular mediator of the anabolic, restorative phase of the hormetic response [56]. These receptor-specific dynamics recapitulate the broader catabolic–anabolic oscillation seen across hormetic systems and align with the neuropeptide dichotomy observed in VP and OT signaling.

Moreover, urocortins (UCNs) act at both central and peripheral sites, including the heart, gastrointestinal tract, and immune system, where they modulate inflammation, oxidative stress, and metabolic regulation, features consistent with broader hormetic adaptation [58,59]. UCN2 and UCN3, in particular, have been implicated in enhancing cardiovascular resilience and attenuating stress-induced autonomic dysfunction, thereby serving as potent endogenous hormetins in their own right. Their biphasic effects are tightly regulated by tissue-specific receptor expression and feedback sensitivity, thus rendering them highly responsive to stress intensity, duration, and context.

Importantly, the CRH-UCN system also interfaces with VP and OT pathways to fine-tune behavioral and physiological outcomes. CRH and VP act synergistically at the pituitary to enhance ACTH release, amplifying glucocorticoid output during acute stress, while urocortin-driven CRHR2 activation counterbalances this effect by dampening HPA axis activity and promoting parasympathetic recovery, mechanistically resembling OT-mediated stress resolution [57,60]. These intersecting networks underscore a hierarchically organized stress-adaptation system, wherein CRH family peptides initiate and regulate the transition between the mobilization and restitution phases of hormesis.

The CRH-UCN axis represents a critical neuroendocrine interface between environmental challenges and systemic adaptation. Its inclusion in the hormetic model enriches our understanding of how organisms maintain stability, growth and restoration through change, using precisely timed, receptor-specific neuropeptide signaling to convert stress exposure into long-term physiological gains. Future research into selective CRHR2 agonists or UCN mimetics may offer novel therapeutic avenues for enhancing resilience, particularly in individuals with dysregulated HPA axis dynamics or impaired stress recovery.

Hormesis has particular relevance in the context of neuroprotection and cognitive resilience, where mild cognitive challenges and psychological stressors can stimulate neurogenesis, synaptic remodeling, and increased resistance to neurodegenerative processes. These forms of behavioral hormesis engage molecular mechanisms such as brain-derived neurotrophic factor (BDNF) upregulation, mitochondrial adaptation, and glutamatergic signaling balance [61,62].

Importantly, stress resilience is not merely a function of exposure magnitude, but of precise timing, context, and neuroendocrine feedback, features central to the hormetic framework.

Within this adaptive schema, neuropeptides such as VP and OT [10] have emerged as critical regulators of systemic stress responses. VP predominantly facilitates catabolic mobilization, acting via HPA axis activation, sympathetic upregulation, and vasoconstrictive pathways. In contrast, OT is associated with anabolic recovery, vagal activation, and anti-inflammatory signaling [63,64]. Together, these neuropeptides mediate a bi-phasic oscillation between challenge and recovery, positioning the components of the VP-OT system as hormonal drivers of stress-response hormesis [65].

Integrating VP and OT into the broader framework of hormesis reveals a neuroendocrine model of adaptation, in which the strategic alternation between mobilization and restitution shapes both physiological and behavioral outcomes. This paradigm has wide-ranging implications, from cardiometabolic adaptation and neuropsychiatric resilience to interventions in aging and neurodegenerative disease, particularly those that exploit cyclic stress exposures, such as intermittent fasting, cold/heat stress, high-intensity interval training, and controlled psychosocial engagement [47,66,67].

Hormesis is also closely linked to stress resilience and adaptation, particularly in the nervous system. Exposure to mild psychological stress can enhance cognitive function and emotional resilience, a phenomenon described as behavioral hormesis [46]. In this context, neuropeptides such as VP and OT play a pivotal role in stress adaptation, helping to regulate the balance between the challenge phase (VP-driven) and the recovery phase (OT-driven) [47] Figure 1.

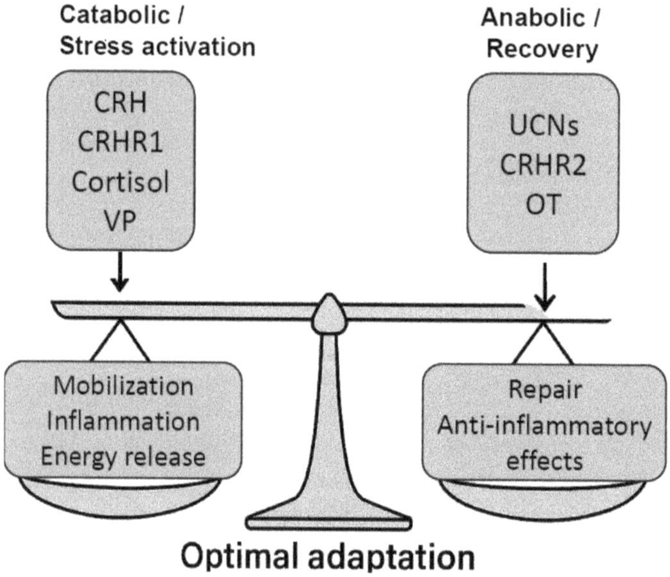

Figure 1. Hormesis represents the dynamic balance between catabolic stress activation and anabolic recovery, orchestrating optimal physiological adaptation. During the catabolic phase, mediators such as cortisol, CRH, VP, and CRHR1 signaling promote mobilization, inflammation, and energy release necessary for immediate stress responses. In contrast, the anabolic recovery phase is governed by OT, UCNs, and CRHR2 pathways, which facilitate tissue repair and exert anti-inflammatory effects. Maintaining equilibrium between these opposing forces enables the organism to maximize resilience and adaptability, highlighting hormesis as a critical framework for health optimization.

By incorporating VP and OT into the hormetic framework, a novel perspective emerges in which these neuropeptides contribute to stress-response hormesis by dynamically modulating physiological and behavioral adaptation processes Figure 1. The dynamic involvement of VP and OT as dual-phase modulators expands this framework, offering a translational lens through which to reframe therapeutic strategies in stress-related pathology, age-associated decline, and chronic disease prevention.

3. Catabolic–Anabolic Cycling Hormesis Model: A Dynamic Framework for Adaptive Health

The CACH model represents an emerging conceptual framework that extends classical hormesis into a temporal and phase-dependent paradigm, wherein biological systems alternate between stress-induced catabolic activation and compensatory anabolic recovery. Unlike linear models of stress adaptation, CACH emphasizes the rhythmic and cyclical nature of organismal resilience, anchoring the health-promoting benefits of transient perturbations in their capacity to trigger restorative, regenerative, and homeostatic processes.

3.1. Theoretical Foundations and Examples

At its core, CACH integrates the well-established biphasic dose–response curve of hormesis with biological oscillation theory, proposing that optimal adaptation requires not just the right magnitude of stimulus, but also precise timing, duration, and recovery dynamics [66,68]. The catabolic phase initiates cellular stress responses, e.g., via ROS signaling, AMP/ATP ratio sensing, or glucocorticoid mobilization, activating defense pathways such as autophagy, mitochondrial uncoupling, and immune priming. This is followed by an anabolic phase, where growth factors, anti-inflammatory mediators, and mitochondrial biogenesis coordinate structural repair and functional recalibration. The coordination between these opposing yet complementary forces allows for efficient adaptation to environmental stressors without tipping into maladaptation or exhaustion Table 3.

3.2. CRH, Urocortins, and the Neuroendocrine Modulation of CACH

CRH and UCNs play a foundational role in orchestrating the initiation, regulation, and resolution of stress responses within the CACH framework. CRH acts as the principal neuropeptide initiating the catabolic phase of stress adaptation. It stimulates ACTH release, thereby activating the HPA axis and promoting systemic glucocorticoid release [55]. This axis supports energy mobilization, immune vigilance, and increased arousal, all of which align with early-phase hormetic stress activation.

Table 3. Biphasic catabolic → anabolic cycles in key physiological systems. This table shows where catabolic-to-anabolic cycling appears across whole-organism systems. It summarizes how diverse biological systems engage in hormetic stress cycles, where an initial catabolic phase triggered by challenge is followed by an anabolic phase of restoration and adaptation. They reflect the CACH model, highlighting the evolutionary advantage of alternating between stress and recovery. Phase specific molecular drivers of CACH. This table outlines the key molecular regulators that operate during the catabolic and anabolic phases of hormesis. The listed signaling pathways function in a phase-specific manner to coordinate energy use, cellular defense, and tissue regeneration, supporting systemic resilience and adaptive plasticity under fluctuating stress conditions.

System	Catabolic Phase ("Challenge")	Anabolic Phase ("Restoration")	Primary References
Skeletal muscle physiology	• Myofibrillar micro-trauma produced by resistance exercise • ↑ AMPK & ROS signaling; temporary protein breakdown	• Fiber hypertrophy; satellite–cell fusion • Mitochondrial biogenesis (↑ PGC-1α, mTORC1)	[53]
Metabolic regulation	• Lipolysis, ketogenesis, autophagy during caloric restriction/intermittent fasting • Activation of SIRT1, FOXO, glucocorticoids	• Re-feeding restores glycogen; ↑ insulin sensitivity • Protein and lipid synthesis via IGF-1 → Akt → mTORC1	[51]
Cognitive performance/brain	• Controlled cognitive load or acute psychological stress → HPA-axis and catecholamine surge • Transient synaptic destabilization	• Parasympathetic rebound; sleep-dependent synaptic remodeling • Neurogenesis and BDNF-mediated circuit strengthening	[61]
Phase	**Key Signaling Nodes/Mediators**	**Core Functions**	**Primary References**
Catabolic (Energy-Mobilizing)	• AMPK, SIRT1, FOXO—energy conservation, autophagy, antioxidant defense • Glucocorticoids, catecholamines, vasopressin—substrate mobilization, cardiovascular tone • Sub-toxic ROS → Nrf2, HSPs, UCPs—stress-sensor activation, mitochondrial uncoupling	Maintain vital function under challenge; initiate cellular "clean-up"	[51,53,69]
Anabolic (Recovery/Growth)	• mTORC1, IGF-1, Akt—protein synthesis, mitochondrial biogenesis, membrane repair • Oxytocin, IL-10, BDNF—anti-inflammatory resolution, synaptic plasticity, tissue re-growth	Restore and enhance structure/function; build resilience to future stressors	[32,33,64]

However, the CRH system does not act in isolation. Its paralogs, the UCNs, modulate stress responsiveness in a receptor- and phase-specific manner, aligning tightly with the cyclical logic of CACH. UCN1 binds to both CRHR1 and CRHR2 and plays a role in initiating the early stress response, while UCN2 and UCN3 exhibit preferential binding to CRHR2, a receptor system associated with recovery, cardiovascular stability, and anxiolytic signaling [56,70]. This receptor divergence reflects a mechanism for transitioning from catabolic stress mobilization to an anabolic resolution, which is a hallmark of CACH's temporal organization.

CRHR2 activation, in particular, is implicated in parasympathetic rebound, cardiac repair, immune rebalancing, and behavioral recovery, functions similar to OT's role in the anabolic phase [70]. UCNs, therefore, act as endogenous hormonal switches, complementing VP and OT by refining the amplitude, duration, and systemic scope of stress-phase transitions. These peptides are expressed not only in the brain but also in peripheral tissues such as the heart, gut, and vasculature, where they modulate inflammatory tone, vascular remodeling, and redox homeostasis [58,59].

Moreover, CRH and VP synergistically potentiate ACTH release during acute stress, amplifying the mobilization of energy substrates and behavioral arousal [43]. This cooperation ensures that the catabolic phase is robustly activated during high-salience stress exposures. In contrast, OT and CRHR2 agonism (via UCN2/3) serve to terminate the stress response, engaging vagal pathways and promoting anti-inflammatory, neurotrophic, and tissue-regenerative processes [46]. The opposing yet coordinated actions of these peptide systems provide molecular substantiation for the phase-specific dynamics of the CACH model.

Taken together, the CRH–UCNs axis functions as an upstream timing mechanism, activating early catabolic defenses and coordinating their resolution through receptor-specific feedback loops. When viewed through the lens of the CACH framework, this neuroendocrine system reinforces the cyclical nature of adaptive physiology, mediating transitions between stress engagement and reparative recalibration. Integrating CRH and UCNs alongside VP and OT enhances our understanding of the molecular choreography underlying resilience, and provides a more nuanced platform for designing phase-targeted interventions across diverse clinical domains.

4. Vasopressin and Oxytocin as Endocrine Drivers of CACH

In addition to CRH and UCNs, VP and OT play an important role in orchestrating the catabolic–anabolic axis of stress adaptation. VP is predominantly active during the catabolic stress phase, enhancing sympathetic tone, vasoconstriction, and HPA axis activity. In contrast, OT becomes prominent during the anabolic recovery phase, supporting vagal activation, anti-inflammatory signaling, endothelial repair, and mitochondrial stabilization [63,65]. This bidirectional neuroendocrine rhythm aligns with autonomic cycling (sympathetic vs. parasympathetic), and can be entrained or modulated via behavioral, environmental, or pharmacologic stimuli.

5. Stress

Stress constitutes a multidimensional construct encompassing psychological, physiological, and environmental challenges to an organism's homeostasis. Within the framework of hormesis, stress functions not solely as a pathological insult, but as a potentially salutogenic stimulus capable of inducing adaptive biological responses when administered within an optimal dose–response window. The biphasic nature of the hormetic curve, with beneficial effects at low-to-moderate intensities and deleterious consequences at high exposures, underscores the critical importance of allostatic regulation, neuroendocrine coordination, and intracellular signaling in mediating organismal resilience [71,72].

5.1. Neuroendocrine Architecture of the Stress Response

The stress response is orchestrated by the neuroendocrine system, prominently featuring the HPA axis and what is sometimes described as the sympatho-adreno-medullary (SAM) system. Upon perception of a stressor, the PVN of the hypothalamus is activated, leading to the co-secretion of CRH and VP that act synergistically on corticotroph cells in the anterior pituitary to promote ACTH synthesis and release, culminating in the secretion from the adrenal cortex of glucocorticoids, primarily cortisol in humans and corticosterone in rodents [55]. Concurrently, the stress-induced activation of the locus coeruleus–norepinephrine (LC-NE) system and the adrenal medulla facilitates a rapid catecholaminergic response, driving cardiovascular, metabolic, and attentional adaptations necessary for immediate survival [73].

5.2. Vasopressin and Oxytocin in the Central Stress Network

Both VP and OT, synthesized predominantly by the magnocellular and parvocellular neurons of the PVN and supraoptic nucleus (SON), modulate the stress response through complex feedback and feedforward loops. VP, acting through V1b receptors in the anterior pituitary, enhances CRH-induced ACTH secretion, particularly during chronic or repeated stress, thereby sustaining HPA axis activity and potentially contributing to hypercortisolemia [46,55].

OT, acting through its G-protein coupled receptor expressed in limbic and hypothalamic circuits, exerts anxiolytic and stress-attenuating effects, partly by dampening amygdala reactivity and modulating central autonomic output [74]. OT's regulatory effects on the autonomic nervous system can also contribute to long-term social bonding and adaptive recovery following stress [6,75–77]. VP and OT not only influence endocrine outputs but also serve as neuromodulators, dynamically regulating synaptic plasticity, emotional processing, and social cognition. Their sometimes antagonistic but functionally complementary effects reveals a calibration of social and environmental adaptability, with VP facilitating vigilance and defensive behaviors, and OT promoting affiliative bonding and the social buffering of stress [35].

6. Hormetic Stress: Cellular Adaptation and Systems Resilience

The concept of hormetic stress challenges the traditional dichotomy of stress as either "good" or "bad," instead proposing a dose-dependent continuum wherein subtoxic exposures to stressors (e.g., heat shock, hypoxia, oxidative stress, caloric restriction, physical exercise, or psychosocial novelty) activate cellular defense pathways. These include the Nrf2 antioxidant system, HSPs, AMPK, sirtuins (e.g., SIRT1), and autophagic machinery, which collectively confer cytoprotection, neuroplasticity, and metabolic efficiency [71,78].

Stress-induced neuropeptides play integral roles in these hormetic adaptations. For instance, VP has been implicated in energy homeostasis and osmoregulation, linking neuroendocrine stress responses to mitochondrial efficiency and metabolic hormesis [79]. OT, conversely, has been shown to reduce ROS production and inflammation while enhancing parasympathetic tone and social reinforcement learning, positioning it as a neuromodulator of stress resilience and organismal recovery [6,80]. It was shown that OT evolved as a molecular safeguard against oxidative stress, acting at both systemic and mitochondrial levels to preserve cellular resilience [81].

6.1. Chronic Stress, Allostatic Load, and Pathophysiological Transition

When stress becomes chronic, cumulative, or dysregulated, the physiological benefits of hormesis are replaced by allostatic load, the wear and tear resulting from the prolonged activation of adaptive systems [82]. Under such conditions, VP expression is upregulated, often independently of CRH, sustaining elevated HPA axis activity and glucocorticoid output, which can lead to hippocampal atrophy, insulin resistance, immunosuppression, and mood dysregulation [60,83]. Simultaneously, chronic stress blunts the oxytocinergic system, impairing social functioning and increasing vulnerability to affective disorders such as depression, anxiety, and PTSD [84]. These impairments may result from the downregulation of oxytocinergic tone and social buffering mechanisms, which have been described as critical mediators of resilience under adversity [48].

Notably, the differential epigenetic regulation of VP and OT genes in response to early-life adversity or trauma further compounds stress susceptibility across the lifespan, supporting a developmental programming model of stress neurobiology [84]. This underscores the clinical relevance of these neuropeptides not only in acute stress modulation, but also in long-term neuropsychiatric trajectories. Although oxytocin supports healthy aging processes, its concentration generally decreases with age due to its association with reproductive function.

6.2. Integration into the Hormetic Framework: Oxytocin and Vasopressin as Bidirectional Modulators of Allostatic Flexibility

Integrating VP and OT into a hormetic model of stress physiology requires recognizing their reciprocal roles in regulating allostatic flexibility, the organism's capacity to mount, modulate, and recover from neuroendocrine and behavioral responses across varying intensities of environmental demands. Hormesis, characterized by a biphasic dose–response pattern wherein moderate stress induces adaptive resilience while excessive stress elicits pathology, hinges on temporally coordinated activity across central and peripheral signaling networks [85].

VP contributes primarily to the reactive phase of allostasis by potentiating HPA axis activity. Via V1b receptor activation, VP synergizes with CRH to promote ACTH secretion and downstream glucocorticoid release, thus enhancing metabolic readiness and cardiovascular tone [86]. However, chronic VP signaling can be associated with allostatic overload, contributing to neuroendocrine rigidity, glucocorticoid resistance, and behavioral inflexibility, hallmarks of maladaptive stress reactivity seen in affective and anxiety disorders [46].

Conversely, OT is preferentially engaged during the resolution and recovery phases of the stress cycle. Central OT receptor (OTR) activation in limbic and hypothalamic circuits, such as the amygdala and ventral hippocampus, attenuates CRH neuron activity, dampens amygdalar hyperexcitability, and re-establishes parasympathetic dominance [46]. OT also facilitates social buffering, reduces neuroinflammation, and enhances hippocampal plasticity and neurogenesis, contributing to the restoration of homeodynamic equilibrium [46,87].

Together, VP and OT define a bidirectional regulatory axis that orchestrates both the mobilization and resolution of stress responses. While VP facilitates immediate systemic robustness during threat exposure, OT supports plasticity, affiliative behavior, and physiological recalibration. The dysregulation of this axis, particularly VP overactivation coupled with insufficient OT tone, has been implicated in the pathophysiology of post-traumatic stress disorder (PTSD), depression, and social dysfunctions [46,88]. This bidirectional modulation supports the conceptualization of OT and VP as evolutionary complements [48], balancing threat detection with social healing and neuroendocrine flexibility [89].

Therefore, the balance between VP and OT signaling constitutes a central mechanism of neuroendocrine hormesis, modulating not only acute stress reactivity but also long-term resilience and adaptive capacity. Recent findings suggest that therapeutic interventions targeting this axis, such as intranasal OT, VP antagonists, or neuromodulation strategies, may enhance psychological recovery, reduce stress-related morbidity, and mitigate cognitive decline in aging populations [90]. However, individual differences in this system are common, and in part may reflect the cumulative properties of hormesis across the lifespan.

6.3. Vasopressin and Oxytocin as Hormetic Modulators

Beyond VP and OT's classical roles in osmoregulation and parturition, they function as modulators of the stress response, exhibiting properties consistent with hormetic agents. We have previously discussed the integrative roles of OT and VP in social attachment and their capacity to modulate stress responses [91]. Their effects are highly context-dependent, influenced by factors such as dosage, exposure duration, receptor subtype activation, and the specific neural circuits involved. We have shown that the behavioral effects of OT and VP vary depending on the perceived emotional context and individual history, reflecting their complex roles in social behavior [89]. At optimal levels, VP and OT can promote adaptive recovery and resilience; however, dysregulation, whether through excessive or insufficient signaling, may lead to maladaptive outcomes.

VP has been implicated in the modulation of social behaviors and stress responses. For example, our prior research in prairie voles indicated that VP plays a crucial role in social bonding and attachment, with its effects being modulated by early social experiences [48]. Studies indicate that VP can influence aggression and social bonding, with effects varying based on administration route and receptor interaction. In another example, in socially isolated mice, VP administration dose-dependently inhibited heightened aggression, suggesting a nuanced role in social behavior modulation [57]. Conversely, chronic elevated VP levels have been associated with increased stress reactivity and anxiety-like behaviors, highlighting the importance of balanced VP signaling.

OT facilitates social bonding and reduces stress. Its administration has been shown to decrease nonsocial risk-based decision-making, indicating a potential role in enhancing prosocial behaviors [92]. OT's actions are influenced by emotional context and individual experiences, contributing to its role in social attachment and stress regulation [93]. However, the effects of OT are complex and can be influenced by factors such as sex, context, and individual differences. For example, OT has been found to exert sex-specific effects on social behaviors, with variations observed in aggression and pair bonding [94]. Additionally, excessive OT stimulation may lead to receptor desensitization or disrupt synaptic homeostasis, underscoring the importance of dosage context and individual histories in its therapeutic application.

The cellular interplay between VP and OT systems adds yet another layer of complexity. Both peptides can bind to each other's receptors, leading to potential crosstalk and varied behavioral outcomes [89]. The interactions between OT and VP are crucial for understanding their roles in social behaviors, as their effects can be paradoxical and context-dependent. The activation of OTRs has been associated with both pro-social and, in certain contexts, pro-aggressive behaviors, potentially mediated by endogenous OT. In contrast, the administration of synthetic OT or VP has been reported to reduce aggression, likely through the activation of vasopressin V1a receptors [95]. This apparent bidirectional modulation underscores the importance of context, dosage, and receptor distribution, suggesting that the behavioral effects of OT and VP depend on a finely tuned balance between their signaling pathways.

In therapeutic contexts, understanding the hormetic properties of VP and OT is essential. For example, the intranasal administration of OT has been explored as a potential treatment for social deficits in neuropsychiatric disorders. The potential for targeting OT and VP pathways for therapeutic interventions in social deficits is associated with neuropsychiatric disorders, highlighting the importance of individualized approaches [48]. Outcomes of interventions, such as intranasal OT, are often variable, and factors such as dosage, individual sensitivity, and the specific behavioral context play significant roles in determining efficacy [96]. Similarly, VP receptor antagonists are being investigated for their potential to mitigate stress-related disorders, but careful titration is necessary to avoid disrupting essential physiological functions. In addition, because VP and its receptors appear to be calibrated in early life by adversity or nurturing, the effects of manipulations of the VP system may appear paradoxical. These unexpected or undesired outcomes could possibly be explained by cross-talk between OT, VP and their receptors, as well as differential sensitivity of these receptors.

Taken together, the current literature on VP and OT suggests that these are integral components of the neuroendocrine system that exhibit hormetic properties, modulating stress responses and social behaviors in a context-dependent manner. Their dualistic nature, as facilitators of both adaptation and potential maladaptation, highlights the importance of precise regulation and the need for nuanced therapeutic approaches that consider individual variability and environmental context.

6.4. Vasopressin's Role in Neuroendocrine Plasticity

VP is increasingly recognized not only as a classical endocrine hormone involved in fluid balance and vasoconstriction, but also as a central neuromodulator with profound effects on neuroendocrine plasticity. Its ability to reshape neural circuits, tune stress reactivity, and mediate behavioral outcomes reflects its integration into the dynamic, hormetic architecture of the stress system. Within this framework, VP supports both immediate adaptation to challenges and long-term recalibration of the HPA axis and limbic processing through activity-dependent, epigenetic, and receptor-specific mechanisms.

6.4.1. Structural and Functional Plasticity of Vasopressin Neurons

VP is synthesized primarily by magnocellular neurons in the PVN and SON of the hypothalamus, as well as parvocellular neurosecretory cells that co-release CRH. These neurons exhibit notable structural plasticity in response to environmental stimuli such as osmotic imbalance or stress exposure. For example, dendritic hypertrophy, increased synaptic input, and glial remodeling have been observed in VP neurons during dehydration and chronic stress, enhancing neurosecretory efficiency [97]. This plasticity supports the concept of neuroendocrine metaplasticity, the long-term modulation of hormone release dynamics based on prior experiences [98]. At the synaptic level, VP neuron excitability is modulated by glutamatergic, GABAergic, nitric oxide, and astrocytic signaling, each of which contributes to flexible hormonal output [99]. This responsiveness allows VP networks to act as finely tuned transducers of internal and external signals, consistent with hormetic regulation.

6.4.2. Epigenetic Regulation and Early-Life Programming

VP gene expression is subject to epigenetic modifications, which can program long-lasting changes in stress physiology. As one example, early-life adversity, such as maternal separation, leads to the hypomethylation of VP promoter/enhancer regions in the PVN, resulting in the persistent upregulation of VP and heightened HPA axis reactivity in adulthood. These molecular changes are part of an adaptive developmental plasticity system that may confer survival advantages in stressful environments but increase vulnerability to anxiety or depressive disorders if environmental predictability is low [84]. VP-sensitive circuits in the bed nucleus of the stria terminalis (BNST), central amygdala, and lateral septum also undergo experience-dependent changes in connectivity and receptor expression, contributing to the modulation of aggression, fear, and social memory [100]. Developmental studies in prairie voles have provided compelling evidence that early-life manipulations of VP pathways, including neonatal handling or pharmacological exposure, can program long-lasting changes in VP receptor expression and aggression, reflecting a form of early-life neuroendocrine imprinting that parallels epigenetic modulation [101].

6.4.3. Receptor-Specific Signaling and Behavioral Adaptation

VP acts on three G-protein-coupled receptors—V1aR, V1bR, and V2R—of which V1aR and V1bR are predominantly expressed in the brain. The activation of V1aR in the amygdala and septum facilitates territorial aggression, social memory, and pair bonding, while V1bR activation in the anterior pituitary stimulates ACTH release and augments HPA axis tone [35,102]. Recent optogenetic and chemogenetic studies have shown that the temporal patterning of VP neuron activity in the extended amygdala can bidirectionally control anxiety-like behavior, suggesting that VP's behavioral impact is tightly linked to network dynamics and circuit state [95,103]. Consistent with these findings, it was shown in prairie voles that developmental VP exposure can lead to increased adult aggression, highlighting the enduring impact of receptor-mediated signaling during critical periods of brain plasticity [104].

6.4.4. Integration of Vasopressin into Hormetic Stress Networks

VP's role in hormetic adaptation lies in its capacity to prime the stress response system under moderate stress while enabling structural remodeling for future responsiveness. Subthreshold VP release can enhance neurotrophic signaling, glucose metabolism, and attentional resources, thereby optimizing brain function under load [105]. However, when VP signaling becomes chronic or dysregulated, particularly in the absence of counter-regulatory oxytocinergic tone, it may contribute to allostatic overload, anxiety disorders, or metabolic dysregulation [46,86]. These outcomes support the general hypothesis that OT and VP act as neurochemical pivots between adaptive cooperation and defensive reactivity, depending on environmental cues and internal states [89]. Thus, VP exemplifies a bidirectional neuroendocrine effector, capable of mediating resilience or pathology depending on dosage, timing, receptor specificity, and environmental context. Its capacity to regulate plasticity at molecular, cellular, and network levels situates it as a central modulator within the hormetic stress architecture.

6.5. Oxytocin's Behavioral and Cellular Roles in Hormetic Regulation

OT is not simply as a "prosocial hormone" but also serves as a context-sensitive regulator of homeostasis. The actions of OT align closely with several benefits attributed to hormesis, and especially the adaptive calibration of biological systems in response to mild or moderate challenges. We specifically hypothesize here that OT may play a pivotal role in what has been called stress-response hormesis.

6.5.1. Context-Dependent Modulation of Stress and Behavior

OT's effects on stress and social behavior are profoundly context-dependent, shaped by internal state, early-life experiences, social environment, and OTR expression patterns [106]. Under moderate, predictable psychosocial stress, endogenous OT release facilitates social engagement, affiliative behavior, and stress attenuation. These effects are supported by evidence showing OT-induced amygdala deactivation and enhanced functional connectivity between the amygdala and prefrontal cortical regions, promoting top-down emotional regulation [69,87].

Such outcomes are consistent with a hormetic interpretation—low to moderate adversity stimulates adaptive neurobehavioral responses that increase the organism's resilience to future stressors [107]. However, in contexts involving unpredictable social threat or early-life adversity, OT may actually enhance social vigilance, ingroup bias, or even anxiogenesis. This outcome has been interpreted to indicate that OT's principal role is to amplify the salience of social stimuli rather than universally promote prosociality [108]. This dual action reflects OT's function as a salience modulator, calibrating behavioral responses in a dose- and context-sensitive manner that is emblematic of hormetic signaling. Although oxytocin has been associated with stress attenuation and recovery, it is not inherently a calming hormone. Instead, OT release often follows stimulating challenges such as endurance exercise, parturition, cold exposure, fasting, or thermoregulatory stress, reflecting its role as a dynamic modulator of adaptive responses rather than a simple stress buffer.

6.5.2. Cellular Mechanisms of Oxytocin-Mediated Hormesis

At the cellular level, OT promotes neuroprotection, synaptic plasticity, and oxidative resilience, hallmarks of cellular hormesis. OTRs are broadly expressed across the central nervous system, immune tissues, and cardiovascular system, enabling OT to exert cytoprotective and anti-inflammatory actions [46]. OT may have evolved as a molecular mechanism for regulating oxidative stress and mitochondrial health, supporting its role in cellular resilience and inflammation control [81]. In neurons, OT enhances mitochondrial efficiency, modulates intracellular calcium dynamics, and reduces oxidative stress through the upregulation of antioxidant enzymes such as superoxide dismutase (SOD) and glutathione peroxidase [109].

OT also facilitates hippocampal neurogenesis and synaptic remodeling, particularly under conditions of intermittent stress exposure, via pathways that overlap with classic hormetic mediators such as BDNF, MAPK/ERK, and PI3K/Akt signaling cascades [110]. These actions not only support cognitive flexibility and emotional regulation, but also protect against neurodegenerative processes. Furthermore, OT suppresses NF-κB activation, reducing pro-inflammatory gene expression and promoting immune tolerance under low-grade immune challenges [110].

6.5.3. Social Hormesis: Affiliative Behavior as Adaptive Stressor

Recent theories have proposed the notion of social hormesis, whereby moderate social demands, such as cooperation, novelty, or conflict resolution, act as adaptive stressors that strengthen emotional and cognitive resilience [48,111]. OT plays a central role in this process: its release in response to social touch, eye contact, and shared goal pursuit not only enhances social bonding, but also initiates stress-buffering physiological cascades [109]. These mechanisms enable the organism to remain engaged in complex social environments without succumbing to chronic stress. As our previous work [48] argues, OT-mediated affiliative behaviors are evolutionarily conserved mechanisms of biobehavioral regulation, wherein moderate social challenges enhance cooperative behavior, stress tolerance, and long-term adaptive capacity. Oxytocin's evolutionary role includes shifting behavioral priorities from food intake toward reproductive behaviors and mate-seeking, which helps explain its diverse behavioral effects.

Importantly, low-dose exogenous OT administration in clinical and experimental contexts has been shown to reduce cortisol levels, increase parasympathetic tone, and enhance threat appraisal accuracy, which are outcomes that reflect hormetic optimization of the stress axis [96]. However, sustained or high-dose administration may lead to OTR desensitization or network dysregulation, again underscoring the hormetic principle that "more is not necessarily better", and that timing, context, and dose are critical determinants of function.

For instance, music-based social interactions such as group singing or rhythmic coordination have been shown to stimulate oxytocinergic activity and enhance prosocial bonding [112], supporting the idea of "social hormesis" through shared sensory and emotional experience.

OT exemplifies a bi-directional hormetic modulator, operating across behavioral and cellular domains to optimize organismal function under moderate challenge. Its ability to both buffer stress and amplify social salience, depending on the environmental and internal context, makes it a precision regulator of adaptive capacity, with broad implications for neuropsychiatric resilience, aging, and immuno-metabolic health.

7. Therapeutical Implications and Translational Opportunities

Framing OT and VP within a hormetic framework repositions these peptides not as binary stress mediators but as precision regulators of stress-response plasticity. Their therapeutic utility lies in the capacity to modulate neuroendocrine tone, behavior, and systemic function in a dose-, context-, and state-dependent manner.

7.1. Precision Neuropeptide Modulation

OT-based therapeutics, particularly intranasal administration, have demonstrated promise in treating autism spectrum disorder (ASD), schizophrenia, borderline personality disorder, and social anxiety disorder, via modulation of social salience and affective processing [90]. However, clinical heterogeneity remains a major challenge. Efficacy may depend on OTR gene polymorphisms, attachment style, and early-life adversity, suggesting the need for personalized interventions. Trials incorporating functional neuroimaging, genetic markers, and hormetic biomarkers (e.g., stress reactivity profiles) could help stratify individuals based on their position on the hormetic response curve [87,96]. This aligns with the proposal that individual responsiveness to OT may depend on the developmental calibration of receptor systems, which could serve as biomarkers for tailored intervention strategies [48].

Similarly, V1bR antagonists have been investigated for the treatment of major depressive disorder (MDD), generalized anxiety disorder, and alcohol dependence, owing to their ability to attenuate hyperactive HPA axis signaling [113]. In contrast, perhaps selective VP agonists could be advantageous under conditions of cognitive fatigue or circadian disruption, as VP promotes wakefulness, working memory, and social vigilance, particularly when administered in short-term, hormetic dosing paradigms [46,105].

7.2. Hormetic Interventions and Lifestyle Medicine

Integrating neuropeptidergic modulation with lifestyle-based hormetic stressors offers synergistic translational potential. Lifestyle interventions, such as exercise, intermittent fasting, cold exposure, and cognitive novelty, are now recognized as controlled stressors that engage hormetic pathways via mitochondrial, anti-inflammatory, and neuroplastic mechanisms [61]. The following are examples.

Physical activity increases endogenous OT release, enhances hippocampal neurogenesis, and upregulates BDNF, suggesting a synergistic mechanism that could be optimized in rehabilitation protocols for PTSD, MDD, and neurodegenerative conditions [87].

Mindfulness-based interventions, attachment therapy, and breathwork have demonstrated effects on OT and VP regulation, recalibrating HPA axis tone and reducing allostatic load, especially in populations with trauma exposure or early-life stress [114]. Such interventions may work in part by engaging conserved oxytocinergic circuits that evolved to buffer oxidative and psychosocial stress, providing a physiological rationale for combining behavioral and biochemical strategies [81].

8. CRH, Urocortins, and the Hormetic Stress Response—Setting the Stage for Oxytocin and Vasopressin

While OT and VP are central to social, emotional, and metabolic adaptations to stress, their functional integration within a hormetic model is best understood against the broader landscape of the HPA axis. At the heart of this system lies the peptides, CRH and UCNs, which act as frontline mediators of stress and adaptive recalibration.

8.1. CRH as a Hormetic Initiator

Hormesis, characterized by a biphasic dose response, describes how exposure to mild or moderate stress can activate adaptive processes, whereas excessive stress becomes detrimental. The HPA axis, regulated primarily by CRH, is a central mediator of these adaptive and maladaptive outcomes. Within the broader context of hormesis, CRH is not simply a stress-activating molecule, it is a conditioning signal, shaping both immediate responses and long-term adaptive plasticity. However, within the framework of hormesis, where low-level stress enhances long-term resilience, CRH's role becomes more nuanced. It is not merely a trigger for downstream activation; CRH also serves as a neuroendocrine "primer" that sets the stage for both acute adaptation and long-term stress conditioning. In other words, CRH operates as a molecular threshold detector, helping the brain and body assess the severity of a challenge and calibrate the physiological response accordingly [107].

8.1.1. Catabolic Phase Activation: Metabolic and Circadian Modulation

CRH, released from the hypothalamic PVN, drives ACTH release from the pituitary and subsequent cortisol/corticosterone secretion. These hormones mobilize metabolic fuels, glucose, fatty acids, and amino acids, supporting energy-intensive processes during stress [115]. The acute elevation of cortisol facilitates gluconeogenesis, lipolysis, and amino acid catabolism, while transient insulin resistance conserves glucose for vital organs [116]. Fasting and exercise-induced CRH release also promotes autophagy, growth hormone secretion, and BDNF expression for neuroplastic benefits [61].

CRH regulates brown adipose tissue (BAT) thermogenesis via sympathetic drive and Uncoupling protein 1 activation, enhancing energy expenditure and cold-induced resilience [117,118]. This is complemented by thyroid and cortisol synergy, sex differences in estrogen-mediated BAT recruitment, and therapeutic implications for metabolic disease [119]. CRH also interacts with the suprachiasmatic nucleus to align cortisol rhythms with circadian timing, modulating glucocorticoid receptor sensitivity [120]. Circadian disruption via sleep loss or light pollution impairs this synchrony, increasing disease risk [121]. Paradoxically and in contrast, mild circadian challenge, like intermittent fasting, may improve resilience through CRH-mediated entrainment [122].

8.1.2. Epigenetic Reprogramming and Intergenerational Stress Signatures

CRH signaling induces epigenetic remodeling, DNA methylation, histone modification, and microRNA regulation, which persistently alters stress responsivity [123]. These modifications can be inherited, shaping behavioral traits and HPA tone across generations. Crucially, these epigenetic effects are reversible. Exercise, social support, and mindfulness interventions restore adaptive CRH regulation [124]. Our studies and others showed that OT and VP also interact with CRH-modulated epigenetic pathways, influencing bonding, emotional resilience, and social cognition [48,125,126].

8.2. CRH as a Conditioning Signal

8.2.1. Immediate Response Calibration

In the short term, CRH helps orchestrate a precisely scaled response. Rather than triggering an all-or-nothing alarm, CRH levels rise in proportion to the intensity and duration of the threat, thereby controlling the magnitude of cortisol release, sympathetic tone, arousal levels, and immune modulation. This titration reflects a hormetic principle, mobilizing just enough stress response resources to overcome the threat while avoiding unnecessary damage from overactivation [115,127].

8.2.2. Neuroplastic Priming

On a longer timescale, repeated or intermittent exposure to low or moderate levels of CRH appears to condition neural circuits involved in emotion regulation, memory, and autonomic control. This has been demonstrated in the prefrontal cortex, amygdala, and hippocampus, where CRH signaling influences synaptic plasticity, dendritic remodeling, and epigenetic programming [99,128,129]. For example, mild, intermittent CRH signaling enhances cognitive flexibility and emotional regulation, fostering the development of a more resilient brain capable of withstanding future stressors, a core outcome of hormetic adaptation. In contrast, chronic or excessive CRH exposure (as seen in toxic stress or trauma) disrupts this balance, leading to maladaptive plasticity, HPA axis dysregulation, and increased vulnerability to anxiety, depression, and metabolic disease [129].

9. CRH, Vasopressin and Oxytocin Interact in Developmental and Behavioral Conditioning

CRH also acts as a developmental cue, particularly in early life, where brief stress exposures can "tune" the HPA axis to become either more resilient or more sensitized, depending on the timing, intensity, and context of CRH activity. We demonstrated that early-life exposure to CRH and OT pathway manipulation causes sex-specific alterations in neuropeptide receptor expression and adult stress reactivity [101], aligning with the idea of CRH as a developmental calibrator of neuroendocrine tone. This is a form of stress inoculation, a hormetic phenomenon whereby early exposure to manageable stress can enhance stress resistance later in life [130]. Moreover, several studies showed that CRH influences behavioral plasticity by modulating fear memory, risk evaluation, and stress coping strategies [128]. It helps encode whether a stressor is novel or familiar, and whether prior responses were successful, thereby conditioning the nervous system's future behavior.

CRH should not be viewed solely as a reactive hormone for acute stress, but as a hormetic regulator and conditioning molecule. Its role may span the following: acute stress orchestration, signal scaling and titration, long-term neuroplastic adaptation, behavioral programming, and the developmental priming of stress resilience. Therefore, CRH lies at the core of a finely tuned conditioning network, shaping both immediate survival responses and the capacity for long-term recovery, learning, and adaptation.

Furthermore, CRH acts as a behavioral conditioner, shaping coping strategies, novelty detection, and arousal modulation. It helps encode contextual threat and adaptive responses in real time, contributing to an organism's future behavioral flexibility. This function is especially relevant in intermittent stress paradigms like exposure therapy, fear extinction, and resilience training, where controlled CRH signaling facilitates memory updating and extinction learning [129].

Importantly, CRH also interfaces with cardiovascular adaptation. CRH not only activates the sympathetic nervous system via central autonomic pathways, but also modulates vascular tone and heart rate variability. In hormetic contexts, these effects can enhance cardiovascular efficiency and stress responsiveness. However, persistent CRH elevation, as in PTSD or chronic anxiety, leads to sympathetic overdrive and cardiovascular pathology [127].

Beyond its intrinsic signaling capacity, CRH's effects are intricately modulated by its interactions with VP and OT, two neuropeptides that share overlapping release pathways and physiological roles. In acute stress states, CRH and VP often act synergistically to enhance ACTH release and drive catabolic mobilization [92]; this synergism between CRH and VP facilitates metabolic mobilization and vigilance [89], while OT functions as an opposing force that downregulates CRH-driven sympathetic arousal and restores affiliative behavior post-threat. This CRH-VP coordination increases sympathetic activity, vascular resistance, and metabolic readiness, aligning with the initial phase of a hormetic response.

In contrast, OT frequently rises in the aftermath of stress, counteracting the effects of both CRH and VP. OT suppresses HPA axis hyperactivity, promotes parasympathetic tone, and modulates limbic circuitry to reduce anxiety and promote social recovery behaviors [6, 64,92]. These effects are well supported by our previous study [6], in which we proposed that OT acts as a neurochemical "recovery signal" following CRH-driven arousal, restoring homeostasis and modulating cardiovascular and emotional responses through prefrontal-limbic circuitry. As such, the CRH-OT dynamic can shape both the magnitude of the initial stress response and the quality of post-stress recovery and integration.

These interrelated dynamics suggest that CRH, VP, and OT act not as isolated hormones, but as neuroendocrine instructors in a broader hormetic conditioning network. Their coordinated oscillation fine-tunes immediate responsiveness and longer-term plasticity, enabling physiological systems to better predict, endure, and adapt to environmental fluctuation.

Therefore, CRH is not merely a stress response initiator, but a conditioning signal that calibrates neuroendocrine sensitivity, behavioral adaptation, cardiovascular reactivity, and long-term resilience. Through its interactions with VP and OT, it anchors a hormetic feedback loop that balances catabolic mobilization with anabolic repair and neuroplastic adaptation.

9.1. CRH as the Gateway to Neuroendocrine Hormesis

CRH initiates a cascade of metabolic, mitochondrial, behavioral, and epigenetic adaptations that exemplify the hormetic principle. While overactivation is pathological, controlled activation through mild physiological stress enhances resilience, flexibility, and energy efficiency [131]. CRH prepares the system not only for challenge but also for recovery, priming the body and brain for the buffering effects of VP and OT, which complete the hormetic cycle by fostering social connection, emotional stability, and long-term homeostasis [6].

9.2. Urocortins in Hormetic Stress Modulation—Bridging Catabolic Initiation and Adaptive Recovery

Urocortins play crucial roles in modulating adaptive responses to physiological stress. Functioning via CRH receptors (CRHR1 and CRHR2), UCNs integrate neuroendocrine signaling with metabolic resilience, mitochondrial adaptation, and cardiovascular protection. Within the hormesis framework, UCNs operate downstream of CRH to fine-tune stress responses, amplifying cellular defenses and promoting recovery after transient metabolic or psychosocial challenges.

9.2.1. Urocortin Subtypes and Functional Roles

UCN 1 binds both CRHR1 and CRHR2, and is implicated in mitochondrial biogenesis, neuroprotection, and metabolic flexibility. It activates PGC-1α, enhancing mitochondrial proliferation and oxidative metabolism [70], and supports glucose uptake and fatty acid oxidation under conditions of fasting or caloric restriction [132]. UCN1 also attenuates oxidative stress and apoptosis in the hippocampus, underscoring its neuroprotective capacity [133].

UCN 2 and 3 preferentially bind CRHR2 and mediate anti-inflammatory, cardioprotective, and metabolic functions. They mitigate myocardial ischemic injury and enhance cardiac performance under stress, suppress pro-inflammatory cytokines, and regulate insulin sensitivity and glucose metabolism, guarding against obesity-linked insulin resistance [134]. Additionally, UCN2/3 selectively bind CRHR2, promote adaptive thermogenesis by stimulating brown adipose tissue (BAT) activity, attenuate myocardial injury, enhance cardiac output, and improve insulin sensitivity [92,111,135].

9.2.2. Hormetic Mechanisms of Urocortins Action

Low-level Urocortins activation enhances mitochondrial function through PGC-1α upregulation and promotes ATP production. UCNs also reduce oxidative stress by upregulating antioxidant enzymes like SOD and glutathione peroxidase [92,132,136], provide vascular protection via eNOS activation, and improve systemic metabolic resilience through enhanced lipid and glucose utilization [134]. However, excessive UCNs signaling can disrupt mitochondrial homeostasis, elevate ROS, and impair electron transport chain efficiency. Chronic overactivation contributes to inflammation, cardiomyocyte hypertrophy, maladaptive remodeling, and HPA axis dysregulation, features linked to depression, anxiety, and metabolic syndrome [70,137].

9.2.3. Urocortins and the Transition to Anabolic Recovery

UCNs, particularly UCN2 and UCN3, complement CRH activity by promoting metabolic and emotional recovery. They activate CRHR2, fostering anti-inflammatory, cardioprotective, and neuroregenerative effects [70]. This mirrors the transition from a catabolic (stressor) to an anabolic (recovery) phase central to the CACH model. Importantly, this recovery phase primes the neuroendocrine environment for the action of OT and VP, which further refine stress responses through social and affiliative buffering mechanisms [6,48,57].

9.3. Integration into the Catabolic-Anabolic Cycling Hormesis Model

UCNs function as pivotal mediators of the CACH model. During the catabolic phase, UCNs alongside CRH activate the HPA axis, promote AMPK signaling [138], and facilitate energy mobilization through gluconeogenesis, fatty acid oxidation, and autophagy [61]. These actions build stress resistance and initiate epigenetic reprogramming toward metabolic flexibility [139].

In the anabolic phase, after CRH subsides, mTOR signaling is upregulated to drive cellular repair, glycogen storage, and protein synthesis [140]. Vasopressin and OT emerge during this recovery window to restore homeostasis, supporting emotional resilience, cardiovascular regulation, and social bonding [48,57,81]. UCNs' activity in this phase also enhances insulin sensitivity, promotes neuroplasticity, and reinforces long-term cognitive and physiological adaptation [131].

9.4. Urocortins as Bidirectional Stress Regulators

UCNs bridge the initial catabolic stress activation with anabolic recovery and resilience, demonstrating their key role in the hormetic stress response (Figure 2). Their ability to activate protective pathways, including AMPK, mitochondrial biogenesis, and anti-inflammatory signaling, positions them as essential players in the adaptive stress response [92,136]. However, the same mechanisms can become pathologic under chronic or excessive stimulation, leading to metabolic and neuropsychiatric dysregulation [70,137].

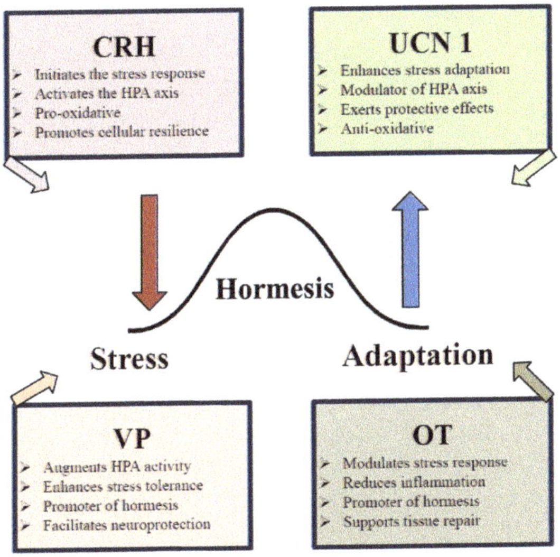

Figure 2. The regulation of hormesis is orchestrated by the dynamic interplay of CRH, UCN1, VP, and OT. CRH initiates the stress response through HPA axis activation, promoting a pro-oxidative state that primes cellular resilience. In contrast, UCN1 enhances stress adaptation by exerting protective, anti-oxidative effects and modulating the HPA axis. VP acts as a potent promoter of hormesis, augmenting stress tolerance, neuroprotection, and HPA axis activity. Meanwhile, OT facilitates recovery by modulating the stress response, reducing inflammation, and supporting tissue repair.

Framing UCNs within the CACH model provides a useful framework for understanding their dual role in stress biology. Their interactions with OT and VP further extend their influence into social, emotional, and intergenerational domains of adaptation [6,46]. Future research may aim to harness these pathways therapeutically, including through lifestyle interventions like exercise and fasting or targeted pharmacologic modulation, to optimize health span, metabolic performance, and resilience to psychological stress [136].

10. Preparing the Stage for Oxytocin and Vasopressin

10.1. Recovery Adaptation Modulators

OT and VP play crucial, complementary roles during stress recovery. OT dampens CRH-induced amygdala activation [49,141], enhances bonding, and promotes parasympathetic tone [49,81], facilitating emotional healing and social reengagement [142,143]. VP, though typically pro-stress during acute threat [43,55], supports circadian rhythm and social hierarchy maintenance under moderate stress [57,100,144]. Together with CRH and UCNs, the VP-OT system forms a feedback loop guiding transition from arousal to recovery [46,57,78]. OT counters CRH-driven anxiety by reducing amygdala activity [141,144] and enhancing vagal tone [49]. This transition supports healing, social approach, and resilience [46,142]. VP amplifies HPA output during acute stress [43,55] but facilitates adaptive responses like energy conservation, circadian alignment, and social vigilance in familiar or moderate stress [57,100,143]. CRH and UCNs initiate the stress response [55,138], while OT and in some cases VP orchestrate recovery [46,57], emphasizing the hypothesis that resilience can depend more on the repair phase than initial stress activation [78]. Adaptive outcomes depend on stressor intensity, timing, and the OT/VP-mediated recovery [46,57,78,145]. Resilience is reconceptualized as the capacity for adaptive recalibration, not just resistance [48,142].

10.2. Integration of Oxytocin and Vasopressin Function

VP initially mobilizes energy and alertness [36,43]; OT later facilitates recovery and plasticity [57,135]. The functions of both are shaped by epigenetics [101,144] and modulate neuroimmune and autonomic functions [6,81,113]. Early-life OT/VP exposure programs lifelong stress responsivity in a sex-specific manner [101,146]. VP promotes ACTH release and HPA activation [43,55,102]. Moderate levels enhance vigilance and metabolism [60], while chronic excess contributes to dysfunction, mood disorders, and gut issues [36,44,84,147–149]. The role of VP is biphasic—beneficial under mild stress or in the face of immediate demands, but harmful when prolonged [55,60]. OT blunts prolonged HPA activation [46,57,67], protects cognition [116], and supports stress recovery [46,48]. It preserves hippocampal structure [49,116], enhances social buffering [135], and modulates inflammation and immunity [135,150,151]. OT's actions span neuroendocrine, immune, and gastrointestinal domains [44,46,152,153], reinforcing its role in resilience. OT is not a panacea, but a neuropeptide with evolutionarily conserved and context-specific physiological effects, comparable to other hypothalamic hormones such as CRH and VP.

10.3. Dynamic OT-VP Interplay

OT and VP act in sequence; VP dominates during acute arousal [48,74], OT during recovery [49,135]. Their feedback loop calibrates stress responses; VP primes OT release [135,154], and OT suppresses excessive VP signaling [57,155]. Sex differences are prominent—males show stronger VP responses [156], while females benefit more from OT buffering [81,157,158].

10.4. Targeted Clinical Applications of OT–VP Modulation

VP–OT balance has therapeutic implications for PTSD, anxiety, depression, and gut–brain disorders. OT enhances recovery [48,135,142], while VP antagonists reduce hyperarousal [159,160]. OT also promotes metabolic health [161,162], cognitive longevity [135,163], and exercise adaptation [161,164].

10.5. Hormesis Blueprint for Resilience

Framing OT and VP within a hormetic framework repositions these peptides not as binary stress mediators but as precision regulators of stress-response plasticity. Their therapeutic utility lies in their capacity to modulate neuroendocrine tone, behavior, and systemic function in a dose-, context-, and state-dependent manner. This paradigm supports the need for a dynamic alternative to traditional fixed-dosing approaches in psychiatry and behavioral medicine, emphasizing individual trajectories, timing, and physiological thresholds. As we have emphasized, OT and VP evolved not merely as stress mediators, but as adaptive neurochemical systems capable of regulating resilience, attachment, and autonomic flexibility across varying stress loads [6]. Together, OT and VP constitute a biphasic stress-regulatory system, in which VP primarily facilitates mobilization and OT promotes repair and recovery [46,135]. When well-regulated, this dynamic pairing enables organisms to convert stress into growth and adaptation, a process that exemplifies hormesis [135,142]. Conversely, the dysregulation of this balance is associated with psychiatric, metabolic, and neurodegenerative pathologies [57,116]. A hormetic blueprint that integrates OT and VP thus redefines stress adaptation as a dynamic, recovery-driven process. The targeted modulation of these peptides, taking into account timing, dose, individual differences, and physiological thresholds, may offer therapeutic benefit across a spectrum of conditions including PTSD, anxiety, depression, gut–brain disorders, metabolic syndrome, cognitive decline, and impaired exercise recovery [81,135,159–166]. This framework invites further investigation into how precision neuroendocrine strategies can optimize resilience and systemic health by leveraging the bidirectional and state-dependent properties of OT–VP signaling Table 4.

Table 4. Comparative roles of oxytocin and vasopressin in hormetic stress regulation. This table summarizes the distinct yet complementary functions of OT and VP across key dimensions of the stress response within the hormetic framework. Each row highlights a specific physiological or behavioral domain, detailing the contributions of OT and VP to stress initiation, adaptation, and recovery. OT predominantly modulates the anabolic recovery phase, promoting emotional, autonomic, immune, and social recalibration, while VP primarily drives the catabolic activation phase, enhancing arousal, vigilance, and metabolic mobilization. Their dynamic interplay orchestrates resilience, with phase-specific and context-dependent modulation shaping adaptive versus maladaptive outcomes.

Aspect	Oxytocin (OT)	Vasopressin (VP)	References
Primary Role in Stress Response	Promotes recovery, emotional regulation, social bonding, resilience	Initiates acute stress response, vigilance, energy mobilization	[6,46,48,49,81,91,142]
Timing of Activation	Activated during post-stress recovery (delayed response)	Activated immediately during stress (early response)	[46,74,135]
Interaction with CRH	Inhibits CRH-induced amygdala activation; suppresses CRH, ACTH	Potentiates CRH activity and ACTH release	[46,57,141]
HPA Axis Effects	Negative feedback on HPA axis; enhances glucocorticoid receptor sensitivity	Stimulates HPA axis and increases glucocorticoid output	[46,47,57,67,135]
Social Behavior Modulation	Enhances prosocial behavior, trust, and social bonding	Supports dominance, aggression, territorial behavior	[48,81,135,142]
Cognitive Effects	Improves cognitive flexibility, prevents hippocampal atrophy	Enhances memory consolidation, but excessive levels cause anxiety and PTSD	[49,116]
Autonomic Regulation	Enhances parasympathetic tone, reduces sympathetic arousal	Increases sympathetic arousal; may disrupt recovery	[49,81,135]
Gastrointestinal Function	Restores vagal motility, reduces inflammation, supports gut integrity	Regulates motility (via V1a/V1b); excess leads to GI dysfunction	[44,46,152,153]
Immune Modulation	Suppresses pro-inflammatory cytokines; activates T-regs and M2 macrophages	Amplifies inflammation under chronic stress	[135,150,151]
Neuroendocrine Plasticity	Programs stress resilience, especially during development	Modulates HPA tone, stress coping (context-dependent)	[88,101,143]
Sex Differences	Stronger response in females; enhances cardiovascular and social resilience	Stronger in males; linked to aggression and prolonged HPA activation	[81,114,135,157]
Therapeutic Potential	Used in PTSD, depression, GI disorders, aging, and metabolic recovery	Targeted in PTSD, anxiety, schizophrenia; VP antagonists under study	[44,45,81,135,161,164–166]

11. Mechanistic Foundations

11.1. Hormetic Biphasic Signaling of Vasopressin and Oxytocin

In summary, both VP and OT demonstrate classic biphasic, dose-dependent properties consistent with the hormetic model, in which low to moderate levels of exposure confer adaptive benefits, while higher or chronic exposure results in adverse effects. These peptides are centrally involved in neuroendocrine coordination during stress and recovery phases, and modulate diverse physiological systems, including emotional regulation, cardiovascular control, mitochondrial function, and immune modulation. At low physiological concentrations, OT and VP enhance stress resilience, promote prosocial behavior, and regulate energy balance. However, at high doses or with prolonged activation, both neuropeptides can shift their effects toward promoting allostatic overload, inflammation, and maladaptive neurobehavioral outcomes [46,167].

OT, in particular, exhibits anxiolytic and anti-inflammatory effects at low doses by modulating HPA axis reactivity and enhancing parasympathetic tone. In contrast, repeated or high-dose administration may cross-activate V1a receptors and increase anxiety, promote aggression, or induce receptor desensitization. Similarly, VP promotes acute cognitive and cardiovascular responsiveness at low levels but exacerbates glucocorticoid resistance, sympathetic overdrive, and systemic oxidative stress when overexpressed or persistently elevated. These patterns of biphasic response align with the principles of hormesis, where moderate stress-induced signaling promotes long-term adaptive capacity, but excess burden tips physiological systems into maladaptation [85].

11.2. Neuroendocrine Integration

The coordination of the catabolic–anabolic shift through the dynamic interplay of CRH, UCNs, VP, and OT represents a temporally regulated and receptor-specific orchestration of stress response systems. The HPA axis initiates the catabolic phase, while the HNS (hypothalamic–neurohypophyseal system) participates in both stress amplification and its resolution. During acute stress, CRH is secreted rapidly (within minutes) from parvocellular neurons in the PVN, activating CRHR1 in the anterior pituitary to stimulate ACTH release and consequent adrenal glucocorticoid secretion [139]. Simultaneously, VP, co-released from adjacent PVN neurons, potentiates ACTH release via V1b receptors, especially under sustained or repeated stress. This synergistic CRH–VP action intensifies the early catabolic response, mobilizing energy stores, increasing cardiovascular output, and modulating immune and behavioral systems [83].

The temporal orchestration of stress responses by CRH, VP, UCNs, and OT follows a highly regulated sequence that reflects distinct yet overlapping neuroendocrine roles during the phases of stress initiation, maintenance, and resolution (Figure 3). As the stressor abates, UCNs and OT assume key roles in initiating the anabolic recovery phase. UCNs are expressed later in the temporal sequence and signal predominantly via CRHR2, which is expressed in the heart, gut, brainstem, and limbic circuits. The activation of CRHR2 facilitates the termination of HPA output, enhances parasympathetic activity, promotes cardiovascular and immune recovery temporal oscillation between activation (CRH/VP) and restoration (UCNs/OT), and reflects the core temporal structure of the CACH model (Figure 4), wherein controlled stress exposure followed by recovery supports adaptive homeostasis and stress resilience.

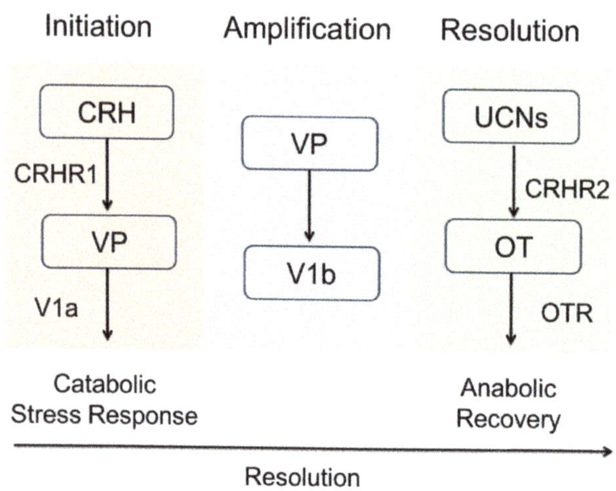

Figure 3. The temporal orchestration of stress responses by CRH, VP, UCNs, and OT follows a highly regulated sequence that reflects distinct yet overlapping neuroendocrine roles during the phases of stress initiation, maintenance, and resolution. CRH, UCNs, VP, and OT interact dynamically to coordinate the shift from catabolic stress responses to anabolic recovery.

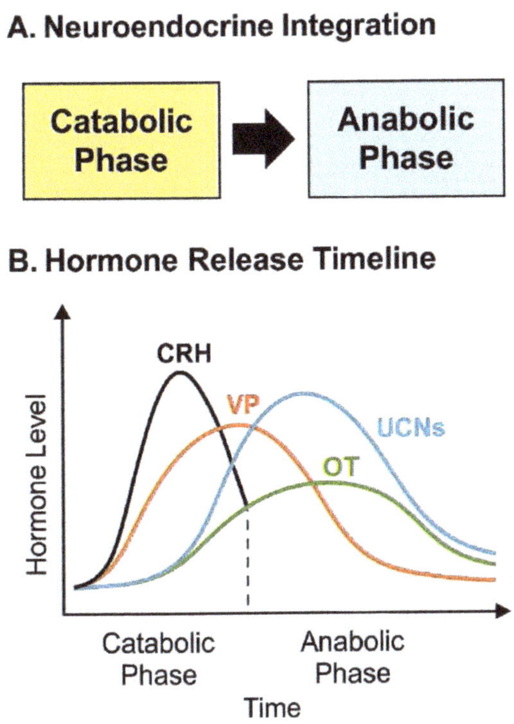

Figure 4. Neuroendocrine integration across stress phases. (**A**) Schematic representation of the biphasic neuroendocrine response to stress, transitioning from a catabolic phase (yellow) to an anabolic phase (blue), reflecting dynamic endocrine and metabolic adaptation. (**B**) Hormone release timeline showing relative levels of CRH (black), VP (red), UCNs (blue), and OT (green) across stress phases. CRH peaks early during the catabolic phase, followed by VP, which bridges both phases. UCNs and OT rise predominantly in the anabolic phase, supporting recovery, regeneration, and recalibration. Hormonal transitions are temporally aligned with physiological shifts from mobilization to restoration.

Interestingly, in all mammals, the OT and VP genes are positioned in close proximity in an inverted orientation on the same chromosome. This genomic architecture allows for mutually exclusive transcription within a single nucleus, a feature that may contribute to the reciprocal regulation observed at the functional level. Moreover, gene conversion events between the second exons of the OT and VP genes are frequent, limiting sequence divergence and possibly preserving complementary functions. This suggests strong evolutionary pressure to maintain this hormetic dyad, despite distinct transcriptional regulation.

11.3. The Temporal Role of Each Peptide in Initiating or Resolving Stress

The temporal orchestration of stress responses by CRH, VP, UCNs, and OT follows a highly regulated sequence that reflects distinct yet overlapping neuroendocrine roles during the phases of stress initiation, maintenance, and resolution (Figure 3). CRH is the earliest and primary hypothalamic peptide released in response to perceived threat, acting via CRHR1 receptors in the anterior pituitary to initiate ACTH secretion and consequent glucocorticoid release from the adrenal cortex [168]. This acute response rapidly mobilizes energy stores, suppresses non-essential functions, and enhances arousal. VP, co-released with CRH from parvocellular neurons in the PVN, sustains and amplifies ACTH secretion through V1b receptor activation, particularly under chronic or repeated stressors, thereby contributing to HPA axis sensitization [83]. This co-activation phase defines the catabolic window, characterized by elevated glucocorticoids, sympathetic drive, and metabolic breakdown. Further insights into the differential dynamics of acute versus chronic HPA axis activation are provided in [169].

In contrast, UCNs, particularly UCN2 and UCN3, emerge later in the stress sequence, and predominantly act via CRHR2 receptors to facilitate stress resolution and promote physiological recovery. CRHR2 signaling counteracts the excitatory CRHR1-driven responses by dampening HPA axis output, restoring cardiovascular tone, and exerting neuroprotective and anti-inflammatory actions [170,171]. UCN1, which has an affinity for both CRHR1 and CRHR2, appears to function as a regulatory buffer during the transition from stress engagement to recovery, modulating emotional responses in limbic circuits. Meanwhile, OT is released from magnocellular neurons of the hypothalamus, and is temporally aligned with the resolution and anabolic phase of the stress cycle (Figure 4). Acting through OT receptors, OT enhances vagal tone, reduces amygdala activation, promotes prosocial behavior, and stimulates anabolic repair mechanisms such as glucose uptake, anti-inflammatory signaling, and parasympathetic dominance [172,173] (Table 5).

Table 5. Temporal dynamics and functional roles of CRH, VP, UCNs, and OT in stress adaptation. CRH initiates the early catabolic phase within seconds, followed by VP sustaining mid-to-late catabolic responses [35,168,169]. UCNs promote transition to anabolic recovery, while OT supports prolonged anabolic repair, social bonding, and anti-inflammatory effects [170–173].

Hormone	Onset Time	Peak Time	Duration	Phase	Functional Role
CRH	Onset within seconds to 2 min after stress	Peaks at 10–20 min	Returns to baseline by ~60–90 min	Early Catabolic	Initiates HPA axis, stimulates ACTH, mobilizes energy
VP	Onset: 2–10 min	Peaks at 20–40 min	May persist up to 2–4 h, with chronic stress	Mid-to-Late Catabolic	Prolongs ACTH release, enhances cardiovascular and metabolic drive
UCNs (esp. UCN2/UCN3)	Onset: ~30–60 min post-stressor	Peaks at 1–3 h	Sustained up to 6 h	Transition to Anabolic	Dampens HPA activation, promotes neuroprotection and tissue repair
OT	Onset: ~20–60 min post-stressor (delayed)	Peaks at 1–4 h	Sustained release up to 12–24 h (especially in recovery-promoting contexts)	Anabolic	Supports parasympathetic tone, social behavior, anti-inflammation, and regenerative recovery

The specificity of these effects is critically shaped by the distribution and G-protein coupling of their respective receptor subtypes. VP signals through three receptors, V1a, V1b, and V2, each with distinct tissue localization and signaling properties. V1a receptors (coupled to Gq proteins) are widely distributed in the vasculature, brain, and liver, mediating vasoconstriction, aggression, and social memory. V1b receptors (also Gq-coupled) are predominantly located in the anterior pituitary and limbic regions, modulating ACTH release and behavioral reactivity. V2 receptors, in contrast, are found in the renal collecting ducts, where they couple to Gs proteins to regulate water reabsorption via cAMP-PKA signaling [174]. These divergent pathways allow VP to engage acute cardiovascular responses (V1a), neuroendocrine amplification (V1b), and systemic fluid balance (V2), depending on stressor type and physiological context.

Similarly, OTRs, which are primarily Gq-coupled, initiate PLC-IP3-mediated Ca^{2+} mobilization, but can also recruit PI3K/Akt, MAPK, and cAMP/PKA pathways in a context- and cell-type-dependent manner. Importantly, OXTRs exhibit region-specific expression in the hypothalamus, amygdala, brainstem, and peripheral tissues such as the heart and GI tract, where they mediate anti-inflammatory, anxiolytic, and parasympathetic functions [175]. Under high ligand concentrations or prolonged stimulation, OT can cross-activate V1a receptors, leading to vasopressor and potentially anxiogenic effects, especially under stress-primed conditions [176]. The receptor coupling profile thus determines whether VP/OT signaling promotes hormetic adaptation through the regulated, pulsatile engagement of survival and recovery pathways, or contributes to pathological allostatic overload through sustained excitotoxic and inflammatory signaling. Collectively, the neuropeptidergic system operates with temporal and receptor precision, coordinating the initiation, amplification, and resolution of stress through CRHR1/2, V1a/b, V2, and OTRs, with functional outcomes that range from catabolic mobilization to anabolic repair. This biphasic and phase-specific organization not only underpins normal stress adaptation, but also provides a mechanistic framework for therapeutic interventions that mimic or modulate these oscillatory dynamics.

12. Hormesis and Allostasis

VP and OT serve as critical neuromodulators within the neuroendocrine system, mediating both the amplification of acute stress responses and the facilitation of recovery processes, depending on timing, context, receptor engagement, and exposure history. Their dual actions significantly shape the balance between allostatic load, the cumulative physiological burden of chronic or dysregulated stress, and allostatic resilience, the organism's capacity to recover from stress while maintaining or enhancing system integrity [173,177].

VP, primarily acting through V1a and V1b receptors, augments hypothalamic–pituitary–adrenal (HPA) axis activity and sympathetic arousal during the catabolic phase of stress. Under acute, time-limited challenges, this facilitates adaptive responses—increased ACTH and glucocorticoid secretion, vasoconstriction, and glucose mobilization [83]. However, under recurrent or chronic stress, persistent VP signaling contributes to elevated allostatic load through glucocorticoid resistance, excessive sympathetic output, impaired negative feedback, and heightened inflammation [178]. In contrast, intermittent or context-appropriate VP release, as observed in circadian regulation or controlled social engagement, can promote stress inoculation and precondition neuroendocrine systems for future perturbations [179]. This aligns with evidence that VP modulates adaptive neuroplasticity in hippocampal and hypothalamic circuits when exposure is moderate and episodic.

OT, in contrast, plays a central role in buffering the consequences of stress and restoring homeostasis. Acting through OT receptors, OT attenuates amygdala activity, reduces HPA reactivity, and enhances vagal tone, key features of allostatic resilience [172]. Importantly, OT also facilitates mitochondrial repair, anti-inflammatory cytokine production, and emotional regulation, especially in social or affiliative contexts. Repeated intermittent OT release, such as that induced by physical touch, social bonding, or moderate physical activity, promotes the adaptive recalibration of the stress system, and may prevent the pathological accumulation of allostatic burden [180,181]. Thus, while VP amplifies stress responses, OT supports system recovery; both peptides can either exacerbate or mitigate allostatic load depending on dose, timing, and receptor context, forming a bidirectional regulatory axis.

Autonomic Effects of Vasopressin and Oxytocin

A core mechanism through which VP and OT influence resilience lies in their ability to mediate the switch between sympathetic and parasympathetic dominance during the stress recovery cycle. VP supports sympathetic drive, enhancing vasoconstriction and alertness via V1a receptor activity in the central and peripheral nervous system [174]. However, it also contributes to the maintenance of arousal states, which, when unresolved, delay recovery and reinforce chronic stress physiology. OT, conversely, promotes parasympathetic engagement through actions on the nucleus ambiguous and dorsal vagal complex, regulating heart rate variability (HRV), gastrointestinal motility, and cardiovagal tone [182]. Notably, OT's effects are potentiated in environments that provide psychosocial safety cues, such as social warmth or emotional closeness (Figure 5), highlighting its role in context-sensitive recovery initiation [173].

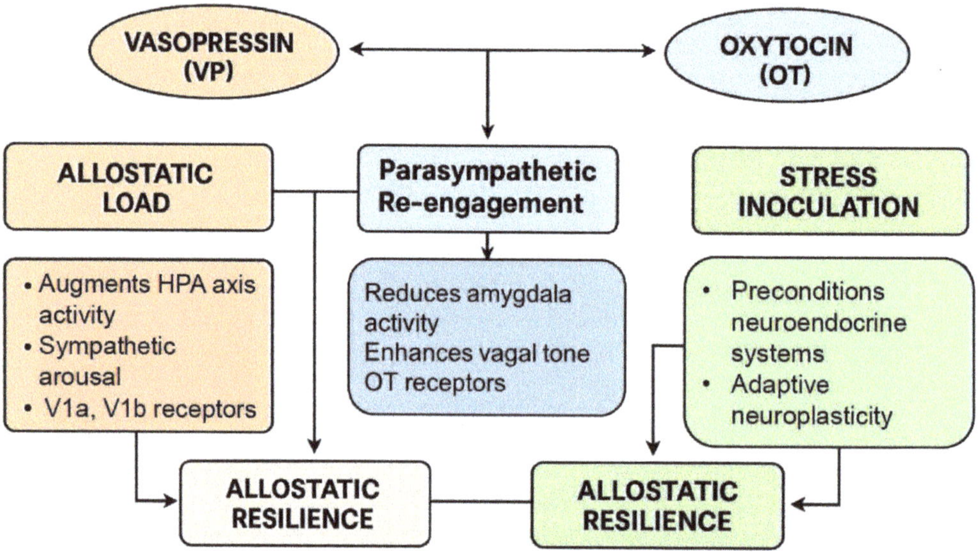

Figure 5. Vasopressin and oxytocin exert distinct but coordinated influences on stress adaptation. VP primarily promotes allostatic load by augmenting HPA axis activity and sympathetic arousal through V1a and V1b receptors, while OT facilitates stress inoculation by enhancing adaptive neuroplasticity and preconditioning neuroendocrine systems. Both peptides converge on parasympathetic re-engagement mechanisms—reducing amygdala activity and enhancing vagal tone—to ultimately promote allostatic resilience.

Moreover, both neuropeptides modulate the timing and extent of autonomic transitions, acting as phase-specific mediators within the CACH framework. During acute stress, VP dominates the autonomic landscape, prolonging sympathetic tone to ensure survival. Upon threat cessation, OT gradually rises, promoting parasympathetic rebound, lowering inflammatory mediators, and facilitating anabolic repair. The dynamic interplay between VP and OT is not merely reciprocal but modulatory, as OT can suppress excessive VP-driven arousal, particularly through the OXTR-mediated inhibition of V1a-induced pathways [176]. Therefore, these neuropeptides do not operate in isolation but as components of a precision-timed feedback network that governs the transition from allostatic activation to resilience-based recalibration.

13. Translational and Clinical Potential

Intermittent intranasal OT and selective VP receptor agonists or antagonists are gaining traction as potential hormetic therapeutics capable of enhancing stress resilience, emotional regulation, and cognitive flexibility in vulnerable or clinical populations. Unlike chronic dosing, which can induce receptor desensitization or paradoxical effects, intermittent administration mimics natural pulsatile peptide release and aligns with the biphasic adaptive principle of hormesis [85]. Clinical trials using 24 IU intranasal OT administered 1–3 times per week in patients with PTSD, social anxiety, or ASD have demonstrated improvements in fear extinction, social engagement, and parasympathetic reactivation, particularly when paired with behavioral therapies and a context of safety [173,183]. Similarly, selective V1a receptor agonists have shown promise in enhancing social cognition and working memory in schizophrenia and depression, while V1b antagonists like ABT-436 have been explored for stress-related mood disorders [184,185].

However, the overuse or continuous administration of OT or VP analogs poses notable risks. High or prolonged doses of OT can lead to OTR downregulation, reduced efficacy, and even anxiogenic or VP-like effects, possibly through V1a receptor cross-activation. This is likely to be most problematic in individuals with heightened baseline stress or specific OXTR polymorphisms [167,176]. Similarly, chronic VP stimulation can exacerbate HPA axis dysregulation, hypertension, and emotional rigidity, increasing rather than reducing allostatic load. This underscores the importance of therapeutic intermittency, allowing for receptor resensitization, adaptive feedback, and the proper sequencing of catabolic-anabolic cycles, akin to physical training or caloric restriction models of hormesis.

To gauge the success of such interventions, several biomarkers have been proposed to track hormetic adaptation mediated by OT and VP. Salivary or plasma OT and VP levels, when sampled longitudinally across stress and recovery periods, can reflect pulsatile neuropeptide dynamics [186]. It is worth noting that methodological differences in peptide quantification (e.g., immunoassays vs. mass spectrometry) can significantly affect data interpretation. It is worth noting that methodological differences in peptide quantification (e.g., immunoassays vs. mass spectrometry) can significantly affect data interpretation. The specificity and sensitivity of oxytocin and vasopressin bioassays remain critical issues in the field, particularly in relation to peripheral sampling and behavioral correlations [187].

In addition, heart rate variability (HRV), a well-established index of parasympathetic activity and vagal tone, is positively associated with OT signaling and successful stress recovery. Increases in high-frequency HRV following OT-enhancing interventions (e.g., therapy, exercise, or dosing) signal autonomic rebalancing and neurovisceral integration [182]. Other informative metrics include diurnal cortisol slope, OT receptor gene methylation status, and functional neuroimaging markers, such as improved amygdala–prefrontal coupling and decreased default mode network overactivity.

Together, these biomarkers allow the real-time and long-term assessment of peptide efficacy and resilience building. Given these properties, OT and VP are strong candidates as mediators of stress inoculation strategies, especially in high-risk populations such as caregivers, first responders, military personnel, and older adults. These groups are particularly vulnerable to cumulative allostatic load, emotional exhaustion, and neurodegenerative stress pathways. Recent insights emphasize that central oxytocin deficiency—particularly arising from hypothalamic–pituitary damage—can result in a spectrum of clinical manifestations, including emotional dysregulation, fatigue, and social withdrawal, further underscoring the therapeutic relevance of oxytocin in stress-related and neuroendocrine disorders [188]. Emerging research suggests that targeted OT interventions, in a safe context and coupled with cognitive training, mindfulness-based stress reduction, or exposure therapy, can enhance resilience, interpersonal connection, and physiological recovery in these cohorts [189]. For example, intranasal OT has been shown to enhance affect regulation and increase HRV in caregivers of patients with dementia, suggesting a role in buffering chronic caregiving stress. Similarly, VP receptor antagonism may offer a novel route to reduce over-reactivity in stress-sensitive professions, although careful considerations of context, dose titration and genetic screening (e.g., AVPR1a and AVPR1b SNPs) may be needed to optimize outcomes [178].

Ultimately, these peptides are not universal solutions but phase-sensitive neuromodulators best applied within adaptive stress frameworks, whether clinical or preventive. When embedded within structured recovery cycles, monitored via physiological biomarkers, and matched to individual receptor genotypes or epigenetic profiles, intermittent OT/VP-based therapies may offer a powerful tool to preempt psychiatric disorders and enhance psychological resilience in the face of chronic adversity.

14. Cytokines as Dynamic Modulators of Hormetic Responses

Cytokines, long regarded primarily as mediators of host defense and inflammation, have emerged as pivotal regulators of stress adaptation across physiological systems. Within the framework of hormesis, cytokines are not merely end-products of injury or infection, but dynamic signaling molecules that initiate, amplify, and resolve adaptive biological responses [190]. Transient cytokine activation, particularly in response to low-dose or moderate stressors, plays a central role in priming tissues for resilience, enhancing cellular repair capacity, and fostering long-term adaptive plasticity.

Recent findings have significantly redefined the traditional pathogenic view of cytokines. Controlled, time-limited inflammatory responses now appear crucial for stress resilience, synaptic remodeling, mitochondrial health, and metabolic regulation [191]. Cytokines such as interleukin-6 (IL-6), tumor necrosis factor-alpha (TNF-α), and interleukin-1 beta (IL-1β) exhibit context-dependent duality; at moderate levels and appropriate timing, they initiate protective pathways including the activation of nuclear factor erythroid 2–related factor 2 (Nrf2), sirtuin signaling, and heat shock protein (HSP) expression [192].

Importantly, the phase-specific orchestration of cytokine activity governs the catabolic-anabolic transition critical for hormetic success. During the early catabolic phase, moderate pro-inflammatory cytokine production mobilizes energy reserves, initiates cellular defense programs, and activates autophagic and proteostatic systems. Subsequently, the timely rise of anti-inflammatory cytokines supports anabolic recovery, tissue repair, neurogenesis, and immunometabolic reprogramming [193,194].

Emerging research further reveals that hormetic cytokine modulation is intimately connected to mitochondrial stress signaling (mitokines), senescence-associated secretory phenotypes (SASP) under controlled conditions, and epigenetic plasticity that primes cells for future stress encounters [195]. Thus, cytokines serve not merely as transient inflammatory triggers, but as dynamic modulators of cellular memory and organismal resilience.

Failure to properly resolve cytokine-mediated stress responses, through excessive magnitude, prolonged duration, or inappropriate spatial spread, results in the breakdown of hormetic processes and the promotion of chronic inflammatory pathologies, such as metabolic syndrome, neurodegeneration, cardiovascular disease, and cancer [196]. Therefore, cytokines represent both a necessary initiator and a potential saboteur of hormetic adaptation, depending on the precision of their regulation.

In sum, cytokines occupy a central node in the hormesis model, bridging immediate stress responses with long-term adaptations by regulating the transition from catabolic mobilization to anabolic recovery. Their temporal, spatial, and quantitative control mechanisms represent essential targets for therapeutic strategies aiming to enhance resilience, optimize stress inoculation, and prevent maladaptive aging processes.

14.1. Pro-Inflammatory Cytokine Surge: The "Alarm Phase" of Hormesis

Upon exposure to low-to-moderate intensity stressors, such as heat shock, physical exertion, ischemic preconditioning, intermittent fasting, or subclinical infection, a rapid but transient surge of pro-inflammatory cytokines initiates the earliest phase of hormesis, often termed the "alarm phase" [190]. Rather than representing maladaptive inflammation, this initial cytokine burst serves as an evolutionarily conserved danger signal that mobilizes broad-spectrum cellular defenses. Key cytokines driving this phase include the following: Tumor necrosis factor-alpha (TNF-α) [197], Interleukin-1 beta [198], and Interleukin-6 [199]. Importantly, the hormetic benefit depends on the magnitude, duration, and resolution of the cytokine surge. Persistent or excessive inflammatory signaling, in contrast, transitions into chronic inflammation, oxidative damage, and tissue pathology, a maladaptive outcome avoided by proper hormetic regulation [200].

Thus, the tightly orchestrated pro-inflammatory cytokine surge during the alarm phase represents a fundamental, beneficial stress-priming mechanism critical for initiating cytoprotective, reparative, and adaptive pathways central to the hormetic response.

14.2. Transition Phase: Activation of Cytoprotective and Repair Programs

Following the initial alarm phase, when stress exposure persists but remains within a hormetic intensity window, a critical transition occurs wherein cytokine signaling shifts from acute inflammatory mobilization to adaptive, cytoprotective reinforcement. This phase is essential for facilitating tissue repair, metabolic recalibration, and long-term resilience, rather than progressing toward pathology [200].

This hormetically tuned cytokine signaling allows stressor to promote resilience rather than degeneration. It bridges the catabolic mobilization of the alarm phase to the anabolic recovery phase, establishing a foundation for adaptive plasticity, metabolic reserve enhancement, and immunological recalibration [201].

Failure to appropriately transition, due to excessive stressor magnitude, persistent cytokine activation, or impaired resolution mechanisms, leads to maladaptive chronic inflammation, mitochondrial dysfunction, and disease progression rather than hormetic benefit.

14.3. Resolution Phase: Anti-Inflammatory and Anabolic Signaling

A critical feature distinguishing beneficial hormetic responses from pathological outcomes is the efficient resolution of the initial inflammatory response. Following the early surge in pro-inflammatory cytokines, a timely shift toward anti-inflammatory and regenerative signaling is indispensable to restore tissue homeostasis, repair damage, and promote adaptation.

14.4. Failure of Resolution: When Hormesis Becomes Pathology

Although transient pro-inflammatory activation is essential for initiating hormetic adaptation, failure to resolve the inflammatory response transforms hormesis from a protective into a pathological process. When cytokine signaling is excessive, prolonged, or dysregulated, multiple maladaptive consequences emerge.

14.5. Biological and Clinical Implications of Cytokine Surges

The success of hormetic adaptation hinges not only on initiating appropriate inflammatory responses, but critically on their timely resolution.

Strategies to modulate cytokine surges, enhance anti-inflammatory pathways, and optimize stress dosing are crucial to prevent hormesis failure and reduce disease burden Figure 6.

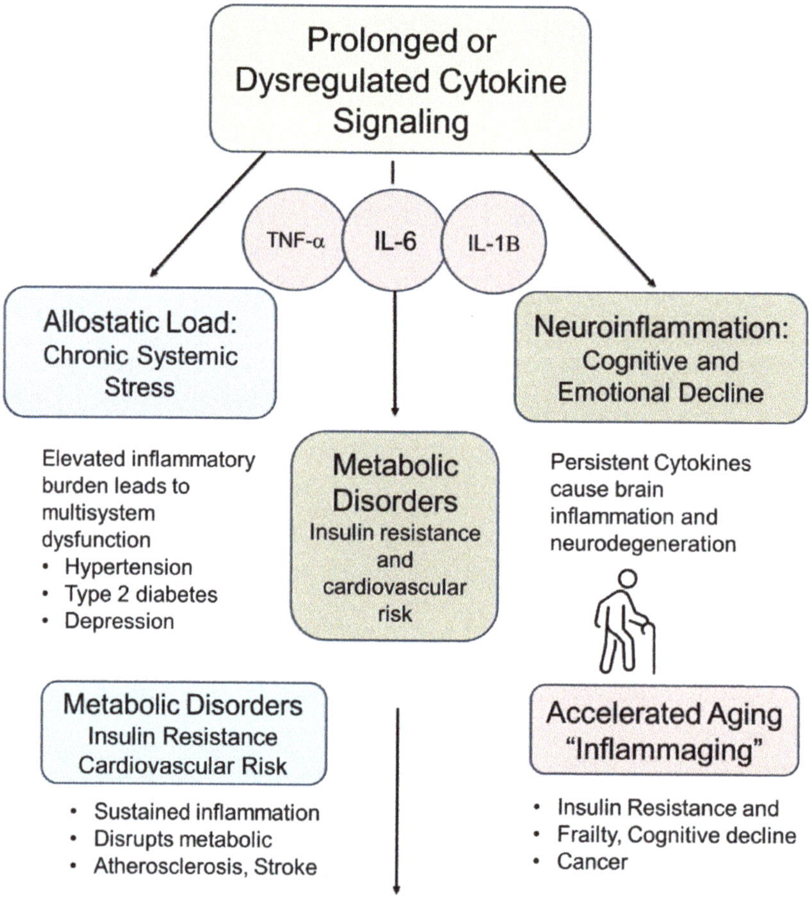

Figure 6. Failure of resolution: transition from hormesis to pathology. This diagram illustrates how prolonged or dysregulated cytokine signaling—specifically chronic elevation of TNF-α, IL-6, and IL-1β—leads to the breakdown of hormetic adaptation and drives pathological outcomes. Failure to resolve early inflammatory activation results in four major maladaptive cascades: (1) Allostatic load, characterized by systemic stress and promoting hypertension, type 2 diabetes, and depression; (2) neuroinflammation, leading to hippocampal atrophy, cognitive decline, and increased neurodegenerative risk; (3) metabolic disorders, including insulin resistance, atherosclerosis, and cardiovascular disease; and (4) accelerated aging ("Inflammaging"), associated with frailty, cognitive impairment, and cancer. Successful hormesis requires the precise regulation and timely resolution of inflammatory responses to prevent these maladaptive trajectories.

14.6. Neuroimmune Interactions: OT, VP, and Cytokine Modulation

The intricate integration between cytokine networks and neuropeptides, particularly OT and VP, has emerged as a pivotal axis in orchestrating successful hormetic adaptation. Rather than acting solely within the nervous or immune systems, these peptides function as neuroimmune regulators, modulating inflammation, stress recovery, and cellular resilience across the catabolic-to-anabolic continuum.

14.6.1. Oxytocin: A Neuroimmune Brake on Inflammation

Direct Inhibition of Pro-inflammatory Cytokines. OT exerts potent anti-inflammatory effects by directly suppressing the secretion of TNF-α, IL-1β, and IL-6 from macrophages, monocytes, and microglia [202]. OT also enhances parasympathetic (Vagal) Tone. OT may strengthen the vagal anti-inflammatory reflex by acting on brainstem nuclei (e.g., dorsal vagal complex), supporting a protective mechanism against excess inflammation.

14.6.2. Neuroprotection via Microglial Modulation

In the CNS, OT reduces microglial activation, inhibits NLRP3 inflammasome formation, and prevents the release of neurotoxic cytokines such as IL-1β and TNF-α [173]. This microglial quiescence protects against stress-induced synaptic loss, hippocampal atrophy, and cognitive dysfunction during and after hormetic stressors (Table 6).

Table 6. Integrated neuroimmune crosstalk between OT and VP in hormetic stress adaptation. This table illustrates the complementary yet opposing roles of oxytocin and vasopressin in regulating cytokine responses, autonomic balance, microglial activation, and stress-phase dominance across the hormetic spectrum. Vasopressin facilitates early stress responses by amplifying pro-inflammatory cytokines, enhancing sympathetic drive, and sustaining glial activation—mechanisms vital for acute mobilization. In contrast, oxytocin supports recovery and resilience by dampening inflammation, promoting vagal tone, and facilitating neuroplasticity. Successful hormesis requires phase-appropriate neuropeptide switching, whereby early VP-driven activation must transition into OT-mediated resolution. Imbalance in this axis, such as prolonged VP signaling or impaired OT release, can disrupt adaptive plasticity and promote chronic pathophysiology including inflammation, affective instability, and premature aging.

Axis	Oxytocin	Vasopressin
Cytokine Modulation	Suppression TNT-a, IL-b, IL-6	Amplifies early pro-inflammatory signals
Parasympathetic/ Sympathetic	Enhances vagal tone, recovery	Boosts sympathetic drive, mobilization
Microglial Activity	Inhibits activation, promotes neuroplasticity	Sustains glial activation if prolonged
Stress Phase	Dominates resolution and recovery phase	Dominates catabolic and alarm phases

15. Integrated Neuroimmune Crosstalk

The balance between OT and VP activity critically tunes cytokine dynamics across the hormetic response. Effective hormetic adaptation depends on phase-appropriate switching: VP-driven mobilization during the alarm and early catabolic stages must be followed by OT-mediated resolution, regeneration, and stress recalibration. Failure to achieve this balance, due to excessive VP or insufficient OT activity, can derail hormesis into chronic inflammation, emotional dysregulation, and accelerated aging. All of these are features of disorders such as severe PTSD, for which an imbalance of VP and OT have recently been documented [18].

16. Future Directions and Research Priorities

Key questions for advancing the clinical application of CACH include the following: What are the optimal oscillation frequencies and recovery intervals to induce maximal benefit without promoting maladaptation? How does biological age, sex, or disease state affect CACH dynamics and responsiveness? Can digital biomarkers (e.g., heart rate variability, mitochondrial respiration, neuropeptide levels) be used to monitor real-time engagement of CACH? Can neuropeptide analogues (e.g., VP/OT modulators) be used therapeutically to entertain beneficial cycles in aging or cardiovascular disease? Can awareness of the hormetic effects of OT and VP be used to index the effectiveness of therapeutic approaches, including those involving either physiology or behavior?

The CACH model provides a powerful, temporally structured lens through which to understand the biological adaptation to stress. By coupling stress-induced activation with recovery-driven repair, this model unifies concepts from mitochondrial biology, neuroendocrinology, and systems physiology. Central to its orchestration are VP and OT, whose cyclical roles in stress and recovery represent a highly conserved mechanism for enhancing resilience, optimizing performance, and preventing chronic disease. As precision medicine shifts toward resilience engineering, CACH offers a conceptual blueprint for designing therapies that harmonize with the body's natural rhythms of adaptation (Table 7).

Table 7. Future directions for advancing CACH research and translational applications. This table presents a strategic roadmap for the future of CACH-oriented research, ranked by translational potential. Longitudinal developmental work is essential for informing early prevention, but is longer-term in scope.

Priority	Future Direction	Focus Area	Potential Impact	Notes
1	Multi-omic profiling to map individual stress-response signatures	Epigenomics, transcriptomics, proteomics	Personalized hormesis models; identification of hormetic thresholds and maladaptive tipping points	Critical for precision medicine and individualized resilience protocols
2	Neuroadaptive technologies for real-time modulation of resilience circuits	Real-time fMRI, tDCS/TMS, closed-loop OT/VP delivery	Enables state-contingent intervention; rapid feedback-based enhancement of stress recovery and cognitive performance	High innovation; bridges neuroscience and technology
3	Integrated lifestyle-based hormetic interventions	Combined use of CR, exercise, social bonding, cognitive stress, and neuropeptide enhancement	Scalable, low-cost strategies to increase population-level resilience and healthspan	High feasibility and translatability to clinical and public health settings
4	Longitudinal developmental studies on neuropeptide plasticity	Developmental biology, early life stress, OT/VP programming	Insights into critical windows, reversibility of early adversity, and preventive strategies	Long timeline; foundational for understanding life-course effects

17. Conclusions

The integration of VP and OT into the hormetic framework provides a paradigm-shifting perspective on how ancient neuropeptides mediate the dynamic oscillation between stress activation and recovery, catabolism and anabolism, vulnerability and resilience. These molecules, previously understood as key modulators of social behavior and homeostasis, are now recognized as core components of a neuroendocrine circuitry that orchestrates systemic adaptation across multiple physiological domains, including immune regulation, metabolism, emotional processing, and cognitive flexibility.

Conceptualized through the lens of CACH, VP and OT serve as temporal gatekeepers of biological rhythm. VP facilitates acute stress reactivity via HPA axis activation, sympathetic tone augmentation, and defensive behaviors, thereby enabling short-term survival and energy mobilization. In contrast, OT mediates reparative processes by downregulating excess arousal and anxiety, enhances parasympathetic balance, modulates inflammation, and promotes social and emotional reintegration. This sequential interplay reflects a fundamental principle of hormesis—beneficial adaptation is contingent upon the successful transition from activation to recovery, and from disruption to recalibration.

Importantly, VP and OT do not act in isolation. They function within a broader network of regulatory peptides, including CRH and UCNs, which initiate, sustain, or resolve stress responses depending on the intensity, duration, and contextual framing of the stressor. CRH and UCNs activate catabolic signaling and neuroendocrine alertness, particularly via CRHR1 and CRHR2, while OT and VP modulate feedback inhibition and downstream repair pathways. The integration of these systems reveals a multistage, homodynamic model in which neuropeptide interactions encode both the threat appraisal and adaptive outcome of a stressor.

Critically, the developmental and sex-specific regulation of these neuropeptide systems must be acknowledged. Early-life adversity, attachment quality, social buffering and perceived safety calibrate lifelong VP and OT tone, with implications for stress susceptibility and resilience. Males and females exhibit differential sensitivity to VP- and OT-mediated effects, suggesting that personalized interventions should account for sex-based neuroendocrine phenotypes and their plasticity across the lifespan.

Therefore, we reemphasize the concept that VP and OT represent more than stress or "anti-stress" hormones. VP and OT serve as evolutionarily conserved mediators of physiological flexibility, social regulation, and adaptive plasticity. It is of interest that CRH, the UCNs and cytokines all existed before the evolution of the modern VP-OT system [14]. We specifically hypothesize here that in mammals, the VP-OT system may have a hierarchical capacity to coordinate these older systems with the behavioral demands of contemporary social species.

The orchestrated, context-dependent, and reciprocal dynamics of these interactive systems offer a unifying framework for understanding and therapeutically harnessing the process identified hormesis. By advancing this integrative neuroendocrine model, we hope to suggest new avenues for precision stress medicine and resilience engineering across mental and physical health domains.

Author Contributions: H.P.N.: Conceptualization, Writing Original Draft. M.A.K., H.L., C.V.D., P.N., J.M.D., E.C.P., S.P.C.: Review & Editing, and C.S.C.: Supervision, Conceptualization, Writing, Review & Editing, Resources. All authors have read and agreed to the published version of the manuscript.

Funding: This research received no external funding.

Conflicts of Interest: The authors declare no conflict of interest.

References

1. Calabrese, E.J.; Mattson, M.P. The Catabolic–Anabolic Cycling Hormesis Model of Health and Resilience. *Ageing Res. Rev.* **2024**, *102*, 102588. [CrossRef]
2. Du Vigneaud, V.; Ressler, C.; Trippett, S. The Sequence of Amino Acids in Oxytocin, with a Proposal for the Structure of Oxytocin. *J. Biol. Chem.* **1953**, *205*, 949–957. [CrossRef]
3. Murphy, M.R.; Seckl, J.R.; Burton, S.; Checkley, S.A.; Lightman, S.L. Changes in Oxytocin and Vasopressin Secretion during Sexual Activity in Men. *J. Clin. Endocrinol. Metab.* **1987**, *65*, 738–741. [CrossRef]
4. Carter, C.S.; Kenkel, W.M.; MacLean, E.L.; Wilson, S.R.; Perkeybile, A.M.; Yee, J.R.; Ferris, C.F.; Nazarloo, H.P.; Porges, S.W.; Davis, J.M.; et al. Is Oxytocin "Nature's Medicine"? *Pharmacol. Rev.* **2020**, *72*, 829–861. [CrossRef] [PubMed]
5. Carter, C.S. The Psychedelic-Peptide Paradox: A Hormetic Hypothesis. *Compr. Psychoneuroendocrinology* **2025**, *23*, 100303. [CrossRef] [PubMed]
6. Carter, C.S. Close encounters with oxytocin. *Compr. Psychoneuroendocrinology* **2023**, *15*, 100189. [CrossRef] [PubMed]
7. Uvnäs-Moberg, K. The Physiology and Pharmacology of Oxytocin in Labor and in the Peripartum Period. *Am. J. Obstet. Gynecol.* **2024**, *230*, S740–S758. [CrossRef]
8. Onaka, T.; Takayanagi, Y. Role of Oxytocin in the Control of Stress and Food Intake. *J. Neuroendocrinol.* **2019**, *31*, e12700. [CrossRef]
9. Landgraf, R.; Neumann, I.D. Vasopressin and Oxytocin Release Within the Brain: A Dynamic Concept of Multiple and Variable Modes of Neuropeptide Communication. *Front. Neuroendocrinol.* **2004**, *25*, 150–176. [CrossRef]
10. Uvnäs-Moberg, K. Oxytocin May Mediate the Benefits of Positive Social Interaction and Emotions. *Psychoneuroendocrinology* **1998**, *23*, 819–835. [CrossRef]
11. Horn, A.J.; Carter, C.S. Love and Longevity: A Social Dependency Hypothesis. *Compr. Psychoneuroendocrinology* **2021**, *8*, 100088. [CrossRef]

12. Szeto, A.; Nation, D.A.; Mendez, A.J.; Dominguez-Bendala, J.; Brooks, L.G.; Schneiderman, N.; McCabe, P.M. Oxytocin Attenuates NADPH-Dependent Superoxide Activity and IL-6 Secretion in Macrophages and Vascular Cells. *Am. J. Physiol. Endocrinol. Metab.* **2008**, *295*, E1495–E1501. [CrossRef] [PubMed]
13. Uvnäs-Moberg, K.; Julius, H.; Handlin, L.; Petersson, M. Editorial: Sensory Stimulation and Oxytocin: Their Roles in Social Interaction and Health Promotion. *Front. Psychol.* **2022**, *13*, 929741. [CrossRef] [PubMed]
14. Carter, C.S.; Kingsbury, M.A. Oxytocin and Oxygen: The Evolution of a Solution to the 'Stress of Life'. Philos. *Trans. R. Soc. Lond. B Biol. Sci.* **2022**, *377*, 1–14. [CrossRef]
15. Theofanopoulou, C.; Gedman, G.; Cahill, J.A.; Boeckx, C.; Jarvis, E.D. Universal Nomenclature for Oxytocin–Vasotocin Ligand and Receptor Families. *Nature* **2021**, *592*, 747–755. [CrossRef]
16. Albers, H.E. The Regulation of Social Recognition, Social Communication and Aggression: Vasopressin in the Social Behavior Neural Network. *Horm. Behav.* **2012**, *61*, 283–292. [CrossRef]
17. Antoni, F.A. Magnocellular Vasopressin and the Mechanism of "Glucocorticoid Escape". *Front. Endocrinol.* **2019**, *10*, 422. [CrossRef]
18. Horn, A.J.; Cole, S.; Nazarloo, N.P.; Nazarloo, P.; Davis, J.M.; Carrier, D.; Bryan, C.; Carter, C.S. Severe PTSD is Marked by Reduced Oxytocin and Elevated Vasopressin. *Compr. Psychoneuroendocrinology* **2024**, *19*, 100236. [CrossRef]
19. Carter, C.S. Neuroendocrine Perspectives on Social Attachment and Love. *Psychoneuroendocrinology* **1998**, *23*, 779–818. [CrossRef]
20. Carter, C.S. Sex, Love and Oxytocin: Two Metaphors and a Molecule. *Neurosci. Biobehav. Rev.* **2022**, *143*, 104948. [CrossRef]
21. Salehi, M.S.; Neumann, I.D.; Jurek, B.; Pandamooz, S. Co-Stimulation of Oxytocin and Arginine-Vasopressin Receptors Affect Hypothalamic Neurospheroid Size. *Int. J. Mol. Sci.* **2021**, *22*, 8464. [CrossRef]
22. Ellis, B.J.; Horn, A.J.; Carter, C.S.; van IJzendoorn, M.H.; Bakermans-Kranenburg, M.J. Developmental Programming of Oxytocin through Variation in Early-Life Stress: Four Meta-Analyses and a Theoretical Reinterpretation. *Clin. Psychol. Rev.* **2021**, *86*, 101985. [CrossRef]
23. Danoff, J.S.; Connelly, J.J.; Morris, J.P.; Perkeybile, A.M. An Epigenetic Rheostat of Experience: DNA Methylation of OXTR as a Mechanism of Early-Life Allostasis. *Compr. Psychoneuroendocrinology* **2021**, *8*, 100098. [CrossRef]
24. Bartz, J.A.; Zaki, J.; Bolger, N.; Ochsner, K.N. Social Effects of Oxytocin in Humans: Context and Person Matter. *Trends Cogn. Sci.* **2011**, *15*, 301–309. [CrossRef] [PubMed]
25. Calabrese, E.J.; Baldwin, L.A. Hormesis: The Dose–Response Revolution. *Annu. Rev. Pharmacol. Toxicol.* **2003**, *43*, 175–197. [CrossRef] [PubMed]
26. Mattson, M.P. Hormesis Defined. *Ageing Res. Rev.* **2008**, *7*, 1–7. [CrossRef] [PubMed]
27. Calabrese, E.J. Hormesis and Medicine. *Br. J. Clin. Pharmacol.* **2008**, *66*, 594–617. [CrossRef]
28. Neumann, I.D.; Landgraf, R. Balance of Brain Oxytocin and Vasopressin: Implications for Anxiety, Depression, and Social Behaviors. *Trends Neurosci.* **2012**, *35*, 649–659. [CrossRef]
29. Calabrese, E.J.; Osakabe, N.; Di Paola, R.; Siracusa, R.; Fusco, R.; D'Amico, R.; Impellizzeri, D.; Cuzzocrea, S.; Fritsch, T.; Abdelhameed, A.S.; et al. Hormesis Defines the Limits of Lifespan. *Ageing Res. Rev.* **2023**, *91*, 102074. [CrossRef]
30. Agathokleous, E.; Calabrese, E.J. Evolution of Hormesis Research: A Bibliometric Analysis. *Arch. Toxicol.* **2024**, *98*, 577–578. [CrossRef]
31. Rattan, S.I.S.; Kyriazis, M. (Eds.) *The Science of Hormesis in Health and Longevity*; Academic Press: Cambridge, MA, USA, 2019; ISBN 9780128142530.
32. Martucci, K.T.; MacNiven, K.H.; Borg, N.; Knutson, B.; Mackey, S. Altered Prefrontal Correlates of Monetary Anticipation and Outcome in Chronic Pain. *Pain.* **2018**, *159*, 1494–1507. [CrossRef]
33. López-Lluch, G.; Navas, P. Coenzyme Q at the Hinge of Health and Metabolic Diseases. *Antioxidants* **2021**, *10*, 1785. [CrossRef] [PubMed]
34. Giona, L.; Musillo, C.; De Cristofaro, G.; Ristow, M.; Zarse, K.; Siems, K.; Tait, S.; Cirulli, F.; Berry, A. Western Diet-Induced Cognitive and Metabolic Dysfunctions in Aged Mice Are Prevented by Rosmarinic Acid in a Sex-Dependent Fashion. *Clin. Nutr.* **2024**, *43*, 2236–2248. [CrossRef] [PubMed]
35. Loth, M.K.; Schmidt, J.C.; Gonzalez, C.A.; Brusman, L.E.; Sadino, J.M.; Winther, K.E.; Protter, D.S.W.; Donaldson, Z.R. Oxytocin and Dopamine Receptor Expression: Cellular Level Implications for Pair Bonding. *bioRxiv* **2025**, preprint. [CrossRef] [PubMed]
36. de Kloet, E.R.; Joëls, M.; Holsboer, F. Stress and the Brain: From Adaptation to Disease. *Nat. Rev. Neurosci.* **2005**, *6*, 463–475. [CrossRef]
37. Mattson, M.P. The Cyclic Metabolic Switching Theory of Intermittent Fasting. *Nat. Metab.* **2025**, *7*, 665–678. [CrossRef]
38. Calabrese, E.J.; Agathokleous, E. Hormesis: Transforming Disciplines That Rely on the Dose Response. *IUBMB Life* **2021**, *73*, 1197–1203. [CrossRef]
39. Radak, Z.; Rattan, S.I.S. Exercise, Hormesis and Ageing: A New Section in Biogerontology. *Biogerontology* **2024**, *26*, 26. [CrossRef]

40. López-Otín, C.; Kroemer, G. Hallmarks of Health. *Cell* **2021**, *184*, 33–63. [CrossRef]
41. Knobloch, H.S.; Grinevich, V. Evolution of Oxytocin Pathways in the Brain of Vertebrates. *Front. Behav. Neurosci.* **2014**, *8*, 31. [CrossRef]
42. Caldwell, H.K.; Petrulis, A. Editorial: The Vasopressin System and Behavior. *Front. Endocrinol.* **2018**, *9*, 438. [CrossRef]
43. Herman, J.P.; McKlveen, J.M.; Ghosal, S.; Kopp, B.; Wulsin, A.; Makinson, R.; Scheimann, J.; Myers, B. Regulation of the Hypothalamic-Pituitary-Adrenocortical Stress Response. *Compr. Psychoneuroendocrinology* **2016**, *6*, 603–621. [CrossRef]
44. Larauche, M.; Moussaoui, N.; Biraud, M.; Bae, W.K.; Duboc, H.; Million, M.; Taché, Y. Brain Corticotropin-Releasing Factor Signaling: Involvement in Acute Stress-Induced Visceral Analgesia in Male Rats. *Neurogastroenterol. Motil.* **2019**, *31*, e13489. [CrossRef] [PubMed]
45. Danhof, H.A.; Lee, J.; Thapa, A.; Britton, R.A.; Di Rienzi, S.C. Microbial Stimulation of Oxytocin Release from the Intestinal Epithelium via Secretin Signaling. *Gut Microbes* **2023**, *15*, 2256043. [CrossRef] [PubMed]
46. Jurek, B.; Neumann, I.D. The Oxytocin Receptor: From Intracellular Signaling to Behavior. *Physiol. Rev.* **2018**, *98*, 1805–1908. [CrossRef]
47. Heinrichs, M.; von Dawans, B.; Domes, G. Oxytocin, Vasopressin, and Human Social Behavior. *Front. Neuroendocrinol.* **2009**, *30*, 548–557. [CrossRef]
48. Carter, C.S. Oxytocin Pathways and the Evolution of Human Behavior. *Annu. Rev. Psychol.* **2014**, *65*, 17–39. [CrossRef]
49. Smith, A.S.; Wang, Z. Hypothalamic Oxytocin Mediates Social Buffering of the Stress Response. *Biol. Psychiatry* **2014**, *76*, 281–288. [CrossRef]
50. Guarente, L.; Sinclair, D.A.; Kroemer, G. Human Trials Exploring Anti-Aging Medicines. *Cell Metab.* **2024**, *36*, 354–376. [CrossRef]
51. Longo, S.K.; Guo, M.G.; Ji, A.L.; Khavari, P.A. Integrating Single-Cell and Spatial Transcriptomics to Elucidate Intercellular Tissue Dynamics. *Nat. Rev. Genet.* **2021**, *22*, 627–644. [CrossRef]
52. Pillon, N.J.; Gabriel, B.M.; Dollet, L.; Smith, J.A.B.; Sardón Puig, L.; Botella, J.; Bishop, D.J.; Krook, A.; Zierath, J.R. Transcriptomic Profiling of Skeletal Muscle Adaptations to Exercise and Inactivity. *Nat. Commun.* **2020**, *11*, 470. [CrossRef] [PubMed]
53. Radak, Z.; Ishihara, K.; Tekus, E.; Varga, C.; Posa, A.; Balogh, L.; Boldogh, I.; Koltai, E. Exercise, Oxidants, and Antioxidants Change the Shape of the Bell-Shaped Hormesis Curve. *Redox Biol.* **2017**, *12*, 285–290. [CrossRef] [PubMed]
54. Mattson, M.P.; Longo, V.D.; Harvie, M. Impact of Intermittent Fasting on Health and Disease Processes. *Ageing Res. Rev.* **2017**, *39*, 46–58. [CrossRef] [PubMed]
55. Aguilera, G. HPA Axis Responsiveness to Stress: Implications for Healthy Aging. *Exp. Gerontol.* **2011**, *46*, 90–95. [CrossRef]
56. Zalachoras, I.; Houtzager, S.W.J.; Servaes, S.; Reul, J.M.H.M. CRH Receptors and Their Role in Neuroendocrine Adaptation to Stress: Implications for Mental Health. *Curr. Opin. Neurobiol.* **2022**, *72*, 134–141. [CrossRef]
57. Fischer, L.; Paschke, B.; Gareis, F.; Schumacher, M.; Liere, P.; Hiergeist, A.; Gessner, A.; Rupprecht, R.; Neumann, I.D.; Bosch, O.J. The Translocator Protein 18 kDa (TSPO) Ligand Etifoxine in an Animal Model of Anxiety: Line- and Sex-Dependent Effects on Emotionality, Stress Reactivity, Spine Density, Oxytocin Receptors, Steroids, and Microbiome Composition. *Neuropharmacology* **2025**, *266*, 110282. [CrossRef]
58. Vetter, C.; Scheer, F.A.J.L.; Shea, S.A. Circadian Biology: Uncoupling Human Body Clocks by Food Timing. *Curr. Biol.* **2017**, *27*, R656–R658. [CrossRef]
59. Brar, B.; Bayoumy, M.; Salama, A.; Henry, A.; Chigurupati, R. A Survey Assessing the Early Effects of the COVID-19 Pandemic on Oral and Maxillofacial Surgery Training Programs. Oral Surg. *Oral. Med. Oral. Pathol. Oral. Radiol.* **2021**, *131*, 27–42. [CrossRef]
60. Winter, J.; Meyer, M.; Berger, I.; Royer, M.; Bianchi, M.; Kuffner, K.; Peters, S.; Stang, S.; Langgartner, D.; Hartmann, F.; et al. Chronic Oxytocin-Driven Alternative Splicing of Crfr2α Induces Anxiety. *Mol. Psychiatry* **2023**, *28*, 4742–4755. [CrossRef]
61. Mattson, M.P.; Arumugam, T.V. Hallmarks of Brain Aging: Adaptive and Pathological Modification by Metabolic States. *Cell Metab.* **2018**, *27*, 1176–1199. [CrossRef]
62. Saleh, M.S.M.; Elsabahy, H.E.; Abdelhamid, Z.N. Information Technology as an Opportunity for Improving Nurses' Work Outcomes in Egypt. *Biosci. Res.* **2021**, *18*, 345–358.
63. Jankowski, P.J.; Sandage, S.J.; Bell, C.A.; Davis, D.E.; Porter, E.; Jessen, M.; Motzny, C.L.; Ross, K.V.; Owen, J. Virtue, Flourishing, and Positive Psychology in Psychotherapy: An Overview and Research Prospectus. *Psychotherapy* **2020**, *57*, 291–309. [CrossRef]
64. Szeto, A.C.H.; Ferreira, A.C.F.; Mannion, J.; Clark, P.A.; Sivasubramaniam, M.; Heycock, M.W.D.; Crisp, A.; Jolin, H.E.; Kozik, P.; Knolle, M.D.; et al. An αvβ3 Integrin Checkpoint Is Critical for Efficient TH2 Cell Cytokine Polarization and Potentiation of Antigen-Specific Immunity. *Nat. Immunol.* **2023**, *24*, 123–135. [CrossRef]
65. Moghadam, M.R.S.; Ghasemnia Arabi, N.; Khoshsima, G. A Review of Case Study Method in Operations Management Research. *Int. J. Qual. Methods* **2021**, *20*, 1–11. [CrossRef]
66. Calabrese, E.J.; Dhawan, G.; Kapoor, R.; Iavicoli, I. Hormesis: Wound Healing and Fibroblasts. *Pharmacol. Res.* **2022**, *182*, 106304. [CrossRef] [PubMed]

67. Lods, M.; Mortessagne, P.; Pacary, E.; Terral, G.; Farrugia, F.; Mazier, W.; Masachs, N.; Charrier, V.; Cota, D.; Ferreira, G.; et al. Chemogenetic Stimulation of Adult Neurogenesis, and Not Neonatal Neurogenesis, Is Sufficient to Improve Long-Term Memory Accuracy. *Prog. Neurobiol.* **2022**, *219*, 102364. [CrossRef] [PubMed]
68. Murphy, M.P.; Dias, G.P. Hormetic Mechanisms in Neuroplasticity: Stress Resilience and Cognitive Adaptation. *Neurosci. Biobehav. Rev.* **2021**, *124*, 304–315. [CrossRef]
69. Zhao, Y.; Wang, M.; Liang, H. FOXO Transcription Factors in Cellular Stress Responses and Aging: Regulation by AMPK and SIRT1. *Mech. Ageing Dev.* **2023**, *211*, 111742. [CrossRef]
70. Perrelli, M.; Goparaju, P.; Postolache, T.T.; Del Bosque-Plata, L.; Gragnoli, C. Stress and the CRH System, Norepinephrine, Depression, and Type 2 Diabetes. *Biomedicines* **2024**, *12*, 1187. [CrossRef]
71. Mattson, M.P. Hormesis and Health: The Integration of Stress Physiology Across Systems. *Trends Endocrinol. Metab.* **2023**, *34*, 9–22. [CrossRef]
72. Calabrese, E.J.; Kozumbo, W.J. The Hormetic Dose-Response Mechanism: Nrf2 Activation. *Pharmacol. Res.* **2021**, *167*, 105526. [CrossRef] [PubMed]
73. Goldstein, D.S.; Kopin, I.J. Linking Stress, Catecholamine Autotoxicity, and Allostatic Load with Neurodegenerative Diseases: A Focused Review in Memory of Richard Kvetnansky. *Cell Mol. Neurobiol.* **2018**, *38*, 13–24. [CrossRef] [PubMed]
74. Neumann, I.D.; Slattery, D.A. Oxytocin in General Anxiety and Social Fear: A Translational Approach. *Biol. Psychiatry* **2016**, *79*, 213–221. [CrossRef] [PubMed]
75. Olff, M.; Frijling, J.L.; Kubzansky, L.D.; Bradley, B.; Ellenbogen, M.A.; Cardoso, C.; van Zuiden, M. The Role of Oxytocin in Social Bonding, Stress Regulation and Mental Health: An Update on the Moderating Effects of Context and Interindividual Differences. *Psychoneuroendocrinology* **2013**, *38*, 1883–1894. [CrossRef]
76. Wang, P.; Yang, H.P.; Tian, S.; Wang, L.; Wang, S.C.; Zhang, F.; Wang, X. Oxytocin-Secreting System: A Major Part of the Neuroendocrine Center Regulating Social Behavior. *J. Neuroendocrinol.* **2019**, *31*, e12792. [CrossRef]
77. Rigney, N.; de Vries, G.J.; Petrulis, A.; Young, L.J. Oxytocin, Vasopressin, and Social Behavior: From Neural Circuits to Clinical Opportunities. *Endocrinology* **2022**, *163*, bqac111. [CrossRef]
78. Giano, Z.; Wheeler, D.L.; Hubach, R.D. Intersectionality and Adverse Childhood Experiences: Racial/Ethnic Differences in the Prevalence of ACEs in a Nationally Representative Sample. *Am. J. Prev. Med.* **2024**, *67*, 328–338. [CrossRef]
79. Szczepańska-Sadowska, E. Interplay of Angiotensin Peptides, Vasopressin, and Insulin in the Heart: Experimental and Clinical Evidence of Altered Interactions in Obesity and Diabetes Mellitus. *Int. J. Mol. Sci.* **2024**, *25*, 1310. [CrossRef]
80. Moslemi, M.; Motamedi, F.; Asadi, S.; Khodagholi, F. Peroxisomal Malfunction Caused by Mitochondrial Toxin 3-NP: Protective Role of Oxytocin. *Iran. J. Pharm. Res.* **2019**, *18*, 296–307. [CrossRef]
81. Carter, C.S.; Porges, S.W. The Biochemistry of Love: An Oxytocin Hypothesis. *EMBO Rep.* **2016**, *17*, 69–75. [CrossRef]
82. Guidi, J.; Lucente, M.; Sonino, N.; Fava, G.A. Allostatic Load and Its Impact on Health: A Systematic Review. *Psychother. Psychosom.* **2021**, *90*, 11–27. [CrossRef]
83. Volpi, S.; Rabadan-Diehl, C.; Aguilera, G. Vasopressinergic Regulation of the Hypothalamic–Pituitary–Adrenal Axis and Stress Adaptation. *Stress* **2004**, *7*, 75–83. [CrossRef]
84. Murgatroyd, C.; Spengler, D. Oxytocin and Stress-Related Disorders: Insights from Epigenetics. *Biol. Psychiatry* **2021**, *89*, 309–317. [CrossRef]
85. Calabrese, E.J.; Mattson, M.P. How Does Hormesis Impact Biology, Toxicology, and Medicine? *NPJ Aging Mech. Dis.* **2017**, *3*, 13. [CrossRef] [PubMed]
86. Engelmann, M.; Ebner, K.; Landgraf, R. Oxytocin and Stress: An Update on the Modulatory Role of Oxytocin in Stress Responses and Its Relevance for Stress-Related Disorders. *Front. Neuroendocrinol.* **2021**, *61*, 100873. [CrossRef]
87. Martins, D.; Brodmann, K.; Veronese, M.; Dipasquale, O.; Mazibuko, N.; Schuschnig, U.; Zelaya, F.; Paloyelis, Y. "Less Is More": A Dose–Response Account of Intranasal Oxytocin Pharmacodynamics in the Human Brain. *Prog. Neurobiol.* **2022**, *211*, 102239. [CrossRef]
88. Zhang, Y.; Li, Q.; Zhuang, Q.; Xu, D.; Yao, S.; Zhao, W.; Becker, B. Oro-Mucosal Administration of Oxytocin Using Medicated Lollipops Alters Social Attention, Similar to Intranasal and Lingual Routes: Implications for Therapeutic Use. *Front. Neurosci.* **2022**, *16*, 1022101. [CrossRef]
89. Carter, C.S. The Oxytocin–Vasopressin Pathway in the Context of Love and Fear. *Front. Endocrinol.* **2017**, *8*, 356. [CrossRef]
90. Leng, G.; Ludwig, M. Oxytocin, A Social Peptide? Deconstructing the Evidence. *Philos. Trans. R. Soc. B Biol. Sci.* **2021**, *376*, 20210055. [CrossRef]
91. Carter, C.S. Oxytocin and love: Myths, metaphors and mysteries. *Compr. Psychoneuroendocrinology* **2021**, *9*, 100107. [CrossRef]
92. Kageyama, K.; Iwasaki, Y.; Daimon, M. Influence of Arginine Vasopressin on the Ultradian Dynamics of the Hypothalamic-Pituitary-Adrenal Axis. *Front. Endocrinol.* **2022**, *13*, 976323. [CrossRef]

93. Carter, C.S.; Perkeybile, A.M. The Monogamy Paradox: What Do Love and Sex Have to Do with It? Front. *Ecol. Evol.* **2018**, *6*, 202. [CrossRef] [PubMed]
94. Sofer, Y.; Zilkha, N.; Gimpel, E.; Wagner, S.; Chuartzman, S.G.; Kimchi, T. Sexually Dimorphic Oxytocin Circuits Drive Intragroup Social Conflict and Aggression in Wild House Mice. *Nat. Neurosci.* **2024**, *27*, 620–630. [CrossRef] [PubMed]
95. Steinman, M.Q.; Duque-Wilckens, N.; Greenberg, G.D.; Hao, R.; Campi, K.L.; Laredo, S.A.; Trainor, B.C. Sex-Specific Effects of Stress on Oxytocin Neurons Correspond with Responses to Intranasal Oxytocin. *Biol. Psychiatry* **2016**, *80*, 406–414. [CrossRef]
96. Yamasue, H.; Okada, T.; Munesue, T.; Kuroda, M.; Fujioka, T.; Uno, Y.; Matsumoto, K.; Kuwabara, H.; Mori, D.; Okamoto, Y.; et al. Effect of Intranasal Oxytocin on the Core Social Symptoms of Autism Spectrum Disorder: A Randomized Clinical Trial. *Mol. Psychiatry* **2020**, *25*, 1849–1858. [CrossRef]
97. Ohbuchi, T.; Haam, J.; Tasker, J.G. Regulation of Neuronal Activity in Hypothalamic Vasopressin Neurons. *Interdiscip. Inf. Sci.* **2015**, *21*, 225–234. [CrossRef]
98. Bains, R.S.; Bhatti, J.S. Oxytocin and Stress Resilience: Neuroprotective Roles, Mechanisms, and Therapeutic Potential. *Neurochem. Int.* **2022**, *157*, 105356. [CrossRef]
99. Duarte-Guterman, P.; Lieblich, S.E.; Qiu, W.; Splinter, J.E.J.; Go, K.A.; Casanueva-Reimon, L.; Galea, L.A.M. Oxytocin Has Sex-Specific Effects on Social Behaviour and Hypothalamic Oxytocin Immunoreactive Cells but Not Hippocampal Neurogenesis in Adult Rats. *Horm. Behav.* **2020**, *122*, 104734. [CrossRef]
100. Caldwell, H.K.; Lee, H.J.; Macbeth, A.H.; Young, W.S. Vasopressin: Behavioral Roles of an "Original" Neuropeptide. *Prog. Neurobiol.* **2008**, *84*, 1–24. [CrossRef]
101. Carter, C.S.; Grippo, A.J.; Pournajafi-Nazarloo, H.; Ruscio, M.G.; Porges, S.W. Oxytocin, Vasopressin and Sociality. *Prog. Brain Res.* **2008**, *170*, 331–336. [CrossRef]
102. Griebel, G.; Holsboer, F. Neuropeptide Receptor Antagonists as Therapeutic Agents: The Vasopressin V1b Receptor Example. *Pharmacol. Ther.* **2021**, *219*, 107709. [CrossRef]
103. Packard, K.; Opendak, M. Rodent Models of Early Adversity: Impacts on Developing Social Behavior Circuitry and Clinical Implications. *Front. Behav. Neurosci.* **2022**, *16*, 918862. [CrossRef] [PubMed]
104. Stribley, J.M.; Carter, C.S. Developmental Exposure to Vasopressin Increases Aggression in Adult Prairie Voles. *Proc. Natl. Acad. Sci. USA* **1999**, *96*, 12601–12604. [CrossRef]
105. Zelena, D.; Pinter, O.; Balazsfi, D.G.; Langnaese, K.; Richter, K.; Landgraf, R.; Makara, G.B.; Engelmann, M. Vasopressin Signaling at Brain Level Controls Stress Hormone Release: The Vasopressin-Deficient Brattleboro Rat as a Model. *Amino Acids* **2015**, *47*, 2245–2253. [CrossRef]
106. Fineberg, S.K.; Ross, D.A. Oxytocin and the Social Brain. *Biol. Psychiatry* **2017**, *81*, e19–e21. [CrossRef]
107. Calabrese, E.J.; Mattson, M.P. Hormesis and Adaptive Plasticity: Implications for Resilience in Neurobiology and Medicine. *Trends Neurosci.* **2023**, *46*, 14–26. [CrossRef]
108. Shamay-Tsoory, S.G.; Abu-Akel, A. The Social Salience Hypothesis of Oxytocin. *Biol. Psychiatry* **2016**, *79*, 194–202. [CrossRef]
109. Kovács, Z.; Szabó, É. Effects of Oxytocin Administration on Oxidative Markers in the Temporal Lobe of Aged Rats. *Neurosci. Behav. Physiol.* **2019**, *49*, 456–462. [CrossRef]
110. Numakawa, T.; Kajihara, R. Involvement of Brain-Derived Neurotrophic Factor Signaling in the Pathogenesis of Stress-Related Brain Diseases. *Front. Mol. Neurosci.* **2023**, *16*, 1247422. [CrossRef]
111. Bordt, E.A.; Smith, C.J.; Demarest, T.G.; Bilbo, S.D.; Kingsbury, M.A. Mitochondria, Oxytocin, and Vasopressin: Unfolding the Inflammatory Protein Response. *Neurotox. Res.* **2019**, *36*, 239–256. [CrossRef]
112. Gebauer, L.; Witek, M.A.G.; Hansen, N.C.; Thomas, J.; Konvalinka, I.; Vuust, P. Oxytocin improves synchronisation in leader-follower interaction. *Sci. Rep.* **2016**, *6*, 38416. [CrossRef] [PubMed]
113. Griebel, G.; Holsboer, F.; Neumann, I.D. Oxytocin and Vasopressin: Signalling, Behavioural Modulation and Potential Therapeutic Targets. *Br. J. Pharmacol.* **2021**, *178*, 1269–1283. [CrossRef]
114. MacDonald, K.; Feifel, D. Oxytocin's role in anxiety: A critical appraisal. *Brain Res.* **2014**, *1580*, 22–56. [CrossRef] [PubMed]
115. Sapolsky, R.M.; Romero, L.M.; Munck, A.U. How Do Glucocorticoids Influence Stress Responses? Integrating Permissive, Suppressive, Stimulatory, and Preparative Actions. *Endocr. Rev.* **2000**, *21*, 55–89. [CrossRef]
116. Lee, S.Y.; Park, S.H.; Chung, C.; Kim, J.J.; Choi, S.Y.; Han, J.S. Oxytocin Protects Hippocampal Memory and Plasticity from Uncontrollable Stress. *Sci. Rep.* **2015**, *5*, 18540. [CrossRef]
117. Morrison, S.F.; Madden, C.J.; Tupone, D. Central Neural Regulation of Brown Adipose Tissue Thermogenesis and Energy Expenditure. *Cell Metab.* **2014**, *19*, 741–756. [CrossRef]
118. Cannon, B.; Nedergaard, J. Stress-Induced Activation of Brown Adipose Tissue Prevents Obesity in Mice. *Cell Rep.* **2015**, *13*, 1523–1530. [CrossRef]

119. Labbé, S.M.; Caron, A.; Bakan, I.; Laplante, M.; Lecomte, R.; Côté, C.H.; Richard, D.; Carpentier, A.C. In Vivo Measurement of Energy Substrate Contribution to Cold-Induced Brown Adipose Tissue Thermogenesis. *FASEB J.* **2015**, *29*, 2046–2058. [CrossRef]
120. Jones, J.R.; Simon, T.; Lones, L.; Herzog, E.D. SCN VIP Neurons Are Essential for Normal Light-Mediated Resetting of Circadian Rhythms and Glucocorticoid Release. *Nat. Commun.* **2021**, *12*, 5280. [CrossRef]
121. Bedrosian, T.A.; Nelson, R.J. Influence of the Modern Light Environment on Mood. *Mol. Psychiatry* **2013**, *18*, 751–757. [CrossRef]
122. Chaix, A.; Zarrinpar, A.; Miu, P.; Panda, S. Time-Restricted Feeding Is a Preventative and Therapeutic Intervention Against Diverse Nutritional Challenges. *Cell Metab.* **2014**, *20*, 991–1005. [CrossRef]
123. Mehta, M.; Liu, Y.; Zhang, P. PGC-1α Controls Mitochondrial Biogenesis and Dynamics in Lead-Induced Neurotoxicity. *Aging* **2015**, *7*, 629–647. [CrossRef]
124. Davidson, R.J.; McEwen, B.S. Social Influences on Neuroplasticity: Stress and Interventions to Promote Well-Being. *Nat. Neurosci.* **2012**, *15*, 689–695. [CrossRef] [PubMed]
125. Furman, D.J.; Chen, M.C.; Gotlib, I.H. Variant in Oxytocin Receptor Gene Is Associated with Amygdala Volume. *Psychoneuroendocrinology* **2011**, *36*, 891–897. [CrossRef] [PubMed]
126. Perkeybile, A.M.; Carter, C.S.; Wroblewski, K.L.; Puglia, M.H.; Kenkel, W.M.; Lillard, T.S.; Karaoli, T.; Gregory, S.G.; Mohammadi, N.; Epstein, L.; et al. Early Nurture Epigenetically Tunes the Oxytocin Receptor. *Psychoneuroendocrinology* **2019**, *99*, 128–136. [CrossRef]
127. Myers, S.S.; Zanobetti, A.; Kloog, I.; Huybers, P.; Leakey, A.D.B.; Bloom, A.J.; Carlisle, E.; Dietterich, L.H.; Fitzgerald, G.; Hasegawa, T.; et al. Increasing CO_2 Threatens Human Nutrition. *Nature* **2014**, *510*, 139–142. [CrossRef]
128. Joëls, M.; Baram, T.Z. The Neuro-Symphony of Stress. *Nat. Rev. Neurosci.* **2009**, *10*, 459–466. [CrossRef]
129. Chen, Y.; Andres, A.L.; Frotscher, M.; Baram, T.Z. Tuning Synaptic Transmission in the Hippocampus by Stress: The CRH System. *Front. Cell. Neurosci.* **2012**, *6*, 13. [CrossRef]
130. Walker, C.M.; Gopnik, A. Discriminating Relational and Perceptual Judgments: Evidence from Human Toddlers. *Cognition* **2017**, *166*, 23–27. [CrossRef]
131. McEwen, B.S.; Akil, H. Revisiting the Stress Concept: Implications for Affective Disorders. *J. Neurosci.* **2020**, *40*, 12–21. [CrossRef]
132. Jamieson, J.P.; Nock, M.K.; Mendes, W.B. Improving Acute Stress Responses: The Power of Reappraisal. *Curr. Dir. Psychol. Sci.* **2013**, *22*, 51–56. [CrossRef]
133. Bale, T.L.; Vale, W.W. CRF and CRF Receptors: Role in Stress Responsivity and Other Behaviors. *Annu. Rev. Pharmacol. Toxicol.* **2004**, *44*, 525–557. [CrossRef] [PubMed]
134. Vandael, D.; Wierda, K.; Vints, K.; Baatsen, P.; de Groef, L.; Moons, L.; Rybakin, V.; Gounko, N.V. Corticotropin-releasing factor induces functional and structural synaptic remodeling in acute stress. *Transl. Psychiatry* **2021**, *11*, 378. [CrossRef]
135. Takayanagi, Y.; Onaka, T. The Oxytocin System and Early-Life Experience-Dependent Plastic Changes. *J. Neuroendocrinol.* **2021**, *33*, e13049. [CrossRef]
136. Kuperman, Y.; Weiss, M.; Dine, J.; Staikin, K.; Golani, O.; Ramot, A.; Nahum, T.; Kühne, C.; Shemesh, Y.; Wurst, W.; et al. CRFR1 in AgRP Neurons Modulates Sympathetic Nervous System Activity to Adapt to Cold Stress and Fasting. *Cell Metab.* **2016**, *23*, 1185–1199. [CrossRef]
137. Li, Y.; Hassett, A.L.; Seng, J.S. Exploring the Mutual Regulation Between Oxytocin and Cortisol as a Marker of Resilience. *Arch. Psychiatr. Nurs.* **2019**, *33*, 164–173. [CrossRef]
138. Roustit, M.M.; Vaughan, J.M.; Jamieson, P.M.; Cleasby, M.E. Urocortin 3 Activates AMPK and AKT Pathways and Enhances Glucose Disposal in Rat Skeletal Muscle. *J. Endocrinol.* **2014**, *223*, 143–154. [CrossRef]
139. Zhang, L.; Li, Y.; Wang, Y. Chronic UCN2 Treatment Desensitizes CRHR2 and Improves Insulin Sensitivity in Skeletal Muscle. *Nat. Commun.* **2023**, *14*, 3953. [CrossRef]
140. Yamada, A.; Chen, X.; Martinez, J.M.; Thompson, B.T. Neural Stimulation Suppresses mTORC1-Mediated Protein Synthesis During Muscle Inactivity. *Sci. Adv.* **2025**, *11*, eadt4955. [CrossRef]
141. Kirsch, P.; Esslinger, C.; Chen, Q.; Mier, D.; Lis, S.; Siddhanti, S.; Gruppe, H.; Mattay, V.S.; Gallhofer, B.; Meyer-Lindenberg, A. Oxytocin Modulates Neural Circuitry for Social Cognition and Fear in Humans. *J. Neurosci.* **2005**, *25*, 11489–11493. [CrossRef]
142. Meyer-Lindenberg, A.; Tost, H. Vasopressin and Post-Traumatic Stress Disorder. *Stress* **2020**, *23*, 631–640. [CrossRef] [PubMed]
143. Bosch, O.J.; Neumann, I.D. Both Oxytocin and Vasopressin Are Mediators of Maternal Care and Aggression in Rodents: From Central Release to Sites of Action. *Horm. Behav.* **2012**, *61*, 293–303. [CrossRef]
144. Spengler, F.B.; Schultz, J.; Scheele, D.; Essel, M.; Maier, W.; Heinrichs, M.; Hurlemann, R. Kinetics and Dose Dependency of Intranasal Oxytocin Effects on Amygdala Reactivity. *Biol. Psychiatry* **2017**, *82*, 885–894. [CrossRef] [PubMed]
145. Miller, A.J.; Slater, M.J.; Turner, M.J. Coach Identity Leadership Behaviours Are Positively Associated with Athlete Resource Appraisals: The Mediating Roles of Relational and Group Identification. *Psychol. Sport. Exerc.* **2020**, *51*, 101755. [CrossRef]

146. Bales, K.L.; Carter, C.S. Developmental Exposure to Oxytocin Facilitates Partner Preference in Male Prairie Voles. *Behav. Neurosci.* **2003**, *117*, 854–859. [CrossRef]
147. Stoop, R. Neuromodulation by Oxytocin and Vasopressin in the Social Brain. *Nat. Rev. Neurosci.* **2022**, *23*, 519–533. [CrossRef]
148. Rigney, N.; de Vries, G.J.; Petrulis, A. Modulation of Social Behavior by Distinct Vasopressin Sources. *Front. Endocrinol.* **2023**, *14*, 1127792. [CrossRef]
149. Yang, X.J.; Jiang, H.M.; Hou, X.H.; Song, J. Anxiety and Depression in Patients with Gastroesophageal Reflux Disease and Their Effect on Quality of Life. *World J. Gastroenterol.* **2021**, *27*, 2985–2995. [CrossRef]
150. Ibayashi, Y.; Hasuzawa, N.; Nomura, S.; Kabashima, M.; Nagayama, A.; Iwata, S.; Kitamura, M.; Ashida, K.; Moriyama, Y.; Yamamoto, K.; et al. Mitochondrial Fission Is Required for Thermogenesis in Brown Adipose Tissue. *PLoS ONE* **2024**, *19*, e0312352. [CrossRef]
151. Szeto, A.; Sun-Suslow, N.; Mendez, A.J.; Hernandez, R.I.; Wagner, K.V.; McCabe, P.M. Regulation of the Macrophage Oxytocin Receptor in Response to Inflammation. *Am. J. Physiol. Endocrinol. Metab.* **2017**, *312*, E183–E189. [CrossRef]
152. O'Malley, D.; Julio-Pieper, M.; Gibney, S.M.; Dinan, T.G.; Cryan, J.F. Oxytocin Inhibits Visceral Hypersensitivity and Colonic Barrier Dysfunction in a Stress-Sensitive Rat Strain. *Neurogastroenterol. Motil.* **2020**, *32*, e13715. [CrossRef]
153. Peters, S.; Slattery, D.A.; Usherwood, T.; Rehn, L. Oxytocin Enhances Resting-State Connectivity Between Amygdala and Medial Frontal Cortex. *Int. J. Neuropsychopharmacol.* **2017**, *20*, 803–809. [CrossRef]
154. Smith, A.S.; Wang, Z. Oxytocin Facilitates Adaptive Fear and Attenuates Anxiety Responses. *Front. Behav. Neurosci.* **2016**, *10*, 92. [CrossRef]
155. Ebner, N.C.; Maura, G.M.; Macdonald, K.; Westberg, L.; Fischer, H. Oxytocin and Socioemotional Aging: Current Knowledge and Future Trends. *Front. Hum. Neurosci.* **2018**, *12*, 1–14. [CrossRef] [PubMed]
156. Steinman, M.Q.; Laredo, S.A.; Lopez, E.M.; Manning, C.E.; Hao, R.C.; Doig, I.E.; Campi, K.L.; Flowers, A.E.; Knight, J.K.; Trainor, B.C. Hypothalamic Vasopressin Systems Are More Sensitive to the Long-Term Effects of Social Defeat in Males versus Females. *Psychoneuroendocrinology* **2015**, *51*, 122–134. [CrossRef]
157. Rigney, N.; Campos-Lira, E.; Kirchner, M.K.; Wei, W.; Belkasim, S.; Beaumont, R.; Singh, S.; de Vries, G.J.; Petrulis, A. A Vasopressin Circuit That Modulates Sex-Specific Social Interest and Anxiety-Like Behavior in Mice. *bioRxiv* **2023**. [CrossRef]
158. Carter, C.S.; Altemus, M. Integrative Functions of Lactational Hormones in Social Behavior and Stress Management. *Ann. N. Y. Acad. Sci.* **1997**, *807*, 164–174. [CrossRef]
159. Ramos, L.; Hicks, C.; Caminer, A.; Narlawar, R.; Kassiou, M.; McGregor, I.S. Oxytocin and Vasopressin Receptor Antagonists as a Novel Pharmacotherapy for Alcohol and Drug Dependence: A Translational Review. *Neurosci. Biobehav. Rev.* **2019**, *106*, 46–68. [CrossRef]
160. Difede, J.; McAleavey, A.A.; Emrich, M.; Jick, A.; Ovalles, A.; Wyka, K.; Spielman, L.; Olden, M.; Peskin, M.; Becket-Davenport, C.; et al. A Proof-of-Concept Randomized Crossover Clinical Trial of a First-in-Class Vasopressin 1a Receptor Antagonist for PTSD. *Contemp. Clin. Trials Commun.* **2023**, *33*, 101116. [CrossRef]
161. Ding, C.; Leow, M.K.-S.; Magkos, F. Oxytocin in Metabolic Homeostasis: Implications for Obesity and Diabetes Management. *Obes. Rev.* **2019**, *20*, 22–40. [CrossRef]
162. Perisic, M.; Woolcock, K.; Hering, A.; Mendel, H.; Muttenthaler, M. Oxytocin and Vasopressin Signaling in Health and Disease. *Trends Biochem. Sci.* **2024**, *49*, 361–377. [CrossRef]
163. Guzmán, Y.F.; Tronson, N.C.; Jovasevic, V.; Sato, K.; Guedea, A.L.; Mizukami, H.; Nishimori, K.; Radulovic, J. Fear-Enhancing Effects of Septal Oxytocin Receptors. *Nat. Neurosci.* **2013**, *16*, 1185–1187. [CrossRef] [PubMed]
164. Chini, B.; Verhage, M.; Grinevich, V. The Action Radius of Oxytocin Release in the Mammalian CNS: From Single Vesicles to Behavior. *Trends Pharmacol. Sci.* **2021**, *42*, 792–806. [CrossRef] [PubMed]
165. Wirobski, G.; Range, F.; Schaebs, F.S.; Palme, R.; Deschner, T.; Marshall-Pescini, S. Life Experience Rather Than Domestication Accounts for Dogs' Increased Oxytocin Release During Social Contact with Humans. *Sci. Rep.* **2021**, *11*, 14423. [CrossRef] [PubMed]
166. Koch, S.B.J.; van Zuiden, M.; Nawijn, L.; Frijling, J.L.; Veltman, D.J.; Olff, M. Intranasal Oxytocin Administration Dampens Amygdala Reactivity Toward Emotional Faces in PTSD Patients. *Neuropsychopharmacology* **2016**, *41*, 1495–1504. [CrossRef]
167. Lefevre, A.; Sirigu, A. The Two-Fold Role of Oxytocin in Social Developmental Disorders: A Cause and a Remedy? *Neurosci. Biobehav. Rev.* **2016**, *63*, 168–176. [CrossRef]
168. Zhang, Z.; Xie, J.; Zhang, H.; Jin, J. CRH Neurons and the Neuroendocrine Regulation of Stress. *Trends Endocrinol. Metab.* **2023**, *34*, 91–106. [CrossRef]
169. Russell, G.M.; Lightman, S.L. The Human Stress Response: Sympathetic Nervous System and HPA Axis. *Neurosci. Lett.* **2019**, *702*, 134–143. [CrossRef]

170. Kageyama, K.; Hanada, K.; Moriyama, T.; Imaizumi, T.; Satoh, K.; Suda, T. Differential Regulation of CREB and ERK Phosphorylation Through Corticotropin-Releasing Factor Receptors Type 1 and 2 in AtT-20 and A7r5 Cells. *Mol. Cell. Endocrinol.* **2007**, *263*, 90–102. [CrossRef]
171. Pérez-García, S.; Juarranz, Y.; Carrión, M.; Gutiérrez-Cañas, I.; Margioris, A.; Pablos, J.L.; Tsatsanis, C.; Gomariz, R.P. Mapping the CRF-Urocortins System in Human Osteoarthritic and Rheumatoid Synovial Fibroblasts: Effect of Vasoactive Intestinal Peptide. *J. Cell. Physiol.* **2011**, *226*, 3261–3269. [CrossRef]
172. Triana-Del Rio, R.; Ranade, S.; Guardado, J.; LeDoux, J.; Klann, E.; Shrestha, P. The Modulation of Emotional and Social Behaviors by Oxytocin Signaling in Limbic Network. *Front. Mol. Neurosci.* **2022**, *15*, 1002846. [CrossRef] [PubMed]
173. Sobota, R.; Mihara, T.; Forrest, A.; Featherstone, R.E.; Siegel, S.J. Oxytocin Reduces Amygdala Activity, Increases Social Interactions, and Reduces Anxiety-Like Behavior Irrespective of NMDAR Antagonism. *Behav. Neurosci.* **2015**, *129*, 389–398. [CrossRef]
174. Barberis, C.; Tribollet, E. Vasopressin and Oxytocin Receptors in the Central Nervous System. *Crit. Rev. Neurobiol.* **1996**, *10*, 119–154. [CrossRef]
175. Tregeagle, D.; Doherty, C.; Callis, T.; Kassiou, M. Non-Peptidic Natural Products That Target and Modulate Oxytocin and Arginine Vasopressin Receptors: Opportunities and Challenges. *J. Pharmacol. Sci.* **2025**, *158*, 238–258. [CrossRef] [PubMed]
176. Song, Z.; Albers, H.E. Cross-Talk Among Oxytocin and Vasopressin Receptors: Relevance for Basic and Clinical Studies of the Brain. *Front. Neuroendocrinol.* **2018**, *51*, 14–24. [CrossRef] [PubMed]
177. McEwen, B.S. Physiology and Neurobiology of Stress and Adaptation: Central Role of the Brain. *Physiol. Rev.* **2007**, *87*, 873–904. [CrossRef]
178. Porges, S.W.; Porges, S. *Our Polyvagal World: How Safety and Trauma Change Us*; W.W. Norton & Company: New York, NY, USA, 2023.
179. Branchi, I. The Double-Edged Sword of Early-Life Adversity: Mechanisms of Adaptive Versus Maladaptive Plasticity. *Dev. Psychobiol.* **2021**, *63*, e22164. [CrossRef]
180. Karelina, K.; Stuller, K.A.; Jarrett, B.; Zhang, N.; Wells, J.; DeVries, A.C. Oxytocin Promotes Resilience to Stress and Inflammation in Rats. *Psychoneuroendocrinology* **2011**, *36*, 1082–1093. [CrossRef]
181. Szabó, P.; Bonet, S.; Hetényi, R.; Hanna, D.; Kovács, Z.; Prisztóka, G.; Križalkovičová, Z.; Szentpéteri, J. Systematic Review: Pain, Cognition, and Cardioprotection—Unpacking Oxytocin's Contributions in a Sport Context. *Front. Physiol.* **2024**, *15*, 1393497. [CrossRef]
182. Quintana, D.S.; Alvares, G.A.; Heathers, J.A.J. Guidelines for Reporting HRV in Research on the Effects of Oxytocin. *Biol. Psychol.* **2016**, *116*, 146–151. [CrossRef]
183. Flanagan, J.C.; Mitchell, J.M.; Baker, N.L.; Woolley, J.; Wangelin, B.; Back, S.E.; McQuaid, J.R.; Neylan, T.C.; Wolfe, W.R.; Brady, K.T. Enhancing Prolonged Exposure Therapy for PTSD among Veterans with Oxytocin: Design of a Multisite Randomized Controlled Trial. *Contemp. Clin. Trials* **2020**, *95*, 106074. [CrossRef] [PubMed]
184. Rubin, L.H.; Rubin, D.B.; Veenstra-VanderWeele, J.; Carter, C.S.; Sweeney, J.A. Sex Differences and Vasopressin System Dysfunction in Schizophrenia and Related Disorders. *Biol. Psychiatry* **2023**, *94*, 183–192. [CrossRef]
185. Mahableshwarkar, A.R.; Jacobsen, P.L.; Chen, Y.; Serenko, M. A Randomized, Placebo-Controlled, Phase 2 Trial of ABT-436, a Novel V1B Receptor Antagonist, for Major Depressive Disorder. *Neuropsychopharmacology* **2015**, *40*, 415–422. [CrossRef]
186. Kemp, A.H.; Quintana, D.S.; Kuhnert, R.L.; Griffiths, K.; Hickie, I.B.; Guastella, A.J. Impact of Oxytocin on Heart Rate Variability and Neurocardiac Function. *Biol. Psychol.* **2012**, *90*, 243–249. [CrossRef]
187. MacLean, E.L.; Wilson, S.R.; Martin, W.L.; Davis, J.M.; Nazarloo, H.P.; Carter, C.S. Challenges for measuring oxytocin: The blind men and the elephant? *Psychoneuroendocrinology* **2019**, *107*, 225–231. [CrossRef]
188. Aulinas, A.; Lawson, E.A. The Oxytocin System and Implications for Oxytocin Deficiency in Hypothalamic-Pituitary Disease. *Endocr. Rev.* **2025**, *46*, bnaf008. [CrossRef]
189. Giovanna, G.; Damiani, S.; Fusar-Poli, L.; Rocchetti, M.; Brondino, N.; de Cagna, F.; Mori, A.; Politi, P. Intranasal Oxytocin as a Potential Therapeutic Strategy in Post-Traumatic Stress Disorder: A Systematic Review. *Psychoneuroendocrinology* **2020**, *115*, 104605. [CrossRef]
190. Calabrese, V.; Wenzel, U.; Piccoli, T.; Jacob, U.M.; Nicolosi, L.; Fazzolari, G.; Failla, G.; Fritsch, T.; Osakabe, N.; Calabrese, E.J. Investigating Hormesis, Aging, and Neurodegeneration: From Bench to Clinics. *Open Med.* **2024**, *19*, 20240986. [CrossRef]
191. Wang, J.; Hodes, G.E.; Zhang, H.; Zhang, S.; Zhao, W.; Golden, S.A.; Bi, W.; Menard, C.; Kana, V.; Leboeuf, M.; et al. Epigenetic Modulation of Inflammation and Synaptic Plasticity Promotes Resilience Against Stress in Mice. *Nat. Commun.* **2018**, *9*, 477. [CrossRef]

192. Xu, J.; Wakai, M.; Xiong, K.; Yang, Y.; Prabakaran, A.; Wu, S.; Ahrens, D.; Molina-Portela, M.D.P.; Ni, M.; Bai, Y.; et al. The Pro-Inflammatory Cytokine IL6 Suppresses Mitochondrial Function via the gp130-JAK1/STAT1/3-HIF1α/ERRα Axis. *Cell Rep.* **2025**, *44*, 115403. [CrossRef]
193. Chung, H.Y.; Kim, H.J.; Kim, J.W.; Yu, B.P. Cytokines as Regulators of the Hormetic Stress Response. *Ageing Res. Rev.* **2023**, *88*, 101927. [CrossRef]
194. Fulop, T.; Larbi, A.; Dupuis, G.; Witkowski, J.M. Immunosenescence and Inflammaging in Healthy Aging: Myth or Reality? *Curr. Opin. Immunol.* **2023**, *83*, 102300. [CrossRef]
195. Wu, D.; Zhou, B.; Xu, X.; Guo, Z. Hormetic Remodeling of Mitochondrial Function and Immune Signaling: A Future Therapeutic Direction. *Trends Mol. Med.* **2024**, *30*, 31–47. [CrossRef]
196. Liberale, L.; Montecucco, F.; Tardif, J.-C.; Libby, P.; Camici, G.G. Inflamm-Ageing: The Role of Inflammation in Age-Dependent Cardiovascular Disease. *Eur. Heart J.* **2020**, *41*, 2974–2982. [CrossRef]
197. Mittal, M.; Siddiqui, M.R.; Tran, K.; Reddy, S.P.; Malik, A.B. Reactive Oxygen Species in Inflammation and Tissue Injury. *Antioxid. Redox Signal.* **2014**, *20*, 1126–1167. [CrossRef] [PubMed]
198. Franceschi, C.; Garagnani, P.; Parini, P.; Giuliani, C.; Santoro, A. Inflammaging: A New Immune–Metabolic Viewpoint for Age-Related Diseases. *Nat. Rev. Endocrinol.* **2018**, *14*, 576–590. [CrossRef]
199. Sarvas, J.L.; Khaper, N.; Lees, S.J. The IL-6 Paradox: Context Dependent Interplay of SOCS3 and AMPK. *J. Diabetes Metab.* **2013**, *S13*, 003. [CrossRef]
200. Hayes, J.D.; Dinkova-Kostova, A.T. The Nrf2 Regulatory Network Provides an Interface between Redox and Intermediary Metabolism. *Trends Biochem. Sci.* **2014**, *39*, 199–218. [CrossRef]
201. Weavers, H. Biological Resilience in Health and Disease. *Dis. Model. Mech.* **2024**, *17*, dmm050799. [CrossRef]
202. Zeytinoğlu, M.; Çınaroğlu, O.S.; Bora, E.S.; Erbaş, O. Healing with Love: Oxytocin Accelerates Oral Ulcer Recovery by Reducing Inflammation. *J. Clin. Med.* **2025**, *14*, 2667. [CrossRef]

current issues in molecular biology

MDPI

Review

The Role of the Arcuate Nucleus in Regulating Hunger and Satiety in Prader-Willi Syndrome

Charlotte Höybye [1,2,*] and Maria Petersson [1,2]

1 Department of Endocrinology, Karolinska University Hospital, 171 76 Stockholm, Sweden; maria.petersson@ki.se
2 Department of Molecular Medicine and Surgery, Karolinska Institutet, 171 77 Stockholm, Sweden
* Correspondence: charlotte.hoybye@ki.se

Abstract: Prader-Willi syndrome (PWS) is a rare genetic disorder. The main characteristics are muscular hypotonia, failure to thrive and feeding problems in infancy, which switch to hyperphagia in early childhood and continue into adulthood. Due to hyperphagia, the risk of developing morbid obesity is high without treatment. PWS is considered a hypothalamic disease, and within the hypothalamus the arcuate nucleus (AC) is of central importance for controlling metabolism, hunger, and satiety. The AC has been studied in several animal models as well as in humans, including PWS. The function of AC is regulated by several neuropeptides and proteins produced within the central nervous system such as oxytocin, orexin, tachykinins as well as the hypothalamic hormones, regulating the adenohypophyseal hormones, also acting as neurotransmitters. Additionally, there are many peripheral hormones among which insulin, leptin, adiponectin, ghrelin, and glucagon-like peptide (GLP-1) are the most important. High levels of adiponectin and ghrelin have consistently been reported in PWS, but dysregulation and deviating levels of many other factors and hormones have also been demonstrated in both individuals with PWS and in animal models. In this review, we focus on the role of AC and peptides and proteins produced within the central nervous system in the regulation of hunger and satiety in PWS.

Keywords: PWS; arcuate nucleus; hyperphagia; oxytocin; orexin; kisspeptin; tachykinins; BDNF; nesfatin-1

1. Introduction

Prader-Willi syndrome (PWS) is a rare, multi-symptomatic, contiguous genetic disorder [1]. PWS is caused by absence of paternally expressed imprinted genes at 15q11.2-q13 through paternal deletion of this region (65–75% of individuals), maternal uniparental disomy 15 (20–30%), or an imprinting defect (1–3%) [1]. The estimated prevalence is 1/10,000-1/21,000 newborns [2].

PWS is considered a hypothalamic disease and many of the symptoms of the syndrome are related to hypothalamic dysfunction. The syndrome is classically characterized by muscular hypotonia, poor growth, and short stature, with feeding problems in infancy replaced by hyperphagia at approximately 4 years of age leading to a high risk of obesity, unless food intake is supervised and controlled [3]. Body composition is abnormal with more body fat than muscle mass, and metabolism is decreased. Behavioral problems including temper outbursts, anxiety, obsessive-compulsive and autism-like features, as well as cognitive and learning disabilities, are common. Endocrine abnormalities are commonly observed, with insufficient growth hormone (GH) secretion, and hypogonadism being most frequently present. The knowledge of PWS and clinical management has improved considerably over recent decades. Together with the approval of GH treatment of children with PWS from the year 2000, it has led to a marked change in the phenotype, with normal adult height and improved psychomotor functioning and body composition [3].

Several of the genes in the affected region of PWS have been studied, and dysfunction of each of them can be associated with one or several of the symptoms of PWS [4]. However, it has not been possible to relate the symptoms of PWS to any single causative gene, and the symptoms are more likely to result from the entire genetic deletion [4]. RNA sequencing of the hypothalamus from individuals with PWS showed that upregulated genes overlapped with the mouse agouti-related peptide (AgRP) neurons, mainly located within the AC and activated by hunger, while downregulated genes overlapped with the expression profile of proopiomelanocortin (POMC) neurons, which in the AC are activated by feeding and satiety [5].

Magnetic resonance (MR) imaging studies have shown that all hypothalamic nuclei are smaller in adults with PWS compared to both age and gender-matched controls as well as participants with obesity [6]. Also, in children with PWS, MR imaging has shown atrophy in the thalamus, pallidum, hippocampus, amygdala, and hypothalamus compared to obese, age- and sex-matched controls [7]. Further, premeal functional MR imaging demonstrated higher activity in reward/limbic regions (nucleus accumbens, amygdala) and lower activity in the hypothalamus and hippocampus in response to food vs. non-food in individuals with PWS compared to participants with obesity [8]. Post-meal, individuals with PWS exhibited a greater stimulation of food activation centers in the limbic and paralimbic regions (hypothalamus, amygdala, hippocampus) and a lower activation in cortical inhibitory areas (orbitofrontal cortex, medial prefrontal cortex) [9].

Thus, knowledge of the function of the hypothalamus is important for understanding metabolism and the regulation of hunger and satiety in PWS. This review summarizes the central regulation with a focus on the arcuate nucleus (AC) and neuropeptides and proteins produced within the central nervous system (CNS) and especially within the hypothalamus in PWS.

2. The Role of the Hypothalamus in Appetite and Metabolism

The hypothalamus is central in controlling metabolism, appetite, thirst, temperature, diurnal rhythm, and hormonal secretion from the pituitary. It is located in the basal part of the brain between the thalamus and the pituitary and integrates and transmits signals from other parts of the brain, the autonomic nervous system, and from hormones and peptides from the periphery. Although the regulation of hunger and satiety is controlled from more than one hypothalamic nucleus, the AC is often considered the most important. The AC has a reciprocal connection with the other hypothalamic nuclei involved in the regulation of hunger and satiety, such as the paraventricular nucleus (PVN), where for example, oxytocin and corticotrophin-releasing hormone (CRH) are produced; the hypothalamic lateral nucleus, which contains many orexin neurons and is considered especially important for hunger; and the ventromedial nucleus, which is considered especially important for satiety.

The hypothalamus, and in particular the AC, have close connections to the mesolimbic system, where dopamine and serotonin are two of the main neurotransmitters. Studies have shown that individuals with PWS also may have deviations in these pathways and neurotransmitters [4,10].

2.1. The Role of the Arcuate Nucleus in Appetite and Metabolism

The AC is located in the mediobasal part of the hypothalamus close to the third ventricle and the median eminence. This nucleus contains two specific neuron populations: the neuropeptide Y (NPY) and AgRP neurons and the POMC cocaine and amphetamine-regulated transcript (CART) neurons. The NPY/AgRP neurons stimulate appetite and the POMC/CART-neurons are stimulated by feeding/satiety and inhibit appetite.

The two neuron populations also interact with each other; for example, there is γ-aminobutyric acid (GABA)-ergic inhibition of the POMC/CART neurons from the NPY/AgRP neurons of the POMC. The NPY/AgRP and POMC/CART neurons both express receptors for many important hormones and neurotransmitters that regulate their activity. For example, leptin, produced mainly in the white adipose tissue, decreases the activity in the NPY/AgRP neurons and increases the activity in the POMC/CART neurons. Similar effects are induced by insulin. Thus, both leptin and insulin stimulate satiety. However, leptin and insulin resistance are common in obesity, where the effects of leptin and insulin within the AC, and especially in the POMC/CART neurons, become less pronounced and sometimes these effects may even be absent [11,12]. However, in this context, it should be noted that individuals with PWS have less insulin resistance and lower insulin levels compared to obese individuals without PWS [13]. One mechanism behind this might be the higher adiponectin levels that have been reported in PWS [14].

Other hormones important for the activity of these neurons in the AC are ghrelin, GH, peptide YY (PYY), orexin, GLP-1, oxytocin, kisspeptin, brain-derived neutrophic factor (BDNF), asprosin, as well as the hypothalamic-pituitary-adrenal (HPA)-axis.

Some of the most important gastrointestinal and adipose tissue hormones regulating neurons in the AC are shown in Figure 1.

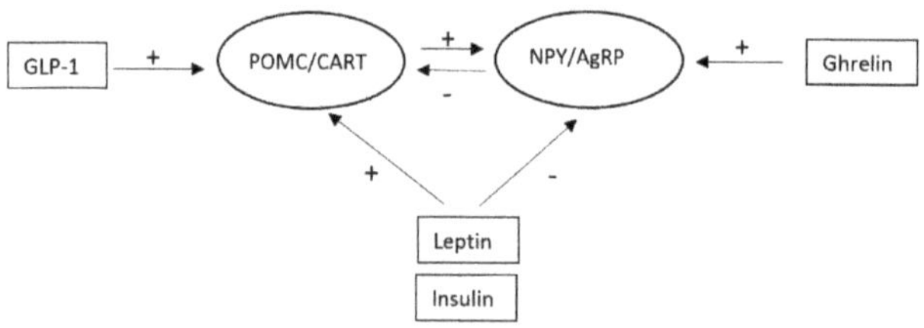

Figure 1. Some of the most important gastrointestinal and adipose tissue hormones regulating neurons in the arcuate nucleus (+ indicates an increased activity in neurons, − indicates a decreased activity in the neurons). Abbreviations: AgRP: agouti-related peptide, CART: cocaine and amphetamine-regulated transcript, GLP-1: glucagon-like peptide 1, NPY: neuropeptide Y, POMC: proopiomelanocortin.

2.2. POMC/CART Neurons

POMC undergoes cleavage into α-MSH that binds to melanocortin (MC) receptors, preferably MC4 receptors, thereby activating satiety. The POMC/CART neurons are regulated primarily by leptin, insulin, and glucose levels but also by GLP-1, which all suppress food intake. Stimulation of POMC/CART neurons in mice reduces food intake and body weight [15]. POMC knockout mice develop hyperphagia, insulin resistance, and obesity. In addition, CART knockout mice exhibit the same characteristics [12].

2.3. NPY/AgRP Neurons

The NPY/AgRP neurons are stimulated by ghrelin and inhibited by leptin, insulin, and glucose acting through the stimulation of NPY receptors and inhibition of MCRs (for reviews see [11,12]). In addition, CRH may both directly, and indirectly through other peptides and neurotransmitters, influence the NPY/AgRP neurons and has been suggested to have a role in stress-related eating behavior [16,17]. Both NPY and AgRP knockout mice have unchanged weight and food intake. In contrast, transgenic mice overexpressing AgRP develop obesity. So do NPY receptor knockouts, although food intake seems to be unaltered [18–21].

2.4. Other Neurons in the Arcuate Nucleus

The NPY/AGRP and the POMC/CART neurons are not the only neurons in the AC. There are, for example, neurons producing GABA and dopamine, where dopamine is released both as a neurotransmitter and as a neurocrine regulator of prolactin. Additionally, several neuropeptides and proteins such as GHRH, kisspeptin, tachykinins, somatostatin, and GnRH are also synthesized within the arcuate nucleus [11].

3. The Arcuate Nucleus in PWS

Dysfunction of the AC is involved in several states and disorders related to overeating, weight and growth diversities, metabolism, and obesity. Therefore, the AC is highly interesting in PWS; and as mentioned above, individuals with PWS generally have a smaller AC [6].

The 15q11-13 region contains both protein-coding DNA genes and differentially imprinted genes. Mutations or deletions of genes in this region (for example *MAGEL2*, *NECDIN*, and *SNORD116*) are in mouse models associated with hypotonia, developmental delay, hypogonadism, hyperphagia, and impaired social skills [22].

The *MAGEL2* gene is expressed in the POMC neurons in the AC of the hypothalamus; knockout of the *MAGEL2* gene disrupts POMC neuronal circuits and functions in rodents and they become obese. A change in the sensitivity and stability of leptin receptors has been suggested to be one of the mechanisms behind this effect [23,24]. In a study of leptin sensitivity in *MAGEL2* gene-knockouts, there was a gradual decrease in leptin sensitivity, such that POMC neurons responded normally to leptin in newborn rats up to 4 weeks of age, whereas the response was reduced at the age of 6 weeks [25].

Contrary to what is expected, the Necdin knockout mice do not usually exhibit obesity. Studies have instead shown that these mice often have lower body weight and a reduced amount of adipose tissue [26].

Neither do *SNORD116* knockout mice typically exhibit obesity. These mice are often born smaller and remain smaller than their wild-type counterparts throughout their lives. However, when *SNORD116* is deleted in adult mice, they do display hyperphagia (increased food intake). Interestingly, only some of these adult knockout mice become obese, and the reason behind this is to our knowledge not known [27].

In PWS, the expression of the transcription factor nescient helix-loop-helix 2 (*NHLH2*) is reduced and *NHLH2*-knockout mice are obese. Indeed, these mice have reduced mRNA levels of both POMC and CART and increased levels of NPY and AGRP [28]. NHLH2 is also associated with kisspeptin and the *NHLH2* knockouts exhibit hypogonadism [29].

In a transgenic mouse model of PWS, which displayed failure to thrive during the neonatal period, the expression of AgRP mRNA was decreased while the mRNA expression of POMC was upregulated in PWS-mice neonates. Since AgRP stimulates appetite and the POMC-derived peptide, α-MSH, stimulates satiety, these changes together will decrease feeding and may thus contribute to the failure to thrive in PWS neonatal mice [30]. These changes are consistent with the findings in Magel 2 knockouts described above and also with the different periods in PWS, in which feeding difficulties in infants change to overeating after a couple of years.

In a study of hypothalamic tissue from individuals with PWS, it was demonstrated that upregulated genes overlapped with rodent RNA-sequencing data from the AgRP/NPYneurons, whereas downregulated genes overlapped with rodent RNA corresponding to the POMC/CART neurons [5]. However, in another study on post-mortem hypothalamic tissues, Goldstone et al. reported unchanged expression of AgRP as well as NPY in PWS [31].

In children with PWS, serum levels of α-MSH were significantly lower compared to obese and lean controls. No difference in AgRP levels was seen [32]. In contrast to what would be expected, the number of NPY neurons are reduced in PWS, maybe because of increased levels of leptin and insulin [4]. NPY in plasma is mainly derived from the sympathetic nervous system, the adrenal gland, and other tissues in the periphery, and does not reflect NPY levels within the brain. Usually, there is no change in plasma levels of NPY in individuals with PWS, although data on NPY are conflicting [13]. Ghrelin levels are increased in individuals with PWS compared to both lean controls and obese subjects [4,33], and as discussed above, ghrelin increases the release of NPY and AgRP from the AC. Snord116 knockout mice have increased ghrelin levels as well [34].

Leptin levels in PWS relate to the amount of adipose tissue in the same way as in obese non-PWS individuals [35], whereas the prevalence of impaired glucose intolerance, hyperinsulinemia, and type 2 diabetes seems to be lower compared to that in obese non-PWS controls [20]. Adiponectin, a peptide that also is produced within the adipose tissue, is mainly associated with positive metabolic effects such as decreased insulin resistance.

Adiponectin receptors are expressed both in the POMC/CART and the NPY/AgRP neurons and increased levels are mainly associated with satiety. Indeed, PWS individuals have been demonstrated to have higher adiponectin levels compared to obese non-PWS controls.

It is important to recognize that there are different genetic mechanisms that can be responsible for PWS, and it is not surprising that there are conflicting findings, since the genotype in the study cohorts were probably not the same type. Even then, other modulatory genes exist so that not everyone would exhibit the same symptoms in every aspect.

4. How Peptides and Proteins Produced Within the Brain Affect the Arcuate Nucleus—And Their Potential Roles in PWS

4.1. Oxytocin

Oxytocin is produced within two hypothalamic nuclei, the PVN and the supraoptical nucleus. Oxytocin is released as a hormone from the neurohypophysis and is mainly associated with its hormonal effects during parturition and breastfeeding. However, it is also released within the brain and involved in, for example, behavior and appetite regulation. Oxytocin has anxiolytic-like effects and it may both increase and decrease food intake depending on physiological context, other hormones, and if in a fed or fasting state [36,37]. There are oxytocin receptors on both the POMC/CART neurons and the NPY/AgRP neurons. Analyses of tissue from the PVN of adults with PWS showed a reduction of both the volume, the cell number, and the immunoreactivity of oxytocin neurons, and a reduction of oxytocin has been suggested to be one of the mechanisms behind the hunger in PWS [38,39]. In contrast, analyses of oxytocin in cerebrospinal fluid showed higher concentrations in adolescents and adults with PWS compared to the controls [40]. In line with this, plasma levels of oxytocin in 23 children with PWS were compared with those of 18 healthy unrelated siblings matched for age and with a similar gender ratio and BMI. The children with PWS were found to have levels of plasma oxytocin more than twice as high compared to those of unrelated siblings [41]. However, analyses of serum oxytocin in adults with PWS showed similar concentrations as in controls but, in relation to their obesity, the concentrations were low [35]. Interestingly, oxytocin is released in response to vagal stimulation [42] and clinical trials of vagal stimulation in PWS individuals have been conducted with positive results [43]. A pronounced dysregulation of the signaling pathways for oxytocin has been shown in the Magel2-knockout mouse model of PWS. Early postnatal treatment with oxytocin restored behavioral changes that have been observed in these mice [44]. In addition, the initial feeding difficulties in the newborn mice were prevented [45]. Consistent with these observations, studies with oxytocin treatment in individuals with PWS have shown improvements in anxiety, compulsiveness, and hyperphagia. However, the degree of oxytocin's effects depends on situation and context and seems to be more pronounced if administered early in life [46].

4.2. Brain-Derived Neurotrophic Factor

The neurotrophin brain-derived neurotrophic factor (BDNF), synthesized in several brain areas, is besides regulating growth involved in the regulation of energy homeostasis, and may directly affect the NPY/AgRP neurons and inhibit their activity, thus acting as a satiety signal. In addition, BDNF injected intracerebroventricularly as well as directly into the PVN decreased food intake and weight gain in rats [47]. In line with this, a study of a *MAGEL2*-null mouse model of PWS assessed the translational potential of hypothalamic adeno-associated virus (AAV)-BDNF gene therapy and the authors found that BDNF gene therapy improved glucose metabolism, insulin sensitivity, circulating adipokine levels, and body composition [48]. Furthermore, in hypothalamic tissue from deceased individuals with PWS, the number of neurons in the hypothalamus expressing BDNF and *NTRK2*, the gene coding for one of its receptors, were found to be lower [5]. Reduced basal circulating BDNF levels and a diminished postprandial peak in comparison to weight-matched controls have also been demonstrated in individuals with PWS [49].

4.3. Orexin

Orexin, which is produced within the hypothalamus, is involved in appetite regulation, arousal, sleep and wakefulness, metabolism and energy expenditure, pleasure and reward-seeking behavior, as well as stress response. There are 2 forms, orexin A and orexin B, which both are important for appetite regulation, but their affinity for the two orexin receptors is different. Changes in peripheral metabolic signals (i.e., glucose, ghrelin, leptin, and blood pH) impact the activity of orexin A neurons in the lateral hypothalamus, which increases eating behaviors through binding to orexin receptors in the hypothalamus, hippocampus, locus coeruleus, and limbic centers. Both the POMC/CART neurons and the NPY/AgRP neurons are provided with orexin receptors. The *MAGEL2*-null mouse has reduced orexin levels as well as orexin neurons [50]. In a study of 23 children with PWS, circulating orexin A levels were found to be higher compared to age-matched controls, suggesting that dysregulation of orexin A signaling might contribute to behavioral problems and hyperphagia in PWS [51]. However, another study of 14 individual with PWS aged 8 to 37 years showed moderately decreased orexin levels in cerebrospinal fluid [52]. There was no correlation between orexin and BMI. The discrepancy between the findings in the two studies might be due to measurements in different fluids and different ages of the patients suggesting involvement of other hormones, such as leptin and ghrelin [53].

4.4. Kisspeptin

Kisspeptin is produced within the hypothalamus, including the AC. It is mainly associated with the onset of puberty and the release of GnRH and is thought to play an important role in reproductive function [54]. However, it has also been demonstrated to modulate the release of GH as well as the activity within the AC, where it stimulates POMC/CART neurons, and it may also inhibit the NPY/AGRP neurons. These effects in the AC both decrease appetite. NHLH2, which is reduced in PWS, takes part in the regulation of the effects of kisspeptin levels. Due to the frequently present hypogonadism in PWS, kisspeptin levels would be expected to be low. Unexpectedly, kisspeptin levels have been demonstrated to be higher in individuals with PWS but are normalized in response to GH treatment [52], which might indicate that the raised kisspeptin levels mainly are due to the GH deficiency.

4.5. Tachykinins

Tachykinins are a large group of neuropeptides including neurokinin A and B, and substance P. Tachykinins are produced both in the periphery and in the brain and are, for example, involved in pain, vascular tone, and behavior. Neurokinin synthesis is also colocalized with the synthesis of kisspeptin in the so-called KNDy (kisspeptin, neurokinin, dynorphin) neurons within the AC. These neurons are of central importance for the regulation of GnRH. However, neurokinins and substance P also modulate the POMC/CART neurons and tachykinin receptors have been demonstrated in these neurons.

In PWS individuals, plasma levels of substance P as well as beta-endorphin (which is cleaved from POMC) have been found to be increased. These changes have been suggested to contribute to hunger in PWS [55].

4.6. Nesfatin-1

Nesfatin-1 is a peptide originating from the precursor molecule nucleobindin-2 and produced in various brain areas including the hypothalamus and the AC. It is also synthesized, for example, within the adipose tissue, the gastrointestinal canal, and the pancreas and it crosses the blood–brain barrier. Serum levels increase in response to feeding and have been demonstrated to inhibit food intake and in addition directly inhibit NPY/AgRP neurons within the AC. Nesfatin-1 has also been demonstrated to increase both oxytocin and CRH (for a review, see [56]). In a recent study of non-obese children with PWS treated with GH, nesfatin-1 levels were higher when compared to age-matched healthy controls [57]. Thus, a change in nesfatin-1 levels might also contribute to the metabolic effects of GH treatment in PWS.

4.7. CRH

CRH is mostly produced within the hypothalamic nucleus PVN and is the main regulator of the HPA-axis. CRH is also produced in the amygdala, and similarly to oxytocin CRH, both from the PVN and the amygdala, is released within the brain, acting as a neurotransmitter. CRH is mainly associated with satiety but during certain circumstances it may also increase the activity of NPY/AgRP neurons, thereby increasing hunger. However, the activity within the HPA-axis seems not to be increased in PWS. Instead, a delayed stress response in children with PWS has been demonstrated [58]. However, this does not exclude deviations within the brain CRH release and pathways. In a study with MRI, an increased activation of the amygdala in response to food odors was seen in adult individuals with PWS [59].

4.8. TRH

TRH is also mainly produced within the PVN and it is the main regulatory hormone in the thyroid axis. Similarly to CRH, it also acts as a neurotransmitter. During starvation and/or long-term illness, pathways from the NPY/AgRP neurons in the AC decrease the release of TRH. There is also a TRH pathway from the PVN to the AC and TRH receptors in the NPY/AgRP neurons, which have been suggested to be excitatory, and thus it would be expected that they should increase hunger [60]. Hypothyreosis is common in PWS and the most common cause is central hypothyroidism, probably caused by a deficiency or dysregulation of TRH [3].

4.9. GHRH

GHRH is produced within the AC and is necessary for the stimulation of growth hormone (GH) release from the pituitary. This release of GH may then modulate the POMC/CART and NPY/AGRP neurons, which both express GH receptors. GH is also released by ghrelin. Individuals with PWS have deficient GH secretion, and leptin levels decrease and body composition improves in response to GH treatment. Worth mentioning is that adiponectin levels do not change in response to GH treatment [35]. GH treatment of individuals with PWS might, through negative feedback, induce the already low GHRH release to become even lower. However, although GHRH is produced within the AC, there are, to our knowledge, no GHRH receptors on the POMC/CART or NPY/AgRP neurons.

5. Discussion

The regulation of hunger and satiety is very complex, and several genes and transcription factors, as well as central and peripheral secreted hormones, neuropeptides, and proteins are involved. In PWS, the dysfunction of the hypothalamus, hyperphagia, abnormal body composition, and obesity make it even more complicated. Within the hypothalamus, the AC is of central importance for metabolism, hunger, and satiety. The focus of our review was the effects of brain-derived peptides and proteins on the AC, and in summary, we found that several of these substances have the potential to disturb the function of the NPY/AgRP and POMC/CART circuits in PWS, resulting in unbalanced systems.

In the present review, we have discussed oxytocin, BDNF, orexin, kisspeptin, tachykinins, nesfatin-1, as well as the hypothalamic hormones and neurotransmitters CRH and TRH, and all of them seem to be dysregulated in individuals with PWS (Table 1). In addition, the release of GHRH is decreased, thus mediating a lower release of GH, which in turn may affect the regulation of the POMC/CART and NPY/AgRP neurons. Indeed, there are improvements in leptin, insulin, and kisspeptin levels in response to GH treatment. The AC is, of course, not the only nucleus within the hypothalamus that regulates appetite and satiety. The PVN, where, for example, oxytocin, CRH, and TRH are produced, also participates in this integrated regulation, as does the ventromedial hypothalamic nucleus (VMH), sometimes considered as the hypothalamic satiety center and the lateral hypothalamic nucleus (LH), sometimes considered as the hypothalamic hunger center. All these nuclei communicate bidirectionally with each other.

Table 1. Changes in peptides and proteins produced within the hypothalamus in animal models or individuals with PWS.

Peptide/Protein	Difference	Type of Study	References
Oxytocin	Usually low	Autopsy material, human blood, and CSF as well as animals	[35,38–41,44]
Orexin	High	Human blood and anjmals	[50,51]
Kisspeptin	High	Human blood	[52]
BDNF	Low	Autopsy material, animals, and human blood	[5,48,49]
Tachykinins	Substance P high	Human blood	[55]
Nesfatin-1	High	Human blood	[57]
CRH	Changed activity	Human blood, MR	[58,59]
TRH	Low	Human blood	[3]
GHRH	Low	Human blood	[3]

Abbreviations: BDNF, brain-derived neurotrophic factor; CRH, corticotrophin releasing factor; GHRH, growth hormone releasing hormone; TRH, thyrotropin releasing hormone.

There are other factors besides the peptides and proteins discussed in the present paper, including the classical neurotransmitters and also lipids. Endocannabinoids are produced in many peripheral organs and tissues as well as the brain and the hypothalamus. There are cannabinoid receptors in the AC and in general they are considered as appetite stimulators.

Indeed, the activity of cannabinoids has been demonstrated to be increased in PWS. The *MAGEL2* knockout mice discussed earlier in this manuscript have been found to have increased activity within their endocannabinoid systems as well. In clinical studies where a cannabinoid receptor 1-antagonist was given to PWS individuals, weight was reduced [61]. However, clinical trials with cannabinoid receptor antagonists have been terminated because of unwanted side effects.

Besides these centrally released factors, there are several hormones produced in the adipose tissue, the pancreas, and the gastrointestinal tract, which also participate in the regulation of the AC in addition to the previously mentioned hormones: leptin, insulin, GLP-1, and ghrelin. Asprosin was discovered in 2016 and is a protein produced in the adipose tissue. Asprosin participates in the regulation of blood glucose levels through stimulating liver glucose production and appetite [62]. In the AC, asprosin may inhibit the POMC/CART neurons, which is thought to be one of the main mechanisms behind its appetite stimulating effect. In a recent study by Faienza et al. [63], obese individuals with PWS had higher asprosin levels when compared to PWS with normal weight. Asprosin levels were also higher in adults compared to adolescents. Interestingly, asprosin levels were significantly higher in patients with deletion versus patients with uniparental disomy. However, no correlation between serum levels of asprosin and appetite in individuals with PWS could be demonstrated. Other peripheral hormones with receptors in the brain may also contribute to changed appetite and eating behavior. Both cholecystokinin (CCK) and PYY levels are lower in individuals with PWS, and these hormones are associated with satiety and increase in response to feeding. In contrast, fasting and postprandial GLP-1 levels have been reported to be similar between individuals with PWS and obese and lean controls [64].

As already mentioned, ghrelin levels are increased in PWS individuals. Animal studies indicate that ghrelin is produced not only in the gastrointestinal tract and islets of Langerhans but also within the hypothalamus [65,66]. Prior to secretion, ghrelin is acetylated (AG), and in this form, it activates the GH secretagogue receptor 1a (GHSR1a), and the release of GH. Adrenocorticotropic hormone (ACTH), cortisol, prolactin, and glucose are also released in response to AG and AG affects feeding, growth, fat stores, and glycemic regulation. Unacetylated ghrelin is converted to AG by the enzyme ghrelin O-acyltransferase (GOAT). GLWL-01 is a selective, reversible inhibitor of GOAT, and in a double-blind, placebo-controlled, phase 2 study, GLWL-01 was found to lower the AG levels and the ratio of AG/UAG [67]. However, no effects on hyperphagia or on anthropometric and clinical parameters related to weight excess in patients with PWS were observed, maybe due to a short duration of treatment and a small study cohort.

UAG inhibits AG-induced hunger, reduces fat deposition, and improves insulin resistance. In studies with livoletide, an UAG analog, no significant change was found in hunger and food-related behaviors as measured with the Hyperphagia Questionnaire for Clinical Trials (HQ-CT) scores, or in fat mass, body weight, or waist circumference compared with placebo [68].

Likewise, studies on setmelanotide, a melanocortin (MC)-4 receptor agonist administrated subcutaneously once daily, showed no changes in weight, DEXA measurements or laboratory findings, while the mean hyperphagia questionnaire scores demonstrated a small, not statistically significant reduction from baseline. Further studies with setmelanotide have not been performed [69].

Ghrelin has, besides its direct effects in the AC, also been demonstrated to increase appetite by acting through dopamine, serotonin, opioid, and cannabinoid systems. We have already discussed the cannabinoid system above but the neurotransmitters have mainly been left aside in the present paper. Worth mentioning here, however, is that serotonin, and especially the 5HT2c receptors, are central in hypothalamic appetite regulation including the AC. Alternate splicing of the serotonin (5-hydroxytryptamine; 5-HT) 2C receptor (5-HT_{2C}R) pre-RNA is negatively regulated by *Snord115* [70,71], and loss of *Snord115* expression is associated with decreased levels of POMC in the AC [71].

Recently, clinical trials with diazoxide, a drug commonly used to treat hypoglycemia and hyperinsulinemia, have been performed. Diazoxide is a potent activator of the adenosine triphosphate-sensitive potassium (KATP) channel and crosses the blood–brain barrier [72]. Activation of the KATP channel in NPY/AgRP neurons in the hypothalamus reduces the secretion of NPY and AgRP [73]. In addition, the KATP channels in the dorsal motor nucleus of the vagal nerve, pancreatic β-cells, and adipocytes are activated, reducing hyperinsulinemia and body fat, directly or indirectly leading to improved insulin and leptin resistance and satiety [73]. These effects have been confirmed in animal models of hyperphagic obesity including Magel-2 null mouse [73].

Diazoxide choline slow-release (DCCR) treatment has in a phase 2 study in individuals with PWS been shown to reduce hyperphagia, aggressive behavior, and body fat, and increase lean body mass [74]. In a following 13 week phase 3 study comparing DCCR to placebo, DCCR improved body composition and clinician-reported outcomes, but did not in the entire cohort significantly improve hyperphagia in the primary analysis. However, DCCR reduced hyperphagia in those with more severe hyperphagia, and in the subset of data collected before the onset of the COVID-19 pandemic [75]. In the open-label extension of the phase 3 study, 125 participants with PWS received DCCR for up to 52 weeks [76]. This study showed that DCCR improved hyperphagia, with greater improvements in those with more severe baseline hyperphagia. Improvements were also seen in aggression, anxiety, and compulsivity, and there were reductions in leptin, insulin, and insulin resistance, as well as a significant increase in adiponectin.

Although patients with PWS are similar in certain aspects, individuals' manifestations may differ depending on the genotype and other modifier genes. Most well-known is the increased risk of psychosis and behavioral challenges in the UPD subgroup. However, it has also been shown that the frequency and severity of hyperphagia and obesity are more prominent in the deletion group compared to the UPD group [77]. Differentially expressed genes associated with PWS are *MKRN3, MAGEL2, NECDIN,* and small nucleolar RNA genes, and abnormalities such as UPD of maternal alleles result in a lack of paternally expressed genes. *SNORD116*, a small nucleolar RNA gene, is known to be the critical gene for most PWS phenotypes, and deletion of this gene has been demonstrated to cause an imbalance in the neuromodulatory systems of the hypothalamus, and results in hyperphagic behavior and sleep disturbances. Studies generating specific information on the function of the AC in deletion and UPD groups is not yet available but would be expected in the future.

6. Conclusions

In conclusion, the AC plays an important role in the very complex and multifactorial regulation of hunger and satiety. Our knowledge is primarily based on animal studies and measurements of circulating levels of various neuropeptides and hormones, which are usually not parallel with CNS levels. In individuals with PWS and in animal models of PWS, hormones such as oxytocin, ghrelin, and adiponectin seem to be secreted and to circulate in different concentrations compared to those in non-PWS humans or animals. Other factors such as deficiency of adenohypophyseal hormones, as well as an abnormal body composition, less insulin resistance, and comorbidities, also affect the outcome. In line with this, some clinical trials in PWS with specific pharmacological compounds important for the regulation of hunger and satiety have been without significant effects. The regulation appears to be multifactorial, involving a complicated interplay between different factors, making it difficult to identify a specific neuropeptide or hormone as the most crucial in PWS. However, promising results have been obtained with treatments using diazoxide and oxytocin and further investigations of the regulation of hunger and satiety in PWS are ongoing.

Author Contributions: Conceptualization, C.H. and M.P.; methodology, C.H. and M.P.; validation, C.H. and M.P.; formal analysis, C.H. and M.P.; investigation, C.H. and M.P.; data curation, C.H. and M.P.; writing—original draft preparation, C.H. and M.P.; writing—review and editing, C.H. and M.P.; visualization, C.H. and M.P.; project administration, C.H. and M.P. All authors have read and agreed to the published version of the manuscript.

Funding: This research received no external funding.

Conflicts of Interest: The authors declare no conflicts of interest.

References

1. Driscoll, D.J.; Miller, J.L.; Cassidy, S.B. *Prader-Willi Syndrome*; Adam, M.P., Feldman, J., Mirzaa, G.M., Pagon, R.A., Wallace, S.E., Amemiya, A., Eds.; GeneReviews®: Seattle, WA, USA; University of Washington: Seattle, WA, USA, 1998; pp. 1993–2024, [updated 5 December 2024]. [PubMed]
2. Bar, C.; Diene, G.; Molinas, C.; Bieth, E.; Casper, C.; Tauber, M. Early diagnosis and care is achieved but should be improved in infants with Prader-Willi syndrome. *Orphanet J. Rare Dis.* **2017**, *12*, 118. [CrossRef] [PubMed] [PubMed Central]
3. Tauber, M.; Hoybye, C. Endocrine disorders in Prader-Willi syndrome: A model to understand and treat hypothalamic dysfunction. *Lancet Diabetes Endocrinol.* **2021**, *9*, 235–246. [CrossRef] [PubMed]
4. Correa-da-Silva, F.; Fliers, E.; Swaab, D.F.; Yi, C.X. Hypothalamic neuropeptides and neurocircuitries in Prader Willi syndrome. *J. Neuroendocrinol.* **2021**, *33*, e12994. [CrossRef] [PubMed] [PubMed Central]
5. Bochukova, E.G.; Lawler, K.; Croizier, S.; Keogh, J.M.; Patel, N.; Strohbehn, G.; Lo, K.K.; Humphrey, J.; Hokken-Koelega, A.; Damen, L.; et al. A Transcriptomic Signature of the Hypothalamic Response to Fasting and BDNF Deficiency in Prader-Willi Syndrome. *Cell Rep.* **2018**, *22*, 3401–3408. [CrossRef] [PubMed] [PubMed Central]
6. Brown, S.S.G.; Manning, K.E.; Fletcher, P.; Holland, A. In vivo neuroimaging evidence of hypothalamic alteration in Prader-Willi syndrome. *Brain Commun.* **2022**, *4*, fcac229. [CrossRef] [PubMed] [PubMed Central]
7. Wu, N.; Yu, H.; Xu, M. Alteration of brain nuclei in obese children with and without Prader-Willi syndrome. *Front. Neuroinform.* **2022**, *16*, 1032636. [CrossRef] [PubMed] [PubMed Central]
8. Holsen, L.M.; Zarcone, J.R.; Brooks, W.M.; Butler, M.G.; Thompson, T.I.; Ahluwalia, J.S.; Nollen, N.L.; Savage, C.R. Neural mechanisms underlying hyperphagia in Prader–Willi syndrome. *Obesity* **2006**, *14*, 1028–1037. [CrossRef] [PubMed]
9. Holsen, L.M.; Savage, C.R.; Martin, L.E.; Bruce, A.S.; Lepping, R.J.; Ko, E.; Brooks, W.M.; Butler, M.G.; Zarcone, J.R.; Goldstein, J.M. Importance of reward and prefrontal circuitry in hunger and satiety: Prader–Willi syndrome vs simple obesity. *Int. J. Obes.* **2012**, *36*, 638–647. [CrossRef]
10. Akefeldt, A.; Ekman, R.; Gillberg, C.; Månsson, J.E. Cerebrospinal fluid monoamines in Prader-Willi syndrome. *Biol. Psychiatry* **1998**, *44*, 1321–1328. [CrossRef]
11. Song, J.; Choi, S.Y. Arcuate Nucleus of the Hypothalamus: Anatomy, Physiology, and Diseases. *Exp. Neurobiol.* **2023**, *32*, 371–386. [CrossRef]
12. Vohra, M.S.; Benchoula, K.; Serpell, C.J.; Hwa, W.E. AgRP/NPY and POMC neurons in the arcuate nucleus and their potential role in treatment of obesity. *Eur. J. Pharmacol.* **2022**, *915*, 174611. [CrossRef]
13. Crino, A.; Graziano, G. Updated on diabetes mellitus and glucose metabolism alterations in Prader-Willi Syndrome. *Curr. Diab Rep.* **2020**, *20*, 7. [CrossRef] [PubMed]
14. Irizarry, K.A.; Bain, J.; Butler, M.G.; Ilkayeva, O.; Muehlbauer, M.; Haqq, A.M.; Freemark, M. Metabolic profiling in Prader-Willi syndrome and nonsyndromic obesity: Sex differences and the role of growth hormone. *Clin. Endocrinol.* **2015**, *83*, 797–805. [CrossRef] [PubMed] [PubMed Central]
15. Aponte, Y.; Atasoy, D.; Sternson, S.M. AGRP neurons are sufficient to orchestrate feeding behavior rapidly and without training. *Nat. Neurosci.* **2011**, *14*, 351–355. [CrossRef] [PubMed]
16. Carr, J.A. Stress, Neuropeptides, and Feeding Behavior: A Comparative Perspective. *Integr. Comp. Biol.* **2002**, *42*, 582–590. [CrossRef] [PubMed]

17. Campbell, R.E.; Grove, K.L.; Smith, M.S. Distribution of corticotropin releasing hormone receptor immunoreactivity in the rat hypothalamus: Coexpression in neuropeptide Y and dopamine neurons in the arcuate nucleus. *Brain Res.* **2003**, *973*, 223–232. [CrossRef]
18. Marie, L.S.; Luquet, S.; Cole, T.B.; Palmiter, R.D. Modulation of neuropeptide Y expression in adult mice does not affect feeding. *Proc. Natl. Acad. Sci. USA* **2005**, *102*, 18632–18637. [CrossRef]
19. Naveilhan, P.; Hassani, H.; Canals, J.M.; Ekstrand, A.J.; Larfalk, A.; Chhajlani, V.; Arenas, E.; Gedda, K.; Svensson, L.; Thoren, P.; et al. Normal feeding behavior, body weight and leptin response require the neuropeptide Y Y2 receptor. *Nat. Med.* **1999**, *5*, 1188–1193. [CrossRef]
20. Qian, S.; Chen, H.; Weingarth, D.; Trumbauer, M.E.; Novi, D.E.; Guan, X.; Yu, H.; Shen, Z.; Feng, Y.; Frazier, E.; et al. Neither agouti-related protein nor neuropeptide Y is critically required for the regulation of energy homeostasis in mice. *Mol. Cell Biol.* **2002**, *22*, 5027–5035. [CrossRef]
21. Ollmann, M.M.; Wilson, B.D.; Yang, Y.K.; Kerns, J.A.; Chen, Y.; Gantz, I.; Barsh, G.S. Antagonism of central melanocortin receptors in vitro and in vivo by agouti-related protein. *Science* **1997**, *278*, 135–138. [CrossRef]
22. Bervini, S.; Herzog, H. Mouse models of Prader-Willi Syndrome: A systematic review. *Front. Neuroendocrinol.* **2013**, *34*, 107–119. [CrossRef]
23. Choi, Y.; Min, H.Y.; Hwang, J.; Jo, Y.H. *Magel2* knockdown in hypothalamic POMC neurons innervating the medial amygdala reduces susceptibility to diet-induced obesity. *Life Sci. Alliance* **2022**, *5*, e202201502. [CrossRef] [PubMed] [PubMed Central]
24. Wijesuriya, T.M.; De Ceuninck, L.; Masschaele, D.; Sanderson, M.R.; Carias, K.V.; Tavernier, J.; Wevrick, R. The Prader-Willi syndrome proteins MAGEL2 and necdin regulate leptin receptor cell surface abundance through ubiquitination pathways. *Hum. Mol. Genet.* **2017**, *26*, 4215–4230. [CrossRef] [PubMed] [PubMed Central]
25. Pravdivyi, I.; Ballanyi, K.; Colmers, W.F.; Wevrick, R. Progressive postnatal decline in leptin sensitivity of arcuate hypothalamic neurons in the Magel2-null mouse model of Prader-Willi syndrome. *Hum. Mol. Genet.* **2015**, *24*, 4276–4283. [CrossRef]
26. Buch, J.R.; Wevrick, R. Loss of the Prader-Willi obesity syndrome protein necdin promotes adipogeneis. *Gene* **2021**, *497*, 45–51.
27. Polex-Wolf, J.; Lam, B.Y.H.; Larder, R.; Tadross, J.; Rimmington, D.; Bosch, F.; Cenzano, V.J.; Eduard Ayuso, E.; Ma, M.K.L.; Rainbow, K.; et al. Hypothalamic loss of Snord116 recapitulates the hyperphagia of Prader-Willi Syndrome. *J. Clin. Investig.* **2018**, *128*, 960–969. [CrossRef] [PubMed]
28. Carraro, R.S.; Nogueira, G.A.; Sidarta-Oliveira, D.; Gaspar, R.S.; Dragano, N.R.; Morari, J.; Bobbo, V.C.D.; Araujo, E.P.; Mendes, N.F.; Zanesco, A.M.; et al. Arcuate Nucleus Overexpression of NHLH2 Reduces Body Mass and Attenuates Obesity-Associated Anxiety/Depression-like Behavior. *J. Neurosci.* **2021**, *41*, 10004–10022. [CrossRef] [PubMed] [PubMed Central]
29. Topaloglu, A.K.; Simsek, E.; Kocher, M.A.; Mammadova, J.; Bober, E.; Kotan, L.D.; Turan, I.; Celiloglu, C.; Gurbuz, F.; Yuksel, B.; et al. Inactivating NHLH2 variants cause idiopathic hypogonadotropic hypogonadism and obesity in humans. *Hum. Genet.* **2022**, *141*, 295–304. [CrossRef] [PubMed]
30. Ge, Y.; Ohta, T.; Driscoll, D.J.; Nicholls, R.D.; Kalra, S.P. Anorexigenic melanocortin signaling in the hypothalamus is augmented in association with failure-to-thrive in a transgenic mouse model for Prader-Willi syndrome. *Brain Res.* **2002**, *957*, 42–45. [CrossRef]
31. Goldstone, A.P.; Unmehopa, U.A.; Bloom, S.R.; Swaab, D.F. Hypothalamic NPY and agouti-related protein are increased in human illness but not in Prader-Willi syndrome and other obese subjects. *J. Clin. Endocrinol. Metab.* **2002**, *87*, 927–937. [CrossRef] [PubMed]
32. Turkkahraman, D.; Sirazi, E.C.; Aykal, G. Serum alpha-melanocyte-stimulating hormone (α-MSH), brain-derived neurotrophic factor (BDNF), and agouti-related protein (AGRP) levels in children with Prader-Willi or Bardet-Biedl syndromes. *J. Endocrinol. Investig.* **2022**, *45*, 1031–1037. [CrossRef] [PubMed]
33. DelParigi, A.; Tschöp, M.; Heiman, M.L.; Salbe, A.D.; Vozarova, B.; Sell, S.M.; Bunt, J.C.; Tataranni, P.A. High circulating ghrelin: A potential cause for hyperphagia and obesity in Prader-Willi syndrome. *J. Clin. Endocrinol. Metab.* **2002**, *87*, 5461–5464. [CrossRef]
34. Ding, F.; Li, H.H.; Zhang, S.; Solomon, N.M.; Camper, S.A.; Cohen, P.; Francke, U. SnoRNA Snord116 (Pwcr1/MBII-85) deletion causes growth deficiency and hyperphagia in mice. *PLoS ONE* **2008**, *3*, e1709. [CrossRef]
35. Höybye, C.; Barkeling, B.; Espelund, U.; Petersson, M.; Thoren, M. Peptides associated with hyperphagia in adults with Prader–Willi syndrome before and during GH treatment. *Growth Horm. IGF Res.* **2003**, *13*, 322–327. [CrossRef] [PubMed]
36. Spetter, M.S.; Feld, G.B.; Thienel, M.; Preissl, H.; Hege, M.A.; Hallschmid, M. Oxytocin curbs calorie intake via food-specific increases in the activity of brain areas that process reward and establish cognitive control. *Sci. Rep.* **2018**, *8*, 2736. [CrossRef] [PubMed] [PubMed Central]
37. Petersson, M.; Uvnäs-Moberg, K. Interactions of oxytocin and dopamine—Effects on behavior in health and disease. *Biomedicines* **2024**, *12*, 2440. [CrossRef]
38. Swaab, D.F.; Purba, J.S.; Hofman, M.A. Alterations in the hypothalamic paraventricular nucleus and its oxytocin neurons (putative satiety cells) in Prader-Willi syndrome: A study of five cases. *J. Clin. Endocrinol. Metab.* **1995**, *80*, 573–579.
39. Swaab, D.F. Prader-Willi Syndrome and the hypothalamus. *Acta Paediatr.* **1997**, *86* (Suppl. 423), 50–54. [CrossRef]

40. Martin, A.; State, M.; Anderson, G.M.; Kaye, W.M.; Hanchett, J.M.; McConaha, C.W.; North, W.G.; Leckman, J.F. Cerebrospinal fluid levels of oxytocin in Prader–Willi syndrome: A preliminary report. *Biol. Psychiatry* **1998**, *44*, 1349–1352. [CrossRef]
41. Johnson, L.; Manzardo, A.M.; Miller, J.L.; Driscoll, D.J.; Butler, M.G. Elevated plasma oxytocin levels in children with Prader–Willi syndrome compared with healthy unrelated siblings. *Am. J. Med. Genet. Part A* **2016**, *170*, 594–601. [CrossRef] [PubMed]
42. Uvnäs-Moberg, K. Role of efferent and afferent vagal nerve activity during reproduction: Integrating function of oxytocin on metabolism and behaviour. *Psychoneuroendocrinology* **1994**, *19*, 687–695. [CrossRef] [PubMed]
43. Qiu, L.; Chang, A.; Ma, R.; Strong, T.V.; Okun, M.S.; Foote, K.D.; Wexler, A.; Gunduz, A.; Miller, J.L.; Halpern, C.H. Neuromodulation for the treatment of Prader-Willi syndrome—A systematic review. *Neurotherapeutics* **2024**, *21*, e00339. [CrossRef] [PubMed]
44. Meziane, H.; Schaller, F.; Bauer, S.; Villard, C.; Matarazzo, V.; Riet, F.; Guillon, G.; Lafitte, D.; Desarmenien, M.G.; Tauber, M.; et al. An early postnatal oxytocin treatment prevents social and learning deficits in adult mice deficient for Magel2, a gene involved in Prader-Willi Syndrome and autism. *Biol. Psychiatry* **2015**, *78*, 85–94. [CrossRef]
45. Schaller, F.; Watrin, F.; Sturny, R.; Massacrier, A.; Szepetowski, P.; Muscatelli, F. A single postnatal injection of oxytocin rescues the lethal feeding behaviour in mouse newborns deficient for the imprinted Magel2 gene. *Hum. Mol. Genet.* **2010**, *19*, 4895–4905. [CrossRef] [PubMed]
46. Petersson, M.; Höybye, C. Is Oxytocin a Contributor to Behavioral and Metabolic Features in Prader-Willi Syndrome? *Curr. Issues Mol. Biol.* **2024**, *46*, 8767–8779. [CrossRef] [PubMed]
47. Wang, C.F.; Bomberg, E.; Billington, C.; Levine, A.; Kotz, C.M. Brain-derived neurotrophic factor in the hypothalamic paraventricular nucleus reduces energy intake. *Am. J. Physiol. Regul. Integr. Comp. Physiol.* **2007**, *293*, R1003–R1012. [CrossRef]
48. Queen, N.J.; Zou, X.; Anderson, J.M.; Huang, W.; Appana, B.; Komatineni, S.; Wevrick, R.; Cao, L. Hypothalamic AAV-BDNF gene therapy improves metabolic function and behavior in the *Magel2*-null mouse model of Prader-Willi syndrome. *Mol. Ther. Methods Clin. Dev.* **2022**, *27*, 131–148. [CrossRef] [PubMed]
49. Bueno, M.; Esteba-Castillo, S.; Novell, R.; Giménez-Palop, O.; Coronas, R.; Gabau, E.; Corripio, R.; Baena, N.; Viñas-Jornet, M.; Guitart, M.; et al. Lack of Postprandial Peak in Brain-Derived Neurotrophic Factor in Adults with Prader-Willi Syndrome. *PLoS ONE* **2016**, *11*, e0163468. [CrossRef] [PubMed] [PubMed Central]
50. Kozlov, S.V.; Bogenpohl, J.W.; Howell, M.P.; Wevrick, R.; Panda, S.; Hogenesch, J.B.; Muglia, L.J.; Van Gelder, R.N.; Herzog, E.D.; Stewart, C.L. The imprinted gene Magel2 regulates normal circadian output. *Nat. Genet.* **2007**, *39*, 1266–1272. [CrossRef]
51. Manzardo, A.M.; Johnson, L.; Miller, J.L.; Driscoll, D.J.; Butler, M.G. Higher plasma orexin a levels in children with Prader-Willi syndrome compared with healthy unrelated sibling controls. *Am. J. Med. Genet. A* **2016**, *170*, 2328–2333. [CrossRef] [PubMed] [PubMed Central]
52. Gimez-Palop, O.; Casamitjana, L.; Corripio, R.; Estab-Castillo, S.; Pareja, R.; Albinana, N.; Rigla, M.; Caixas, A. Growth Hormone (GH) Treatment Decreases Plasma Kisspeptin Levels in GH-Deficient Adults with Prader-Willi Syndrome. *J. Clin. Med.* **2021**, *10*, 3054. [CrossRef]
53. Omokawa, M.; Ayabe, T.; Nagai, T.; Imanishi, A.; Omokawa, A.; Nishino, S.; Sagawa, Y.; Shimizu, T.; Kanbayashi, T. Decline of CSF orexin (hypocretin) levels in Prader-Willi syndrome. *Am. J. Med. Genet. A* **2016**, *170*, 1181–1186. [CrossRef] [PubMed]
54. Clarke, H.; Dhillo, W.S.; Jayasena, C.N. Comprehensive Review on Kisspeptin and Its Role in Reproductive Disorders. *Endocrinol. Metab.* **2015**, *30*, 124–141. [CrossRef] [PubMed] [PubMed Central]
55. Butler, M.G.; Nelson, T.A.; Driscoll, D.J.; Manzardo, A.M. Evaluation of plasma substance P and beta-endorphin levels in children with Prader.Willi syndrome. *J. Rare Disord.* **2015**, *3*, 1–13. [PubMed]
56. Öztürk Özkan, G. Effects of Nesfatin-1 on Food Intake and Hyperglycemia. *J. Am. Coll. Nutr.* **2020**, *39*, 345–351. [CrossRef] [PubMed]
57. Gajewska, J.; Szamotulska, K.; Klemarczyk, W.; Chełchowska, M.; Strucińska, M.; Ambroszkiewicz, J. Circulating Levels of Nesfatin-1 and Spexin in Children with Prader-Willi Syndrome during Growth Hormone Treatment and Dietary Intervention. *Nutrients* **2023**, *15*, 1240. [CrossRef] [PubMed]
58. Grootjen, L.N.; Kerkhof, G.F.; Juriaans, A.F.; Trueba-Timmermans, D.J.; Hokken-Koelega, A.C.S. Acute stress response of the HPA-axis in children with Prader-Willi syndrome: New insights and consequences for clinical practice. *Front. Endocrinol.* **2023**, *14*, 1146680. [CrossRef] [PubMed]
59. Strelnikov, K.; Debladis, J.; Salles, J.; Valette, M.; Cortadellas, J.; Tauber, M.; Barone, P. Amygdala hyperactivation relates to eating behaviour: A potential indicator of food addiction in Prader-Willi syndrome. *Brain Commun.* **2023**, *5*, fcad138. [CrossRef] [PubMed] [PubMed Central]
60. Krashes, M.J.; Shah, B.P.; Madara, J.C.; Olson, D.P.; Strochlic, D.E.; Garfield, A.S.; Vong, L.; Pei, H.; Watabe-Uchida, M.; Uchida, N.; et al. An excitatory paraventricular nucleus to AgRP neuron circuit that drives hunger. *Nature* **2014**, *507*, 238–242. [CrossRef]
61. Ibrahim, K.; Earley, B.J.; Udi, S.; Nemirovski, A.; Hadar, R.; Gammal, A.; Hirsch, H.J.; Pollak, Y.; Gross, I.; Eldar-Geva, T.; et al. Targeting the endocannabinoid/CB1 receptor system for treating obesity in Prader-Willi syndrome. *Mol. Metab.* **2016**, *5*, 1187–1199. [CrossRef]

62. Yuan, M.; Li, W.; Zhu, Y.; Yu, B.; Wu, J. Asprosin: A Novel Player in Metabolic Diseases. *Front. Endocrinol.* **2020**, *11*, 64. [CrossRef]
63. Faienza, F.M.; Chiarito, M.; Aureli, A.; Buganza, R.; Corica, D.; Delvecchio, M.; De Sanctis, L.; Fintini, D.; Grugni, G.; Licenziati, M.R.; et al. Lack of correlation between asprosin serum levels and hyperphagic behavior in subjects with Prader-Willi Syndrome. *J. Endocrinol. Investig.* **2024**, 1–8. [CrossRef]
64. Purtell, L.; Sze, L.; Loughnan, G.; Smith, E.; Herzog, H.; Sainsbury, A.; Steinbeck, K.; Campbell, L.V.; Viardot, A. In adults with Prader-Willi syndrome, elevated ghrelin levels are more consistent with hyperphagia than high PYY and GLP-1 levels. *Neuropeptides* **2011**, *45*, 301–307. [CrossRef] [PubMed]
65. Cabral, A.; López Soto, E.J.; Epelbaum, J.; Perelló, M. Is Ghrelin Synthesized in the Central Nervous System? *Int. J. Mol. Sci.* **2017**, *18*, 638. [CrossRef]
66. Ferrini, F.; Salio, C.; Lossi, L.; Merighi, A. Ghrelin in central neurons. *Curr. Neuropharmacol.* **2009**, *7*, 37–49. [CrossRef]
67. Miller, J.L.; Lacroix, A.; Bird, L.M.; Shoemaker, A.H.; Haqq, A.; Deal, C.L.; Clark, K.A.; Ames, M.H.; Suico, J.G.; de la Peña, A.; et al. The Efficacy, Safety, and Pharmacology of a Ghrelin O-Acyltransferase Inhibitor for the Treatment of Prader-Willi Syndrome. *J. Clin. Endocrinol. Metab.* **2022**, *107*, e2373–e2380. [CrossRef] [PubMed] [PubMed Central]
68. Millendo Therapeutics SAS. A Phase 2b/3 Study to Evaluate the Safety, Tolerability, and Effects of Livoletide (AZP-531), an Unacylated Ghrelin Analogue, on Food-Related Behaviors in Patients with Prader-Willi Syndrome. Available online: https://clinicaltrials.gov/ct2/show/NCT03790865 (accessed on 10 February 2025).
69. Mahmoud, R.; Kimonis, V.; Butler, M.G. Clinical Trials in Prader-Willi Syndrome: A Review. *Int. J. Mol. Sci.* **2023**, *24*, 2150. [CrossRef] [PubMed]
70. Kishore, S.; Stamm, S. The snoRNA HBII-52 regulates alternative splicing of the serotonin receptor 2C. *Science* **2006**, *311*, 230–232. [CrossRef] [PubMed]
71. Garfield, A.S.; Davies, J.R.; Burke, L.K.; Furby, H.V.; Wilkinson, L.S.; Heisler, L.K.; Isles, A.R. Increased alternate splicing of Htr2c in a mouse model for Prader-Willi syndrome leads disruption of $5HT_{2C}$ receptor mediated appetite. *Mol. Brain* **2016**, *9*, 95. [CrossRef] [PubMed]
72. Kishore, P.; Boucai, L.; Zhang, K.; Li, W.; Koppaka, S.; Kehlenbrink, S.; Schiwek, A.; Esterson, Y.B.; Mehta, D.; Bursheh, S.; et al. Activation of K(ATP) channels suppresses glucose production in humans. *J. Clin. Investig.* **2011**, *121*, 4916–4920. [CrossRef] [PubMed] [PubMed Central]
73. Cowen, N.; Bhatnagar, A. The Potential Role of Activating the ATP-Sensitive Potassium Channel in the Treatment of Hyperphagic Obesity. *Genes* **2020**, *11*, 450. [CrossRef] [PubMed] [PubMed Central]
74. Kimonis, V.; Surampalli, A.; Wencel, M.; Gold, J.A.; Cowen, N.M. A randomized pilot efficacy and safety trial of diazoxide choline controlled-release in patients with Prader-Willi syndrome. *PLoS ONE* **2019**, *14*, e0221615. [CrossRef] [PubMed] [PubMed Central]
75. Miller, J.L.; Gevers, E.; Bridges, N.; Yanovski, J.A.; Salehi, P.; Obrynba, K.S.; Felner, E.I.; Bird, L.M.; Shoemaker, A.H.; Angulo, M.; et al. DESTINY PWS Investigators. Diazoxide Choline Extended-Release Tablet in People With Prader-Willi Syndrome: A Double-Blind, Placebo-Controlled Trial. *J. Clin. Endocrinol. Metab.* **2023**, *108*, 1676–1685. [CrossRef] [PubMed] [PubMed Central]
76. Miller, J.L.; Gevers, E.; Bridges, N.; Yanovski, J.A.; Salehi, P.; Obrynba, K.S.; Felner, E.I.; Bird, L.M.; Shoemaker, A.H.; Angulo, M.; et al. C601/C602 Investigators. Diazoxide choline extended-release tablet in people with Prader-Willi syndrome: Results from long-term open-label study. *Obesity* **2024**, *32*, 252–261. [CrossRef] [PubMed]
77. Dai, Y.L.; Qin, Y.F.; Chao, Y.Q.; Hu, C.X.; Xia, F.L.; Zou, C.C. The genotype and phenotype correlation of Prader-Willi syndrome. *Rare Dis. Orphan Drugs J.* **2024**, *3*, 33. [CrossRef]

current issues in molecular biology

MDPI

Review

The Dynamicity of the Oxytocin Receptor in the Brain May Trigger Sensory Deficits in Autism Spectrum Disorder

Claudia Camerino [1,2]

[1] Department of Precision and Regenerative Medicine, School of Medicine, University of Bari Aldo Moro, P.za G. Cesare 11, 70100 Bari, Italy; ccamerino@libero.it

[2] Department of Physiology and Pharmacology "V. Erspamer", Sapienza University of Rome, P.le Aldo Moro 5, 00185 Rome, Italy

Abstract: Sensory processing abnormalities have been noted since the first clinical description of autism in 1940. However, it was not until the release of the fifth edition of the Diagnostic and Statistical Manual of Mental Disorders (DSM-5) in 2013 that sensory challenges were considered as symptoms of autism spectrum disorder (ASD). Multisensory processing is of paramount importance in building a perceptual and cognitive representation of reality. For this reason, deficits in multisensory integration may be a characteristic of ASD. The neurohormone oxytocin (Oxt) is involved in the etiology of ASD, and there are several ongoing clinical trials regarding Oxt administration in ASD patients. Recent studies indicate that Oxt triggers muscle contraction modulating thermogenesis, while abnormal thermoregulation results in sensory deficits, as in ASD. Activation of the Oxt system through exposure to cold stress regulates the expression of oxytocin receptor (Oxtr) in the brain and circulating Oxt, and if this mechanism is pathologically disrupted, it can lead to sensory processing abnormalities since Oxt acts as a master gene that regulates thermogenesis. This review will describe the sensory deficits characteristic of ASD together with the recent theories regarding how the modulation of Oxt/Oxtr in the brain influences sensory processing in ASD.

Keywords: oxytocin; oxytocin receptor; thermoregulation; autism spectrum disorder; sensory processing; skeletal muscle; Diagnostic and Statistical Manual of Mental Disorders (DSM-5)

1. Introduction

Autism is a complex neurodevelopmental condition and its cause is still unknown. ASD is characterized by two main symptoms: persistent social communication and interaction difficulties and restrictive and repetitive behavior, interests and activities [1]. The majority of research on autism spectrum disorder (ASD) has focused on social, communication and learning/cognitive challenges associated with the disease [2]. However, it is only in the last 10 years that a new diagnostic criterion for ASD has been accepted by the scientific community and included in the Diagnostic and Statistical Manual of Mental Disorders (DSM-5): sensory processing [3]. Indeed, abnormalities in sensory processing can influence behavioral and cognitive experience in ASD patients since alterations in social cognition are marked by a very different perceptual experience of the world. Atypical sensory experience is estimated in about 90% of ASD individuals in every sensory modality, such as taste, touch, audition, smell and vision. The neurobiological alterations that affect processes as diverse as cognition and sensory experience in ASD and the common thread between them are still unclear. In order to begin to understand the disease, it is essential to clarify whether sensory difference represents secondary consequences after reduced social interaction or sensory deficits influence both development and neurobiology. ASD affects human experience from sensation to perception, to motor behavior and cognition. This is why it is important to understand how these diverse domains can be related. Research on sensory symptoms may help in simplifying this complexity by focusing on circuit-level alterations in the brain that may affect cortical processing in ASD and offer translational and pharmacological potential to cure this disease. Indeed, according to the "sensory-first" theory, social/cognitive symptoms may be downstream effects of atypical sensory processing in early development, while according to the "top-down" theory, symptoms relating to sensation and social cognition might co-arise from alteration in attention or prolonged lack of social interaction [2,3]. Perturbations in the inhibitory neurotransmitter gamma-aminobutyric acid (GABA) receptor have been associated with autism, and GABAergic signaling is disrupted in several different mouse models of autism [2,3]. Other neuromodulators such as testosterone and oxytocin (Oxt) modulate GABAergic signaling and are associated with autistic traits [4–8]. Oxt is a neurohypophysial hormone, as previously described [9]. Oxt triggers skeletal muscle contraction through thermoregulation, and cold stress exposure increases oxytocin receptor (Oxtr) expression in the brain and the Soleus muscle (Sol), while it decreases circulating Oxt in mice, leading to what we have called "The oxytonic effect" [10]. Oxt's regulation of thermogenesis is linked to Prader–Willi syndrome (PWS) and Schaaf–Yang syndrome (SYS) since both these pathologies present increased circulating Oxt levels, muscle hypotonicity, decreased Oxtr expression in the brain and frequent episodes of hypothermia and hyperpyrexia with no infection [11,12]. PWS and SYS patients also present a very high incidence of ASD [13] and they often present a history of low body temperature and/or high body temperature with no apparent reason provided on their medical history form [12]. PWS arises from the lack of expression of paternally inherited genes on chromosome 15q11-q13 or maternal uniparental disomy or an imprinting defect. Mage family member L2 (Magel2) is one of the affected genes located on 15q11-q13 and mutations in Magel2 have been found in PWS, SYS and ASD individuals [11,12]. In a mouse model of autism in which Magel2 has been knocked down, Oxtr is downregulated or upregulated according to postnatal Oxt treatment [14]. Magel2-KO mice present neurodevelopmental impairment and autistic-like behavior [12]. In this review, we explain the sensory deficits described in ASD in light of the novel theories that see Oxt/Oxtr expression as being dynamically modulated in the brain, and we hope that these results may be instrumental in designing precisely timed Oxt-based therapeutic strategies that target specific brain regions and can improve the clinical features, both sensorial and social, of ASD.

2. Autism and Sensory Deficits

In this section, we will describe some of the sensory deficits documented in ASD and their manifestations (Table 1). Pediatric patients with ASD suffer hyper- or hyposensitivity to visual, auditory, tactile and olfactory stimuli. ASD individuals also report paradoxical heat sensations when cold is perceived as burning hot, indicating disruption of thermosensory integration since central processing of information seems to be altered in ASD rather than peripheral perceptions [15,16]. The existence of sensory abnormalities in ASD was only recently included in the DSM-5 as diagnostic criteria for ASD (American Psychiatry Association 2013) and have for a long time been overlooked [15]. Central mechanisms during the integration of multisensory stimulus rather than peripheral mechanisms cause sensory deficits in ASD. In the DSM-5, both hyper- and hypo-reactivity to sensory input or unusual interest in sensory aspects or the external milieu are mentioned as diagnostic criteria for ASD. These criteria include somatosensory abnormalities such as indifference to pain related to temperature, adverse response to texture or excessive touching of objects (APA 2013). For example, vibro-tactile stimulus perceived as static and poor vibro-tactile amplitude discrimination have been reported in children with ASD [16]. Hypersensitivity to thermal pain and innocuous thermal stimuli has been found in ASD in adults [17]. Conversely, adolescents with ASD showed normal detection of thermal pain but hyposensitivity to innocuous thermal stimuli [15]. Moreover, in adolescents and adults with ASD, normal pain detection was observed during thermal and electrical stimulation [18], where pressure pain was also lower in children [19]. In a recent study, the somatosensory perception of patients with ASD was investigated, since altered detection of somatosensory stimuli may cause unusual sensory perception in ASD using the quantitative sensory testing (QST) protocol on neuropathic pain. Hyper- or hyposensitivity to sensory inputs is normal in certain parts of the body and abnormal in other parts, in line with the different innervation from the many brain areas with peripheral and central neuronal activity. ASD patients show similar thresholds in detecting innocuous warmth and cool and light pressure on their palm and forearm, but hypersensitivity to vibrations was seen in ASD on the forearm and hypersensitivity to thermal pain at both sites. Interestingly, individuals with ASD exhibit enhanced perception of certain stimulus properties such as vision, where local, circumscribed properties of a visual stimulus are seen instead of the complete picture. This peculiarity of ASD is described as follows: "ASD individuals see the tree but not the forest". A similar pattern of perception is also present in the auditory domain where the change of a single note of a melody is perceived to be stronger than global melodic changes [20]. That sensory deficits in ASD vary according to the age of patients and the area of the body is an important concept consistent with the plasticity of the Oxtr expression in the brain, as we will explain in the following sections [13].

Table 1. In autism spectrum disorder, sensory deficits are present in all domains of sensorial experience.

Sensory Deficits in Autism Spectrum Disorder		
Auditory	Enhanced auditory perception of details at the expense of the global situation. For example, hearing the change of a single note and not the change of the entire melody.	Hyposensitivity and hypersensitivity in all domains.
Vision	Enhanced perception of certain stimulus properties since circumscribed properties of a visual stimulus are seen over the complete picture. For example, seeing the tree and not the forest.	
Tactile	Sensorial experience changes according to the part of the body, as pressure is perceived normally on the palm of the hand and forearm, while hypersensitivity to vibrations is reported in the forearm only. Adverse response to texture is reported where vibro-tactile stimulus was generally perceived as static.	
Thermoregulation	Paradoxical heat sensation is reported where cold is perceived and burning hot and hypo- or hypersensitivity to thermal innocuous thermal stimuli is also reported according to the age of the patients. Prader–Willi syndrome individuals often present autism spectrum disorder and hypothermia and hyperpyrexia with no infection are reported.	

3. Oxytocin and Sensory Deficits in Autism Spectrum Disorder

Oxt/Oxtr$^{-/-}$ mice have impaired thermoregulation, resulting in thermosensory deficits and the inability to maintain a stable temperature [21,22]. Our previous studies indicate that temperature regulation is impaired in PWS individuals since they show significant changes in thermoregulation compared to siblings or controls [19,21–23]. PWS is caused by the loss of expression of a critical genetic region on chromosome 15q11-q13 and a very high percent of PWS individuals also present ASD. Our studies also indicate that Oxt acts as a master gene regulating thermogenesis and influences the manifestation of the PWS phenotype [9,13]. A small but statistically significant proportion of PWS children have been reported to experience persistent or episodic hyperpyrexia/hypothermia [11], whereas fingertip temperature is unaltered [22,24,25]. About 90% of ASD individuals have a sensory abnormality that manifests with hyper- and hypo-reactivity to smell, taste, audition, vision and tactile sensitivity. The sensory deficit influences the perception of the external word and social behavior [26]. Oxt is also released in response to tactile stimuli, and Oxt in the plasma of autistic children is decreased compared to neurotypical individuals [27]. Abnormalities in the gene that encodes for Oxtr [28] as well as in Oxt have been found in ASD [29]. This is why we hypothesize that Oxt regulation of thermogenesis can cause sensory abnormalities in ASD (Table 2). Moreover, children with ASD present altered thermal thresholds with reduced sensitivity to warmth, coolness and heat, but not pain threshold. We speculated that this evidence may be significant since ASD individuals, unlike PWS individuals, do not have hypothalamic syndrome, which can per se cause temperature fluctuations and sensory abnormalities. However, a more comprehensive study of ASD is not within the scope of this article, which regards exclusively the relationship between ASD and sensory deficits. Oxt's regulation of thermogenesis may be responsible for sensorial functionality and body temperature regulation through muscle contraction in healthy individuals, while a dysfunctional Oxt system can cause varying degrees of sensory abnormalities and muscle hypotonicity, as seen in PWS and ASD.

Table 2. Oxytocin and sensory deficits in autism spectrum disorder.

• Oxytocin is a master gene regulating thermogenesis.
• Oxytocin-deficient mice and oxytocin receptor-deficient mice have impaired thermoregulation and inability to maintain a stable temperature.
• Oxytocin is released due to tactile stimuli.
• Autism spectrum disorder children are reported to have a lower oxytocin concentration.
• Autism spectrum disorder individuals are reported to have sensory deficits such as paradoxical heat sensations, altered thermal threshold, and hyper- or hypo-reactivity to smell, taste, audition, vision and tactile sensitivity.
• The altered sensorial perception in ASD individuals can cause the cognitive and social dysfunction typical of this condition.

4. Oxytocin Receptor Modulation Is Dynamic Across Lifespan and Is Sexually Dimorphic

Oxt levels are lower in autistic compared with neurotypical children; however, this difference is not detectable in adults [30]. ASD is characterized by deficits in social interaction and communication and restricted interests together with sensory deficits (APA 2013). These traits are all thought to be mediated by Oxt. Indeed, several components of the Oxt/Oxtr system have been associated with ASD. Variations in these genes can affect Oxt/Oxtr expression and distribution. Some studies found Oxt levels to be inversely correlated with ASD severity; however, the expression of Oxtr in the brain has not been investigated. The fact that Oxt is lower in autistic children highlights an involvement of the Oxt system in the development or manifestation of ASD [28]. From this perspective, the Oxt system can shape the brain for social interaction and sensory perception during a critical period in infancy during which there is probably a critical window when Oxt administration may be effective. However, little is known about the developmental trajectory of Oxt/Oxtr and the causality of these manifestations in ASD, which is why studies on the genetics of Oxt/Oxtr are necessary. In this regard, thermoregulation may modulate Oxtr in the brain and cause sensory deficits in ASD [30]. The experimental manipulation of Oxtr expression levels and Oxtr gene KO can have a significative effect on behavior and physiology [31]. This can lead to the identification of a critical window of Oxtr expression in the brain together with associated genes, which is important for understanding ASD and other psychiatric illness. Oxt/Oxtr expression is dynamic throughout a person's development, with critical periods. Indeed, Oxtr peaks in early childhood and late adulthood and is highly correlated with dopaminergic signaling across a lifespan to adapt to shifting environmental challenges. The peak in Oxtr expression observed during early childhood is probably stronger in males because it is influenced by gonadal steroids. These sex differences may contribute to the reported sex differences in neurodevelopmental disorder diagnoses and is consistent with the sex bias of ASD recurring more often in males than females. Specifically, the early childhood peak of Oxtr binding that was observed in brain tissue of the ventral pallidum tissue from neurotypical donors was absent in the same tissue in autistic donors. Moreover, Oxtr expression and binding are higher in the ventral pallidum and nucleus basalis of Meynert brain tissue of neurotypical donors than the same tissue of autistic donors [31]. Oxtr undergoes epigenetic modifications, and these modifications predict the severity of symptoms in adults with ASD. Indeed, the neural networks involved in reward processing and social capability in ASD involve the Oxt/Oxtr system, which is sensitive to epigenetic processes caused by environmental exposure, and these epigenetic modifications account for the features of autistic traits. Specifically, Oxtr hypermethylation in the intron 1 area of MT2 was related to a less severe developmental phenotype making this site a potential biomarker of adults with ASD with less severe verbal communication deficits [32]. It is worth noting that thermoregulation and thermogenesis can also cause epigenetic modifications of Oxtr.

5. Current Theories Explaining Sensory Deficits in Autism Spectrum Disorder

Individuals with ASD "See the trees but not the forest", which means that ASD individuals see the details of the perceptual world rather than the global picture [30]. To understand autistic sensory experience, the perceptual processing cannot be simply characterized as a talent or a deficit reflecting neither hyposensitivity nor hypersensitivity, but sensory experience in ASD exhibits a bias toward local versus global characteristics of a sensory scene which can be more or less advantageous according to the task [2]. Moreover, these sensory processing abnormalities also impact other domains of ASD such as social communications, worsening their severity [3]. This is why individuals with ASD have enhanced performance in tasks that rely on the analysis of stimulus details but have difficulties when these details need to be integrated into a complete image. Sensory difference is a common issue in ASD and much of our cognitive and social representations are dependent upon sensory inputs. In this regard, there are at least five theories that explain the ASD. The first theory, also called "Theory of mind" [33], suggests that ASD individuals have a decreased ability to understand other people's feelings and this can be caused by abnormalities in sensory processing. Neuroimaging studies reveal that inferior frontal regions including the mirror neuron system and the temporoparietal junction are involved in these tasks. The second theory, named "Weak central coherence" [34], proposes that the meaning of things is built through the integration of information across lower-level sensory and higher-level cognitive processing, and this is abnormal in ASD individuals. The classic illustration of how this theory works is that to understand a complex visual scene, the integration of all details is necessary and not focusing on single details, as seen in the tree and forest example. The third theory is named "Predictive coding hypothesis" and is based on the notion that individuals with ASD do not have a robust historical representation of the world, making it difficult for them to predict upcoming events and limiting interactions with external environments. This explains, for example, the tendency of ASD individuals to engage in repetitive behavior to limit novel sensory inputs in an endlessly novel world [35]. The fourth theory is called "Reduced sensory precision and reliability" and states that changes in the variability of neural response patterns create variability (or less reliability) in their behaviors and perceptions [36]. Finally, the fifth and last theory is based on evidence that GABA functioning is altered in ASD, leading to increased excitation for greater glutamatergic signaling, while inhibition is decreased for less GABAergic signaling. Indeed, glutamate and GABA signaling are involved in cortical function, including sensory processing and social function. In this context, increased excitability may explain hypersensitivity in sensory processing, with sensory input eliciting an abnormally large cortical response that makes the experience overwhelming [4–7,9,36]. This theory has been confirmed by a recent study on animal models [37]. However, here, we propose a novel theory according to which the sensory deficits in ASD can be triggered by the differential modulation of Oxtr in the brain, as we explain in the following sections.

6. Oxytocin and Oxytocin Receptor Expression Modulation in the Brain May Trigger Sensory Deficits in Autism Spectrum Disorder: New Theories

In this section, we will describe the most recent studies regarding the modulation of Oxtr in the brain and how it can impact sensory deficits in ASD. In particular, we will focus on four recent studies focusing on the Oxt/Oxtr system.

6.1. Oxytocin Is a Master Gene That Regulates Thermogenesis, and Cold Stress Modulates Oxytocin Receptor in the Brain and Peripheral Oxytocin

The first study [10] shows that Oxt is anabolic in muscle, and Oxt anti-obesogenic effects are also related to its positive effects on muscle mass. We characterized Oxt and Oxtr expression in different tissues after a cold stress (CS) challenge in mice by an in vivo approach. In this study, a cold stress protocol was elaborated and mice were kept at 4 °C for either 6 h or 5 days. Then, mice were sacrificed and properly dissected. Blood was collected and brains were quickly extracted, snap-frozen and analyzed by histomorphometry as previously shown [10]. Mice initially lost body weight, but after 5 days these mice regained their body weight, indicating that they were in good health. The exposure to cold stress activates shivering thermogenesis. In this regard, we hypothesize that there is a feedback loop between the hypothalamus and muscle that regulates the rate of central Oxt release and Oxtr expression level in the brain together with muscular Oxtr expression. This feedback loop between the brain and muscle is activated in response to situations requiring increased musculoskeletal performance such in shivering thermogenesis [11,13]. We showed in mice that Oxtr expression increases in the Soleus (Sol) muscle and in the brain after the CS challenge, while circulating Oxt decreases [10]. The dynamicity of Oxtr following a thermogenic challenge can trigger sensory deficits in ASD since Oxt regulates thermogenesis (Table 3). The cold stress model is explained in Table 3 and has been previously published [10].

Table 3. Ex vivo study of oxytocin and oxytocin receptor expression in the brain after exposure to 6 h or 5 days of cold stress in wild-type mice and measurement of circulating oxytocin. "−" indicates a significative decrease compared to untreated control; "+" indicates a significative increase compared to untreated control. Cold stress increases Oxtr in Paraventricular and Supraoptical Nuclei of hypothalamus and decreases circulating Oxt in a feed-back/feed-forward loop with the brain [10].

Ex vivo Studies				
	Oxt 6 h	**Oxtr 6 h**	**Oxt 5 d**	**Oxtr 5 d**
Hippocampus	*	*	*	−
Hypothalamus	*	*	*	*
Paraventricular Nucleus	*	+	*	+
Supraoptical Nucleus	*	+	*	+
Circulating Oxytocin	Decreases			

* no changes detected.

6.2. Oxytocin Receptor Expression Is Dynamic and Is Normalized by Oxytocin During a Specific Temporal Window

The second study [14] reports that Magel2 is a gene included in the PWS locus [38], and in addition to many pathological PWS phenotypic traits, the syndrome presents a high incidence of ASD [39]. Magel2-KO mice recapitulate autistic traits and neurodevelopmental impairments. In this study, three to five hours after delivery, pups were subcutaneously injected with saline or Oxt, and this administration was repeated every 2 days. Then, 8 days after birth or 90 days after birth, mice were sacrificed, the brains were extracted and Oxtr was quantified as described [14]. An ultrasonic microphone was used to record the pups' vocalization. The aim of this study was to extend the regional mapping of Oxtr in male and female Magel2-KO brains with or without Oxt treatment. This is important because Oxt is strongly regulated in the first week of life in mice, and understanding the specific site of action of Oxt in the brain and the specific window of time when it is effective will be essential to design a therapy with Oxt for ASD. Moreover, Oxtr is sexually dimorphic, which is why Oxtr was differently modulated according to gender after treatment with Oxt.

Early postnatal Oxt treatment prevents neonatal lethality in these mice and prevents social and learning deficits in adult Magel2-KO mice [40]. Indeed, region-specific alterations in Oxtr expression are present in Magel2-KO mice, and postnatal Oxt treatment could modulate Oxtr in specific brain regions. In Magel2-KO pups characterized by deficient production of hypothalamic Oxt [38], the neuropeptide may be unable to play its role as a mediator of early sensory functions, and the postnatal supplementation of Oxt may restore this function [14]. The action of Oxt in the brain is mediated by Oxt binding to Oxtr [39]. Environmental factors during early infancy can epigenetically modify the Oxtr gene and influence its expression level in adulthood [40]. Early modulation of Oxtr expression is important in neurodevelopmental disorders with social and cognitive deficits, as in ASD. Indeed, several mouse models of neurodevelopmental disorders present deficits in Oxt or Oxtr expression [41]. Oxtr is strongly regulated in the first three weeks of life in mice [42]. A recent study investigated whether Oxt treatment received in the first weeks of life had short- or long-term effects on regional Oxtr expression and if it is gender specific, since ASD had a higher incidence in males than in females [14]. Indeed, the Oxt system is sexually dimorphic [43]; for example, in humans, following a social stress test, a single dose of intranasal Oxt increases distress and anger in women but reduces distress in men [44]. Magel2-KO mice show highly variable region-specific patterns of Oxtr expression that vary according to Oxt administration, age and gender. Indeed, at post-natal day 8 (P8) Magel2-KO mice show a significant reduction in Oxtr levels in the brain compared to controls, indicating a major defect in Oxtr expression in the brain. This defect is present in males and females, and is not removed by the treatment with Oxt [14]. On the other hand, in Magel2-KO mice at P90, Oxt treatment normalizes Oxtr expression, specifically in those regions of the brain where Oxtr is abnormally upregulated compared to the control, such as the amygdala and hippocampus and piriform cortex, indicating an impaired developmental pattern of Oxtr expression. This effect was more evident in male mice than in female mice, which is consistent with the well-recognized sex bias in ASD incidence. One possible explanation is the existence of a temporal window in which these regions of brain are particularly sensitive to Oxt, since during neurodevelopment, brain circuits mature at different times and Oxt modulates sensory inputs and shapes brain circuits and connectivity. Conversely, when the Oxt system is defective, sensory inputs are defective or absent, failing to shape mature brain circuits. A similar time window was observed during a clinical trial with Oxt in PWS infants where Oxt administration prevents the failure to thrive of these infants only before 6 months of age [45]. The same results after Oxt administration were not detectable at any other age considered.

6.3. The Oxytocin System Requires Both Oxytocin/Oxytocin Receptor Expression and Synaptic Excitation/Inhibition Trasmission

The third study [46] shows that in the absence of Magel2, overall ex vivo Oxt neuron activity is suppressed for altered synaptic input profile resulting from decreased excitatory and increased inhibitory currents in mice. This shows that perturbations of the Oxt system include both neuropeptide expression [13,20] and inhibitory/excitatory synaptic transmission [46]. However, while the studies described above [13,20] rely on peptide or mRNA expression level measurement, Oxt is a signaling molecule that functions in a broader circuit of signaling and is dependent upon other hormones such as estrogens. For this reason, in a Magel2-KO mouse model, synaptic and cell autonomous properties of Oxt neurons were investigated. Magel2-KO mice were injected stereotaxically with Oxt and voltage clamp recording was performed as described [46]. Indeed, the Oxt system works through both Oxt expressing neurons and their connection and the expression of Oxt and Oxtr. The combined addition of synaptic blockers for GABAergic and glutamatergic transmission reduced the synaptic activity of Oxt neurons in control but not in Magel2-KO mice, and this indicates a loss of excitatory drive in these mice [46]. These results are consistent with previous studies on the involvement of GABA transmission in sensory deficits of ASD [2,3]. So despite the restoration of Oxt levels in ASD, Oxt neuron defects at circuit level persists, meaning that Magel2 deficiency permanently alters these circuits and the constant presence of Magel2 is essential for Oxt circuit function [37].

6.4. Hypothalamic Gray Matter Volume and Concentration Are Reduced in Autism Spectrum Disorder and Are Positively Associated with Peripheral Oxytocin

Finally, the fourth study [47] describes how the structural characteristics of the hypothalamus produce different results between ASD and neurotypical controls in human patients with a reduced gray matter volume or concentration in ASD children or adolescence compared with healthy controls of corresponding age. The study [47] is particularly relevant since it translates research from a mouse model to human patients. ASD individuals met DSM-5 criteria for autism and were diagnosed in accordance with current guidelines. In this study, Oxt was measured in blood samples obtained exclusively at rest to rule out any effect of Oxt on muscle contraction in ASD individuals [48]. Structural brain scans were obtained and images were processed and analyzed using voxel-based morphometry as described [47]. Notably, healthy carriers of Oxtr variants were associated with an increased likelihood of ASD, displaying a significant decrease in gray matter volume in the hypothalamus [49]. In a recent study, the morphological characteristics of the hypothalamus and their relationship with Oxt in ASD patients were analyzed through a region-of-interest analysis using voxel-based morphometry [37]. The aim of the study was to determine if there are differences in hypothalamic volume between autistic and neurotypical adults, if differences in the Oxt system in ASD were reflected in the hypothalamic structure and if these differences could be attributed to Oxt. To achieve this, the authors compared the gray matter volume between autistic and non-autistic control groups; then, they compared the differences in the relationship between hypothalamic gray matter volume and peripheral Oxt, and last, they analyzed the association between hypothalamic gray matter volume and ASD quotient scores as a measure of autistic traits [37]. Hypothalamic gray matter volume does not change between autistic and non-autistic individuals; however, comparing the group differences in terms of hypothalamic gray matter volume and peripheral Oxt, a positive association was found in the ASD group and a negative association in the non-autistic group. Hypothalamic gray matter volume was also associated with the autistic group [50]. Basal concentration of Oxt is reported to be lower in autistic children [31,48] but this difference disappears in autistic adults because of developmental changes in the Oxt system in ASD. The importance of these developmental effects has also been shown for Oxtr expression patterns [31]. This means that Oxt and Oxtr levels are dynamic and may normalize with age in ASD, translating into normalization of gray matter volume. This also means that the structural properties of the hypothalamus are related to Oxt levels in ASD [47]. This is consistent with the concept of a time window in which the treatment with Oxt is effective [51]. The most important changes in the hypothalamus and Oxt may all be modulated by variations in Oxtr expression [50]. Gray matter volume seems to increase alongside the increase in autistic traits, and this finding is probably caused by abnormalities in the Oxt system and Oxtr expression levels. These results show that the increase in Oxt levels can be caused by variations in Oxtr in the brains of ASD individuals, as seen after the cold stress challenge [13,51]. Overall, the studies described above [13,20,47,48] are consistent with the concept that Oxtr expression in the brain is dynamic and highly regulated by factors such as thermogenic challenge and muscle contraction [10], the temporal window of Oxt administration, sexual dimorphism and reproductive stage [14]. The integrity of synaptic transmission [46] and the feed-back between the brain and circulating Oxt [13,48] are also important. These data, together with the evidence that Oxt is a master gene regulating thermogenesis, made us hypothesize that the dysfunctional Oxt system could trigger sensory deficits in ASD. A role for hypothalamic hormones in sensorial functions has also been recently seen for gonadotropin-releasing hormone (GnRH) for olfactory function, consistent with our data on Oxt [52,53]. Indeed, the integrity of the Oxt system and the expression of Oxtr is essential to maintain the homeostasis of the body. This is evident, for example, in mental illnesses such as eating disorders (ED) where ED-related Oxtr haplotypes alter the relationship between proteins important for ED such as TGF-beta and sterol regulatory element-binding proteins (SREBPs) and Oxtr expression. Such disruptions cause the failure of compensatory Oxt-dependent mechanisms that would protect against starvation and stress contributing to ED [52]. Oxtr expression in the brain is directly regulated by the intake of carbohydrates and lipids [54,55], proving the dynamicity of Oxtr expression in the brain. The most recent theories explaining ASD are shown in Table 4.

Table 4. Most recent theories explaining autism spectrum disorder.

Theory of mind	Suggests that ASD individuals have a decreased ability to understand other people's feelings and this can be caused by abnormalities in sensory processing.
Weak central coherence	Proposes that the meaning of things is built through the integration of information across lower-level sensory and higher-level cognitive processing and this is abnormal in ASD individuals [34].
Predictive coding hypothesis	Is based on the notion that individuals with ASD do not have a robust historical representation of the world, making it difficult for them to predict upcoming events and limiting interactions with external environments.
Reduced sensory precision and reliability	States that changes in variability of neural response pattern create variability (or less reliability) in their behaviors and perceptions.
Altered GABAergic signaling	Theory is based on evidence that the imbalance in excitatory and inhibitory processes is changed in ASD with increased excitation for greater glutamatergic signaling, while inhibition is decreased for less GABAergic signaling.
Dynamicity of oxytocin receptor expression in the brain	The variation in oxytocin receptor expression levels may cause sensory deficits, deficits in social interaction/communication and restrictive interests in autism spectrum disorder.

7. Conclusions

Developmental changes can influence the structure of the hypothalamus and the expression of Oxt/Oxtr, triggering sensory deficits typically seen in autism. The action of Oxt in the brain is mediated by Oxt binding to its specific receptor Oxtr, which is a G-protein-coupled receptor expressed in several areas of the brain. An important feature of Oxtr is the variability of its distribution in the different areas of the brain and it can vary according to gender, age, pathological conditions and environmental factors [11,20]. The distribution of Oxtr in the brain varies in mammals, consistent with social behavior and sexual dimorphism. Moreover, Oxtr can undergo epigenetic modifications since early life experience has long-term effects on the Oxt system and expression of Oxtr. Early life Oxt exposure may influence life-long Oxtr expression, a theory known as "hormonal imprinting", and it is important in neurodevelopmental disorders characterized by social abilities since many animal models of neurodevelopmental disorders present abnormalities in Oxt release and Oxtr distributions [41,44,47]. Oxt regulates thermogenesis [10,13]. Indeed, the Oxt/Oxtr system in a healthy brain is activated after thermogenic challenge, while a dysfunctional Oxt/Oxtr system in the brain may trigger sensory deficits in ASD [11,13]. In this regard, there may exist a time window for Oxt administration in ASD during which Oxt treatment is effective. Future studies are necessary to prove this preliminary evidence. It is worth noting that environmental factors may cause epigenetic modifications of the Oxtr expression in the brain, as described for thermogenic challenge, altering Oxtr expression levels in the brain, which can trigger sensory deficits in ASD. The Oxtr expression is dynamic and influenced by epigenetic factors, reproductive stage and even nutrition. This is why a limitation of this study that needs to be acknowledged is the challenge of translating animal models to human patients. Indeed, progress in the biological detection and pharmacological treatment of ASD has been limited for three main reasons. First, there were difficulties in obtaining brain-related biological samples such as brain and cerebrospinal fluid from patients with ASD. Second, the studies on ASD were performed on an animal model that lacked the sophisticated social and cognitive abilities disrupted in ASD. Third, the primate species used as a model for ASD still have several limitations [1,55,56]. We hope that the novel perspective provided in this article can add new knowledge to the understanding of ASD.

Funding: This research was supported by the Grant "Ateneo 2017" from the University of Bari "Aldo Moro" granted to Prof. Claudia Camerino.

Institutional Review Board Statement: This study was conducted according to the guidelines of the Declaration of Helsinki and approved by the committee Organization for Animal Health O.P.B.A. of the University of Bari on 21 January 2019 and by the Italian Minister of Health aut. N 665/2019 PR.

Conflicts of Interest: The author declares no conflicts of interest.

References

1. Parker, K.J. Leveraging a translational research approach to drive diagnostic and treatment advances for autism. *Mol Psychiatry* **2022**, *27*, 2650–2658. [CrossRef] [PubMed]
2. Robertson, C.E.; Baron-Cohen, S. Sensory perception in autism. *Nat. Rev. Neurosci.* **2017**, *18*, 671–684. [CrossRef]
3. Baum, S.H.; Stevenson, R.A.; Wallace, M.T. Behavioral, perceptual, and neural alterations in sensory and multisensory function in autism spectrum disorder. *Prog. Neurobiol.* **2015**, *134*, 140–160. [CrossRef]
4. Baron-Cohen, S.; Lombardo, M.V.; Auyeung, B.; Ashwin, E.; Chakrabarti, B.; Knickmeyer, R. Why are autism spectrum conditions more prevalent in males? *PLoS Biol.* **2011**, *9*, e1001081. [CrossRef]
5. Gordon, I.; Vander Wyk, B.C.; Bennett, R.H.; Cordeaux, C.; Lucas, M.V.; Eilbott, J.A.; Zagoory-Sharon, O.; Leckman, J.F.; Feldman, R.; Pelphrey, K.A. Oxytocin enhances brain function in children with autism. *Proc. Natl. Acad. Sci. USA* **2013**, *110*, 20953–20958. [CrossRef]
6. Harden, S.W.; Frazier, C.J. Oxytocin depolarizes fast-spiking hilar interneurons and induces GABA release onto mossy cells of the rat dentate gyrus. *Hippocampus* **2016**, *26*, 1124–1139. [CrossRef] [PubMed]
7. Owen, S.F.; Tuncdemir, S.N.; Bader, P.L.; Tirko, N.N.; Fishell, G.; Tsien, R.W. Oxytocin enhances hippocampal spike transmission by modulating fast-spiking interneurons. *Nature* **2013**, *500*, 458–462. [CrossRef] [PubMed]
8. Zhang, L.; Chang, Y.H.; Feldman, A.N.; Ma, W.; Lahjouji, F.; Barker, J.L.; Hu, Q.; Maric, D.; Li, B.S.; Li, W.; et al. The expression of $GABA_A$ receptor α2 subunit is upregulated by testosterone in rat cerebral cortex. *Neurosci. Lett.* **1999**, *265*, 25–28. [CrossRef]
9. Camerino, C. The Long Way of Oxytocin from the Uterus to the Heart in 70 Years from Its Discovery. *Int. J. Mol. Sci.* **2023**, *24*, 2556. [CrossRef] [PubMed]
10. Conte, E.; Romano, A.; De Bellis, M.; de Ceglia, M.; Rosaria Carratù, M.; Gaetani, S.; Maqoud, F.; Tricarico, D.; Camerino, C. Oxtr/TRPV1 expression and acclimation of skeletal muscle to cold-stress in male mice. *J. Endocrinol.* **2021**, *249*, 135–148. [CrossRef]
11. Camerino, C. The Pivotal Role of Oxytocin's Mechanism of Thermoregulation in Prader-Willi Syndrome, Schaaf-Yang Syndrome, and Autism Spectrum Disorder. *Int. J. Mol. Sci.* **2024**, *25*, 2066. [CrossRef]
12. Rose, D.J.; Castner, D.M.; Wilson, K.S.; Rubin, D.A. Examination of sensory reception and integration abilities in children with and without Prader-Willi syndrome. *Res. Dev. Disabil.* **2024**, *149*, 104730. [CrossRef] [PubMed]
13. Camerino, C. Oxytocin's Regulation of Thermogenesis May Be the Link to Prader–Willi Syndrome. *Curr. Issues Mol. Biol.* **2023**, *45*, 4923–4935. [CrossRef]
14. Gigliucci, V.; Busnelli, M.; Santini, F.; Paolini, C.; Bertoni, A.; Schaller, F.; Muscatelli, F.; Chini, B. Oxytocin receptors in the Magel2 mouse model of autism: Specific region, age, sex and oxytocin treatment effects. *Front. Neurosci.* **2023**, *17*, 1026939. [CrossRef]
15. Duerden, E.G.; Taylor, M.J.; Lee, M.; McGrath, P.A.; Davis, K.D.; Roberts, S.W. Decreased sensitivity to thermal stimuli in adolescents with autism spectrum disorder: Relation to symptomatology and cognitive ability. *J. Pain* **2015**, *16*, 463–471. [CrossRef] [PubMed]
16. Puts, N.A.J.; Wodka, E.L.; Tommerdahl, M.; Mostofsky, S.H.; Edden, R.A.E. Impaired tactile processing in children with autism spectrum disorder. *J. Neurophysiol.* **2014**, *111*, 1803–1811. [CrossRef]
17. Cascio, C.; McGlone, F.; Folger, S.; Tannan, V.; Baranek, G.; Pelphrey, K.A.; Essick, G. Tactile perception in adults with autism: A multidimensional psychophysical study. *J. Autism Dev. Disord.* **2008**, *38*, 127–137. [CrossRef]
18. Yasuda, Y.; Hashimoto, R.; Nakae, A.; Kang, H.; Ohi, K.; Yamamori, H.; Fujimoto, M.; Hagihira, S.; Takeda, M. Sensory cognitive abnormalities of pain in autism spectrum disorder: A case-control study. *Ann. Gen. Psychiatry* **2016**, *15*, 8. [CrossRef] [PubMed]

19. Riquelme, I.; Hatem, S.M.; Montoya, P. Abnormal Pressure Pain, Touch Sensitivity, Proprioception, and Manual Dexterity in Children with Autism Spectrum Disorders. *Neural Plast.* **2016**, *2016*, 1723401. [CrossRef]
20. Fründt, O.; Grashorn, W.; Schöttle, D.; Peiker, I.; David, N.; Engel, A.K.; Forkmann, K.; Wrobel, N.; Münchau, A.; Bingel, U. Quantitative Sensory Testing in adults with Autism Spectrum Disorders. *J. Autism Dev. Disord.* **2017**, *47*, 1183–1192. [CrossRef] [PubMed]
21. Takayanagi, Y.; Kasahara, Y.; Onaka, T.; Takahashi, N.; Kawada, T.; Nishimori, K. Oxytocin receptor-deficient mice developed late-onset obesity. *NeuroReport* **2008**, *19*, 951–955. [CrossRef]
22. Williams, M.S.; Rooney, B.L.; Williams, J.; Josephson, K.; Pauli, R. Investigation of thermoregulatory characteristics in patients with Prader-Willi syndrome. *Am. J. Med. Genet.* **1994**, *49*, 302–307. [CrossRef]
23. Camerino, C. Low sympathetic tone and obese phenotype in oxytocin-deficient mice. *Obesity* **2009**, *17*, 980–984. [CrossRef]
24. Ince, E.; Ciftçi, E.; Tekin, M.; Kendirli, T.; Tutar, E.; Dalgiç, N.; Oncel, S.; Dogru, U. Characteristics of hyperthermia and its complications in patients with Prader Willi syndrome. *Pediatr. Int.* **2005**, *47*, 550–553. [CrossRef]
25. DiMario, F.J.; Burleson, J.A. Cutaneous blood flow and thermoregulation in Prader-Willi syndrome patients. *Pediatr. Neurol.* **2002**, *26*, 130–133. [CrossRef]
26. Balasco, L.; Provenzano, G.; Bozzi, Y. Sensory Abnormalities in Autism Spectrum Disorders: A Focus on the Tactile Domain, From Genetic Mouse Models to the Clinic. *Front. Psychiatry* **2019**, *10*, 1016. [CrossRef] [PubMed]
27. Modahl, C.; Green, L.; Fein, D.; Morris, M.; Waterhouse, L.; Feinstein, C.; Levin, H. Plasma oxytocin levels in autistic children. *Biol. Psychiatry* **1998**, *43*, 270–277. [CrossRef]
28. Ebstein, R.P.; Israel, S.; Lerer, E.; Uzefovsky, F.; Shalev, I.; Gritsenko, I.; Riebold, M.; Salomon, S.; Yirmiya, N. Arginine vasopressin and oxytocin modulate human social behavior. *Ann. N. Y. Acad. Sci.* **2009**, *1167*, 87–102. [CrossRef] [PubMed]
29. Campbell, D.B.; Datta, D.; Jones, S.T.; Batey Lee, E.; Sutcliffe, J.S.; Hammock, E.A.D.; Levitt, P. Association of oxytocin receptor (OXTR) gene variants with multiple phenotype domains of autism spectrum disorder. *J. Neurodev. Disord.* **2011**, *3*, 101–112. [CrossRef] [PubMed]
30. John, S.; Jaeggi, A.V. Oxytocin levels tend to be lower in autistic children: A meta-analysis of 31 studies. *Autism* **2021**, *25*, 2152–2161. [CrossRef]
31. Rokicki, J.; Kaufmann, T.; de Lange, A.-M.G.; van der Meer, D.; Bahrami, S.; Sartorius, A.M.; Haukvik, U.K.; Steen, N.E.; Schwarz, E.; Stein, D.J.; et al. Oxytocin receptor expression patterns in the human brain across development. *Neuropsychopharmacology* **2022**, *47*, 1550–1560. [CrossRef] [PubMed]
32. Andari, E.; Nishitani, S.; Kaundinya, G.; Caceres, G.A.; Morrier, M.J.; Ousley, O.; Smith, A.K.; Cubells, J.F.; Young, L.J. Epigenetic modification of the oxytocin receptor gene: Implications for autism symptom severity and brain functional connectivity. *Neuropsychopharmacology* **2020**, *45*, 1150–1158. [CrossRef]
33. Baron-Cohen, S. The autistic child's theory of mind: A case of specific developmental delay. *J. Child Psychol. Psychiatry* **1989**, *30*, 285–297. [CrossRef] [PubMed]
34. Burnette, C.P.; Mundy, P.C.; Meyer, J.A.; Sutton, S.K.; Vaughan, A.E.; Charak, D. Weak central coherence and its relations to theory of mind and anxiety in autism. *J. Autism Dev. Disord.* **2005**, *35*, 63–73. [CrossRef] [PubMed]
35. Sinha, P.; Kjelgaard, M.M.; Gandhi, T.K.; Tsourides, K.; Cardinaux, A.L.; Pantazis, D.; Diamond, S.P.; Held, R.M. Autism as a disorder of prediction. *Proc. Natl. Acad. Sci. USA* **2014**, *111*, 15220–15225. [CrossRef]
36. Pérez Velázquez, J.L.; Galán, R.F. Information gain in the brain's resting state: A new perspective on autism. *Front. Neuroinformatics* **2013**, *7*, 37. [CrossRef]
37. Rubenstein, J.L.R.; Merzenich, M.M. Model of autism: Increased ratio of excitation/inhibition in key neural systems. *Genes Brain Behav.* **2003**, *2*, 255–267. [CrossRef]
38. Oumi, T.; Ukena, K.; Matsushima, O.; Ikeda, T.; Fujita, T.; Minakata, H.; Nomoto, K. Annetocin: An oxytocin-related peptide isolated from the earthworm, Eisenia foetida. *Biochem. Biophys. Res. Commun.* **1994**, *198*, 393–399. [CrossRef]
39. Meziane, H.; Schaller, F.; Bauer, S.; Villard, C.; Matarazzo, V.; Riet, F.; Guillon, G.; Lafitte, D.; Desarmenien, M.G.; Tauber, M.; et al. An Early Postnatal Oxytocin Treatment Prevents Social and Learning Deficits in Adult Mice Deficient for Magel2, a Gene Involved in Prader-Willi Syndrome and Autism. *Biol. Psychiatry* **2015**, *78*, 85–94. [CrossRef] [PubMed]
40. Busnelli, M.; Chini, B. Molecular Basis of Oxytocin Receptor Signalling in the Brain: What We Know and What We Need to Know. *Curr. Top. Behav. Neurosci.* **2018**, *35*, 3–29. [CrossRef]
41. Onaka, T.; Takayanagi, Y. The oxytocin system and early-life experience-dependent plastic changes. *J. Neuroendocrinol.* **2021**, *33*, e13049. [CrossRef] [PubMed]
42. Mayer, A.V.; Wermter, A.-K.; Stroth, S.; Alter, P.; Haberhausen, M.; Stehr, T.; Paulus, F.M.; Krach, S.; Kamp-Becker, I. Randomized clinical trial shows no substantial modulation of empathy-related neural activation by intranasal oxytocin in autism. *Sci. Rep.* **2021**, *11*, 15056. [CrossRef]

43. Newmaster, K.T.; Nolan, Z.T.; Chon, U.; Vanselow, D.J.; Weit, A.R.; Tabbaa, M.; Hidema, S.; Nishimori, K.; Hammock, E.A.D.; Kim, Y. Quantitative cellular-resolution map of the oxytocin receptor in postnatally developing mouse brains. *Nat. Commun.* **2020**, *11*, 1885. [CrossRef]
44. Carter, C.S.; Kenkel, W.M.; MacLean, E.L.; Wilson, S.R.; Perkeybile, A.M.; Yee, J.R.; Ferris, C.F.; Nazarloo, H.P.; Porges, S.W.; Davis, J.M.; et al. Is Oxytocin "Nature's Medicine"? *Pharmacol. Rev.* **2020**, *72*, 829–861. [CrossRef] [PubMed]
45. Kubzansky, L.D.; Gilthorpe, M.S.; Goodman, E. A prospective study of psychological distress and weight status in adolescents/young adults. *Ann. Behav. Med.* **2012**, *43*, 219–228. [CrossRef]
46. Ates, T.; Oncul, M.; Dilsiz, P.; Topcu, I.C.; Civas, C.C.; Alp, M.I.; Aklan, I.; Ates Oz, E.; Yavuz, Y.; Yilmaz, B.; et al. Inactivation of Magel2 suppresses oxytocin neurons through synaptic excitation-inhibition imbalance. *Neurobiol. Dis.* **2019**, *121*, 58–64. [CrossRef] [PubMed]
47. Haaf, R.; Brandi, M.-L.; Albantakis, L.; Lahnakoski, J.M.; Henco, L.; Schilbach, L. Peripheral oxytocin levels are linked to hypothalamic gray matter volume in autistic adults: A cross-sectional secondary data analysis. *Sci. Rep.* **2024**, *14*, 1380. [CrossRef]
48. Albantakis, L.; Brandi, M.-L.; Brückl, T.; Gebert, D.; Auer, M.K.; Kopczak, A.; Stalla, G.K.; Neumann, I.D.; Schilbach, L. Oxytocin and cortisol concentrations in adults with and without autism spectrum disorder in response to physical exercise. *Compr. Psychoneuroendocrinol.* **2021**, *5*, 100027. [CrossRef] [PubMed]
49. Tost, H.; Meyer-Lindenberg, A. Dopamine-glutamate interactions: A neural convergence mechanism of common schizophrenia risk variants. *Biol. Psychiatry* **2011**, *69*, 912–913. [CrossRef]
50. Caria, A.; Ciringione, L.; Falco, S. de Morphofunctional Alterations of the Hypothalamus and Social Behavior in Autism Spectrum Disorders. *Brain Sci.* **2020**, *10*, 435. [CrossRef]
51. Wolfe, F.H.; Auzias, G.; Deruelle, C.; Chaminade, T. Focal atrophy of the hypothalamus associated with third ventricle enlargement in autism spectrum disorder. *NeuroReport* **2015**, *26*, 1017–1022. [CrossRef] [PubMed]
52. Ahmed, I.A.; Liu, J.-J.; Gieniec, K.A.; Bair-Marshall, C.J.; Adewakun, A.B.; Hetzler, B.E.; Arp, C.J.; Khatri, L.; Vanwalleghem, G.C.; Seidenberg, A.T.; et al. Optopharmacological tools for precise spatiotemporal control of oxytocin signaling in the central nervous system and periphery. *Res. Sq.* **2023**. [CrossRef]
53. Manfredi-Lozano, M.; Leysen, V.; Adamo, M.; Paiva, I.; Rovera, R.; Pignat, J.-M.; Timzoura, F.E.; Candlish, M.; Eddarkaoui, S.; Malone, S.A.; et al. GnRH replacement rescues cognition in Down syndrome. *Science* **2022**, *377*, eabq4515. [CrossRef] [PubMed]
54. Davis, C.; Patte, K.; Zai, C.; Kennedy, J.L. Polymorphisms of the oxytocin receptor gene and overeating: The intermediary role of endophenotypic risk factors. *Nutr. Diabetes* **2017**, *7*, e279. [CrossRef] [PubMed]
55. Oztan, O.; Garner, J.P.; Constantino, J.N.; Parker, K.J. Neonatal CSF vasopressin concentration predicts later medical record diagnoses of autism spectrum disorder. *Proc. Natl. Acad. Sci. USA* **2020**, *117*, 10609–10613. [CrossRef]
56. Oztan, O.; Talbot, C.F.; Argilli, E.; Maness, A.C.; Simmons, S.M.; Mohsin, N.; Del Rosso, L.A.; Garner, J.P.; Sherr, E.H.; Capitanio, J.P.; et al. Autism-associated biomarkers: Test-retest reliability and relationship to quantitative social trait variation in rhesus monkeys. *Mol. Autism* **2021**, *12*, 50. [CrossRef]

current issues in molecular biology

MDPI

Review

Oxytocin, the Love Hormone, in Stem Cell Differentiation

Luca Pampanella [1], Giovannamaria Petrocelli [1], Federica Forcellini [1], Sara Cruciani [2], Carlo Ventura [1,3], Provvidenza Maria Abruzzo [1,*], Federica Facchin [1,4,*] and Silvia Canaider [1,4]

1 Department of Medical and Surgical Sciences (DIMEC), University of Bologna, Via Massarenti 9, 40138 Bologna, Italy; luca.pampanella2@unibo.it (L.P.); giovannam.petrocell2@unibo.it (G.P.); federica.forcellini2@studio.unibo.it (F.F.); carlo.ventura@unibo.it (C.V.); silvia.canaider@unibo.it (S.C.)

2 Department of Biomedical Sciences, University of Sassari, Viale San Pietro 43/B, 07100 Sassari, Italy; scruciani@uniss.it

3 National Laboratory of Molecular Biology and Stem Cell Bioengineering, National Institute of Biostructures and Biosystems (NIBB), Via di Corticella 183, 40129 Bologna, Italy

4 IRCCS Azienda Ospedaliero-Universitaria di Bologna, Via Massarenti 9, 40138 Bologna, Italy

* Correspondence: provvidenza.abruzzo2@unibo.it (P.M.A.); federica.facchin2@unibo.it (F.F.); Tel.: +39-051-2094122 (P.M.A.); +39-051-2094114 (F.F.)

Abstract: Oxytocin (OXT) is a neurohypophysial nonapeptide that exerts its effects mainly through the oxytocin receptor (OXTR). Several studies have pointed out the role of OXT in the modulation of stem cell (SC) fate and properties. SCs are undifferentiated cells characterized by a remarkable ability to self-renew and differentiate into various cell types of the body. In this review, we focused on the role of OXT in SC differentiation. Specifically, we summarize and discuss the scientific research examining the effects of OXT on mesodermal SC-derived lineages, including cardiac, myogenic, adipogenic, osteogenic, and chondrogenic differentiation. The available studies related to the effects of OXT on SC differentiation provide little insights about the molecular mechanism mediated by the OXT–OXTR pathway. Further research is needed to fully elucidate these pathways to effectively modulate SC differentiation and develop potential therapeutic applications in regenerative medicine.

Keywords: oxytocin; oxytocin receptor; stem cells; stem cell differentiation; cardiogenesis; myogenesis; adipogenesis; osteogenesis; chondrogenesis; odontogenesis

1. Introduction

Oxytocin (OXT) was the first peptide hormone to be structurally determined and chemically synthesized in a biologically active form [1]. OXT is a nonapeptide containing an internal disulfide bond between its Cys(1) and Cys(6) residues. It is mainly synthesized in the magnocellular neurons of the hypothalamic paraventricular and supraoptic nuclei [2], but it is also produced locally in peripheral tissues where it exerts paracrine and autocrine functions (Figure 1) [3]. In response to specific stimuli, OXT is released into the systemic circulation from the posterior pituitary gland and exerts its function by binding to the oxytocin receptor (OXTR) [4]. OXTR belongs to the heptahelical G protein-coupled receptor family, and it is expressed in many tissues like the myometrium, mammary gland, thymus, heart, ovary, kidney, and brain [5]. Upon binding to its receptor, OXT triggers signal transduction, mainly involving the phosphoinositide pathway; this activation increases the levels of intracellular Ca^{2+}, which plays a key role in several intracellular functions, such as inducing muscle contractions [6]. As a matter of fact, OXT was first identified for its function in stimulating uterine smooth muscle contractions [7]. During childbirth, the mechanical stimulation from the stretching of the cervix and uterus triggers the release of OXT, which induces uterine contractions, thereby facilitating labor [8]. Beyond its role during parturition, multiple functions of OXT have been described in human physiology (Figure 1) [9,10]. For instance, OXT plays a key role in lactation [4]: breastfeeding activates sensory neurons in the nipple, which send signals via the spinal cord to the hypothalamus.

This, in turn, leads to the release of OXT, causing milk ejection [4]. In addition, OXT is often referred to as the "love hormone" due to its association with reproduction, birth and maternal behavior, social bonding, sexual behavior, and emotional well-being [11]. Psychosocial stimuli, such as emotional and social interactions, can also stimulate OXT release and modulate social responses [12]. Furthermore, OXT is involved in other central and peripheral physiological processes [9], including inflammation and immune system regulation [13,14], energy metabolism [15], stress responses [16], pain modulation [4], male reproductive system regulation [17], water homeostasis regulation in the kidney [18], cardioprotection and cardiomyogenesis [19], as well as thermoregulation and body composition regulation [20–23]. Finally, its other less studied functions are related to other organs such as the pancreas, liver, eye, skin, bone, and skeletal musculature [3,17]. Given the numerous functions of OXT, it is unsurprising that altered levels of OXT have been associated with several diseases and mental health disorders [9]. Low levels of OXT were observed in patients with Autism Spectrum Disorder, a neurodevelopmental condition characterized by deficits in social interaction and communication [24]. Additionally, low levels of OXT were associated with depression in both non-pregnant adults [25] and postpartum women [26], as well as in anxiety disorder [27] and schizophrenia [28]. On the other hand, elevated levels of OXT have been observed in several pathological conditions, including Obsessive Compulsive Disorder (OCD) [29,30], maternal aggression [31], and tumor growth [32]. Numerous experimental and clinical studies suggest that pharmacological modulation of OXT could be a promising therapeutic approach for the treatment of diseases characterized by altered OXT expression [27,28,33]. Among the various functions of OXT, we focused specifically on its relationship with stem cell (SC) differentiation.

SCs are undifferentiated cells characterized by their ability to differentiate into specialized cell types and by their capability to self-renew [34]. Due to these biological properties, SCs play a crucial role in many physiological processes, such as embryonic development, tissue homeostasis, and the repair of damaged tissue [35,36].

Moreover, SCs are used in biological research to study developmental processes and disease pathogenesis, and to aid drug discovery [37].

Furthermore, due to their remarkable self-renewing and differentiation abilities, SCs represent an attractive tool for regenerative medicine, a branch of translational research aimed at restoring damaged cells, tissues, or organs through cell-based therapy or by inducing endogenous repair and regeneration processes [38,39]. Different sources of SCs are available for tissue regeneration. Based on their origin, SCs can be classified into embryonic SCs (ESCs), fetal SCs, adult SCs, and induced pluripotent SCs (iPSCs), each differing in their differentiation potential [34,40]. Due to ethical concerns in ESC usage [34] and to the tumorigenicity risk of the employment of promising iPSCs [41], researchers have increasingly turned their attention to adult SCs, such as mesenchymal SCs (MSCs), which are multipotent and naturally present in adult tissues and organs.

Several preclinical and clinical studies demonstrated that SC therapy, particularly using MSCs, can promote tissue repair in injured organs in vivo. This includes applications in bone repair, cutaneous wound healing, pulpitis, and ischemic cardiac tissue through SC differentiation and the secretion of anti-inflammatory molecules [38,42,43].

Therefore, understanding and controlling SC properties, such as proliferation and differentiation, is one of the main goals of researchers developing cell-based therapies to treat a wide range of diseases. In this context, growing evidence has revealed that OXT is an interesting molecule that is able to modulate SC differentiation. OXT stimulates cardiac differentiation in various SCs [19,44] and promotes chondrogenic commitment [45]. In addition, OXT improves muscle and liver regeneration [46,47] and enhances neurogenesis [48]. It also stimulates osteogenic differentiation by inhibiting adipogenesis in both human adipose-derived mesenchymal SCs (ADSCs) and human bone marrow mesenchymal SCs (BMSCs) [49–52].

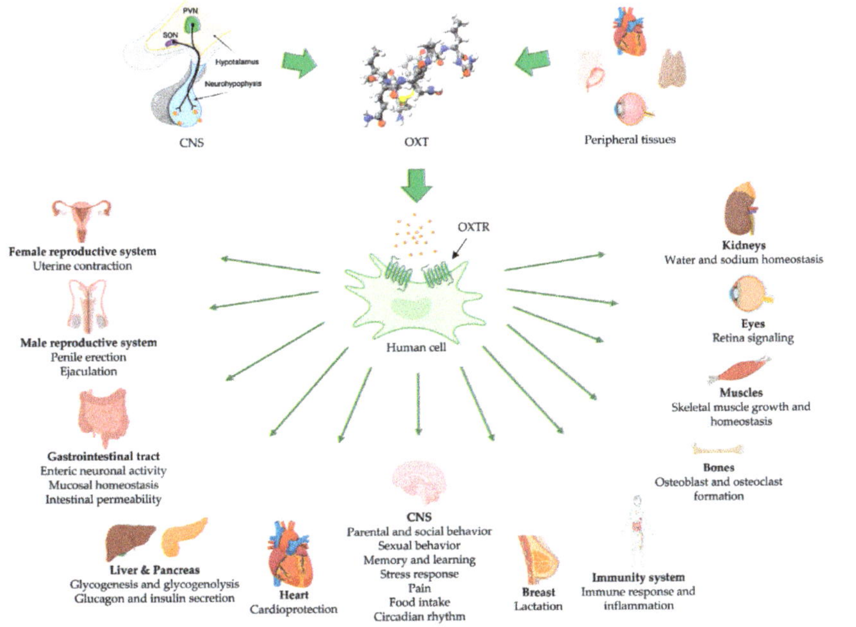

Figure 1. Oxytocin release and its main functions in human tissues and organs. Its 3D molecular structure (National Institute of Health, NIH) is shown as ball-and-stick model using the Jmol variant of Corey–Pauling–Koltun (CPK) coloring: gray = carbon; white = hydrogen; red = oxygen; yellow = sulfur; blue = nitrogen. CNS, central nervous system; OXTR, oxytocin receptor; OXT, oxytocin; PVN, paraventricular nucleus; SON, supraoptic nucleus. Individual images were obtained from BioRender (https://www.biorender.com/, accessed on 23 September 2024).

In this review, we thoroughly examine the current literature discussing the modulatory role of OXT on mesodermal SC differentiation capacity, focusing on cardiac, myogenic, adipogenic, osteogenic, chondrogenic, and odontogenic lineages.

2. Search Strategy

The search was performed in March 2024 using the publicly available database PubMed (National Centre for Biotechnology Information, NCBI, Bethesda, MD, USA). No filters were applied during this process and the search terms, generated using the Medical Subject Headings (MeSH) database, were combined into the search strings reported in Figure 2. After an initial screening using the exclusion criteria reported in Figure 2, original articles relevant to the aim of our review were included. The contributions of the authors are summarized in the text in chronological order, with some exceptions where appropriate. In addition, we included tables that report experimental details, such as the OXT doses, administration methods, and specific results obtained from the studies (see in the next paragraphs). Furthermore, for easier comprehension, Figure 3 summarizes the cell types in which OXT has been shown to positively or negatively modulate specific differentiation processes.

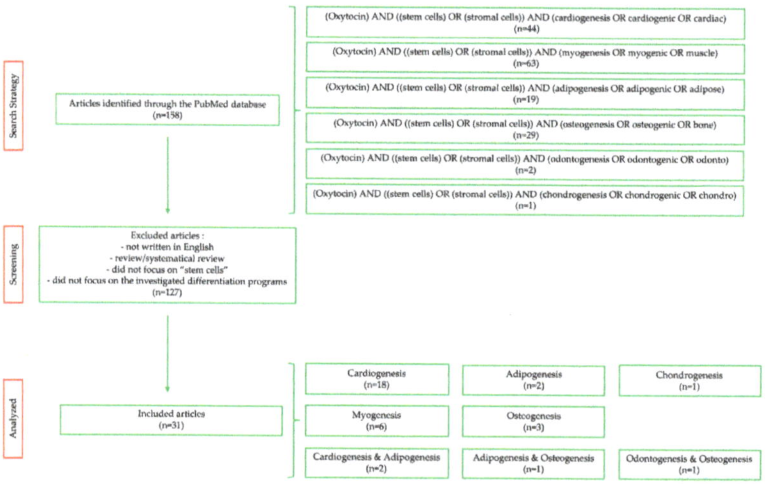

Figure 2. Flow diagram of PubMed search showing number of articles identified (n), according to the declared criteria, and analyzed.

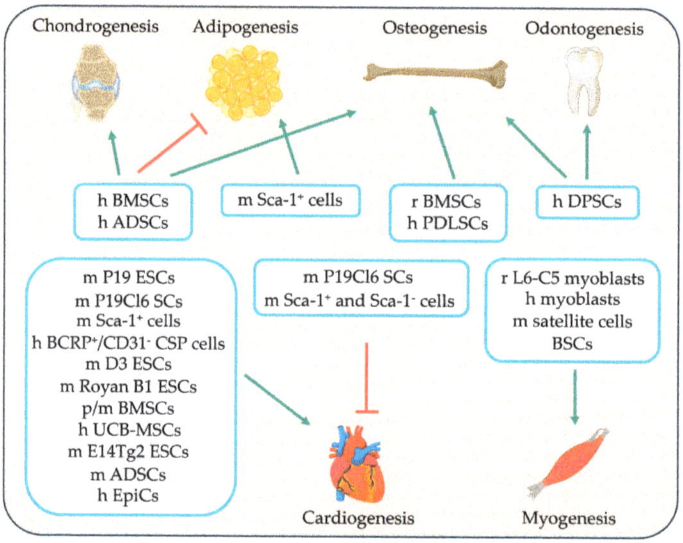

Figure 3. Stem cell types that were treated with oxytocin (OXT) and their differentiation programs. Cells with green arrows are cells prone to differentiation towards the indicated lineages when treated with OXT; cells with red lines are cells that are unable to differentiate into the indicated lineages when treated with OXT. h, human; m, mouse; p, porcine; r, rat. ADSCs, adipose-derived mesenchymal stem cells; $BCRP^+/CD31^-$ CSP cells, breast cancer resistance protein-positive and CD31-negative cardiac side population cells; BMSCs, bone marrow mesenchymal stem cells; BSCs, bovine satellite cells; D3 ESCs, D3 embryonic stem cells; DPSCs, dental pulp stem cells; E14Tg2 ESCs, E14Tg2 embryonic stem cell line; EpiCs, induced pluripotent stem cell-derived epicardial cells; $Sca\text{-}1^+/Sca\text{-}1^-$ cells, Sca-1-positive/negative cells; L6-C5 myoblasts, mononucleated L6 myoblasts, subclone C5; P19 ESCs, P19 embryonic stem cells; P19Cl6 SCs, P19 clone 6 stem cells; PDLSCs, periodontal ligament-derived stem cells; Royan B1 ESCs, Royan B1 embryonic stem cells; UCB-MSCs, umbilical cord-derived mesenchymal stem cells. Individual images were obtained from BioRender (https://www.biorender.com/, accessed on 23 September 2024).

3. Oxytocin in Stem Cell Cardiac Differentiation

In 2022, the first study investigating the potential role of OXT in the differentiation process was published [53]. The elevated OXT levels observed in the fetal and newborn mouse heart, when cardiomyocytes show intense hyperplasia, led the authors to hypothesize a crucial role of this hormone in cardiac differentiation. To test this hypothesis, they used P19 mouse ESCs (P19 ESCs) and observed that OXT stimulated the cells to start beating by day 8 of the differentiation protocol, earlier than the cells treated with dimethyl sulfoxide (DMSO), which is known to promote differentiation into cardiomyocytes by day 12. The cardiomyocyte phenotype was further confirmed by the increase in the expression of *atrial natriuretic peptide* (*ANP*), *sarcomeric myosin heavy chain* (*MHC*), and *dihydropyridine receptor alpha 1* (*DHPR-α1*), as well as by the presence of abundant mitochondria. Notably, OXT-treated P19 ESCs showed an increase in the expression of OXTR, while an OXT antagonist abrogated cardiomyocyte formation, suggesting a key role of OXT in cardiac differentiation [53]. This role was further supported by the presence of high levels of OXT and OXTR proteins in the developing rat hearts at day 21 of gestation and postnatal days 1–4, when cardiomyocytes exhibit intense hyperplasia [54].

We noticed that following this study [53], many researchers used OXT and its differentiation capacity to validate the stemness properties of specific cells employed in their studies. Nonetheless, since these cells had already been confirmed as SCs, the results reported in these scientific articles were included in this review as further evidence of the ability of OXT to induce cardiac differentiation. For example, in a study investigating the stemness of Sca-1-positive (Sca-1$^+$) cells isolated from adult mouse hearts, it was reported that OXT induced Sca-1$^+$ cells to differentiate into beating cardiomyocytes expressing cardiac gene and protein markers [55]. On the contrary, in another study, it was observed that Sca-1$^+$ and Sca-1-negative (Sca-1$^-$) cells isolated from adult mouse skeletal muscle and exposed to OXT did not differentiate into cardiomyocytes [56]. OXT led to a reduction in the expression of two cardiac markers, *NK2 homeobox 5* (*Nkx2.5*) and *GATA-binding protein 4* (*GATA4*); this promoted the differentiation of Sca-1$^+$ and Sca-1$^-$ cells into adipocytes and epithelial cells, respectively [56].

Another research group also performed different experiments to demonstrate the SC properties of the breast cancer resistance protein-positive and CD31-negative (BCRP$^+$/CD31$^-$) cardiac side population (CSP) cells obtained from neonatal rat hearts [57]. Among them, the authors treated these cells with OXT and found that it induced the differentiation of BCRP$^+$/CD31$^-$ CSP cells into beating cardiomyocytes after 3 weeks of treatment. These findings were further confirmed by the increased expression of both cardiac markers, such as *Nkx2.5*, *GATA4*, *myocyte-enhancer factor 2C* (*MEF-2C*), and *ANP*, and of contractile proteins, including myosin light chain 2v (MLC-2v), cardiac troponin T (cTnT), and sarcomeric α-actinin [57].

In 2013, OXT was used to assess the differentiation potential of human BCRP$^+$/CD31$^-$ CSP cells, which represent the cardiac resident progenitor cells. However, these cells did not exhibit spontaneous beating even after 21 days of the differentiation protocol, although they expressed cTnT protein and showed an increase in the levels of *α-MHC* mRNA [58], suggesting that they did not achieve full differentiation.

In 2007, the effect of OXT on murine P19 clone 6 (P19Cl6) cells, a subclone of P19 ESCs, was investigated by several authors, yielding contrasting results. In one research study [59], it was demonstrated that following OXT treatment, P19Cl6 SCs showed a beating phenotype and expressed elevated levels of the mRNA of the cardiomyocyte-specific transcription factor *GATA4* as well as MLC-2v protein.

On the other hand, the ability of OXT to induce cardiac differentiation in P19Cl6 SCs was not replicated in a subsequent study [60], although it was confirmed that OXT induces cardiac differentiation in P19 ESCs. In fact, the modulation of the mRNA and protein expression of cardiac markers, such as *myosin light chain 2a* (*MLC-2a*), *α-MHC*, *β-MHC*, and cardiac troponin I (cTnI), were observed [60]. The cardiac differentiation of OXT-treated P19 ESCs was further validated in 2009 [61]. To understand the different behaviors of

P19 and P19Cl6 SCs during cardiac differentiation, OXTR and its relationship with early cardiac marker genes, particularly *GATA4*, were analyzed [60]. The data revealed that although OXT induced an increase in OXTR levels, it failed to stimulate cardiomyogenic differentiation in mP19Cl6 SCs, probably because it was unable to properly induce the expression of *GATA4*, which is essential for this process [60].

Conversely, in another study, it was demonstrated that OXT promoted spontaneous beating in both mouse P19 ESCs and the subclone P19Cl6 [62]; the results indicated that nitric oxide (NO) plays a key role in OXT-mediated cardiogenesis, as evidenced by the abrogation of OXT effects when NO synthase was inhibited [62]. The role of NO in OXT-mediated cardiogenic differentiation was also confirmed in 2011 in porcine BMSCs [63]. The study emphasized the importance of treating cells with OXT at early culture passages, when they express high levels of *OXTR* transcripts, which increased their responsiveness to OXT stimulation and improved their differentiation potential [64]. The effects of OXT on the P19Cl6 SC model were also investigated in another research study [65] where the authors focused on the cardiac differentiation of P19Cl6 SCs cultured as confluent monolayers or in aggregates (embryoid bodies, EBs). Interestingly, they found that OXT induced P19Cl6 SCs to differentiate into cardiomyocytes only when they grew in aggregates, whereas it was ineffective when the cells were maintained in a monolayer culture. This finding suggests that cell interactions within EBs influence the cell fate of OXT-treated P19Cl6 SCs [65].

The role of OXT in promoting cardiac differentiation was further demonstrated in murine Royan B1 ESCs. OXT promoted the early maturation of ESC-derived cardiomyocytes, as evidenced by enhanced chronotropic responses and increased expression of cardiac markers, such as cTnI. However, no changes in the ultrastructural characteristics of the cardiomyocytes at any stage of development were observed [66].

In 2008, researchers started to study the involvement of C-terminally extended forms of OXT in cardiogenesis. These forms are derived from the processing of OXT–neurophysin precursors in the hypothalamus and act as intermediate prohormones [67]. In particular, the OXT forms OXT-Gly-Lys-Arg (OXT-GKR), OXT-Gly-Lys (OXT-GK), and OXT-Gly (OXT-G) were used to treat mouse D3 ESCs. OXT-GKR increased the number of beating cells on days 5 and 12, which exhibited a ventricular cell phenotype [67]. Similar results were obtained in cells overexpressing OXT-GKR (OXT-GKR$^+$ cells), which showed increased expression of *GATA4* and *MLC-2v* mRNAs. Subsequently, it was demonstrated that OXT-GKR was the dominant form of OXT in newborn rat hearts and it had accumulated concomitantly with OXTR expression in mouse embryos at day 15 [68]. Moreover, OXT-GKR induced contracting cell colonies and more efficiently promoted the expression of ventricular cardiomyocyte markers than OXT and it also reduced the expression of skeletal muscle markers, such as *MEF-2C*, *myogenin* (*MyoG*), and *myogenic differentiation 1* (*MyoD*) mRNAs, in P19 ESCs. These findings led to the hypothesis that the C-terminally extended OXT molecules promote cardiomyocyte differentiation and contribute to heart growth during fetal life [68].

Therefore, in 2014, the same research group treated BCRP$^+$/CD31$^-$ CSP cells with OXT-GKR [69]. OXT-GKR enhanced cell viability, increased both the formation and size of cell aggregates, induced synchronized contraction of cells, and stimulated the expression of cardiomyocyte markers. Moreover, OXT-GKR induced endothelial differentiation by promoting the formation of a network of tubular cells, the expression of von Willebrand factor (vWF), and the creation of Weibel–Palade bodies, the storage granules in endothelial cells [69].

In a study published in 2010, a pro-migratory effect of OXT on umbilical cord blood-derived MSCs (UCB-MSCs) injected into the myocardium of infarcted rats was reported [70]. In 2012, the same research group evaluated the effect of OXT on UCB-MSCs in vitro [71], demonstrating that, when exposed to OXT, these cells differentiated into cardiomyocytes and expressed cardiac gene and protein markers. Moreover, UCB-MSCs co-cultured with neonatal rat cardiomyocytes subjected to a hypoxia/reoxygenation insult (HR-CMs) showed enhanced expression of cardiac proteins such as connexin 43 (CX43), cTnI, and α-sarcomeric actin (α-SA), indicating that direct contact between UCB-MSCs and HR-CMs may reinforce the effects of OXT on cardiac differentiation [71]. Similar effects were observed in mouse ADSCs, which differentiated into cardiomyocytes when treated with OXT, either alone or in combination with relaxin. Notably, the OXT/relaxin combination enhanced the effects of OXT [72]. The cardiogenic property of OXT was further demonstrated in a murine ESC line (E14Tg2) [73]. OXT reduced the expression of the SC marker *POU domain, class 5, transcription factor 1* (*OCT4*), while it increased the levels of the mesoderm marker *mesoderm posterior 1* (*Mesp1*) and of cardiogenic markers during the differentiation process. Interestingly, an upregulation of the expression of the *fibroblast growth factor 1* (*FGF1*) gene was observed, particularly in the *FGF1B* transcript levels. FGF1 is expressed in the human fetal heart and plays a role in enhancing cardiac regeneration. Blocking FGF1 actions or its interaction with its receptor, fibroblast growth factor receptor (FGFR), led to a reduction in the efficiency of beating cell formation and in the mRNA levels of cardiomyocyte markers. Moreover, Lin and colleagues observed that inhibiting the FGF1-FGFR downstream effectors AKT serine/threonine kinase (AKT) or protein kinase C (PKC) (particularly PKC ε) further impaired cardiac differentiation, demonstrating that FGF1 could regulate cardiogenesis through PKC signaling [73].

Recently, in a study published in 2022 [74], it was demonstrated that OXT stimulated epicardial cell proliferation, induced epithelial-to-mesenchymal transition (EMT), and increased the transcriptional activity in human induced pluripotent stem cell-derived epicardial cells (hEpiCs). These findings suggested that OXT induces the activation of epicardial cells to a progenitor-like state (hEpiPCs); these multipotent cells play a crucial role in cardiac regeneration after injury, as they can differentiate into various cardiac lineages [74]. The role of OXT in enhancing epicardial cell function was confirmed in a zebrafish model, where OXT levels increased following cardiac cryoinjury. OXT contributed, through the involvement of the transforming growth factor beta (TGF-β) pathway, to epicardial activation and heart regeneration. Furthermore, the authors demonstrated that OXT signaling plays a crucial role in the development of the epicardium in zebrafish embryos [74].

Overall, the studies mentioned above highlight the role of OXT as a key inducer of cardiogenic differentiation in various SC types (Figure 3 and Table 1).

Table 1. Effects of oxytocin on cardiogenic differentiation.

Oxytocin Formulation/Dose	Species and Cell/Animal Model	Treatment	Cellular Mechanism *	Effect(s) *	Ref.
OXT: 100 nM	m P19 ESCs	4 days, then cells were cultured until day 14 without OXT	↑ spontaneous beating ↑ ANP mRNA ↑ mitochondria number ↑ sarcomeric MHC and DHPR-α1 mRNAs ↑ OXTR mRNA and protein OXT antagonist (vasoticin) inhibited OXT effects	↑ cardiogenesis	[53]
OXT: 100 nM	m Sca-1^{+} cells	72 h	↑ spontaneous beating ↑ sarcomeric structures ↑ Nkx2.5, GATA4, MEF-2C, α-MHC, β-MHC, MLC-2a, MLC-2v, and cardiac α-actin mRNAs ↑ GATA4, ANP, cTnT, MLC-2v, sarcomeric MHC, CX43, and tropomyosin proteins ↑ OXTR mRNA OXT antagonist (vasoticin) inhibited OXT effects	↑ cardiogenesis	[55]
OXT: 100 nM	m P19 ESCs and m P19Cl6 SCs	EBs treated for 4 days, then cells were cultured until day 14 or 16 without OXT	↑ spontaneous beating ↑ GATA4, ANP, Nkx2.5, MEF-2C, α-MHC, and MyoG mRNAs NOS inhibitor (L-NAME) blocked OXT action	↑ cardiogenesis OXT acts through NO signaling	[62]
OXT: 1000 nM	m P19Cl6 SCs	EBs treated for 5 days, then cells were cultured until day 14 without OXT	↑ spontaneous beating ↑ GATA4 mRNA ↑ MLC-2v protein OXTR antagonist (H-9405) reduced MLC-2v protein expression	↑ cardiogenesis	[59]
OXT: 10 nM	m Royan B1 ESCs	EBs treated for 5 days, then cells were cultured until day 30 without OXT	↑ spontaneous beating ↑ cTnI protein expression ↑ β-MHC, ANP, and MLC-2v mRNAs at early stage of development	↑ cardiogenesis	[66]

Table 1. *Cont.*

Oxytocin Formulation/Dose	Species and Cell/Animal Model	Treatment	Cellular Mechanism *	Effect(s) *	Ref.
OXT: 100 nM	r $BCRP^+/CD31^-$ CSP cells	72 h	↑ spontaneous beating ↑ Nkx2.5, GATA4, MEF-2C, β-MHC, and MLC-2c mRNAs ↑ GATA4, ANP, cTnT, MLC-2v, and sarcomeric α-actinin proteins	↑ cardiogenesis	[57]
OXT: 1000, 100 or 10 nM	m P19 ESCs and m P19Cl6 SCs	4 days, then cells were cultured without OXT	↑ spontaneous beating only in P19 cells ↑ MLC-2a, α-MHC, β-MHC, Myf5, MyoD, and MyoG mRNAs (only in P19 cells) ↑ cTnI protein (only in P19 cells) ↑ GATA4 mRNA (only in P19 cells)	↑ cardiogenesis GATA4 upregulation has a key role in cardiogenesis induction	[60]
OXT: 100 nM	m $Sca\text{-}1^+$ and $Sca\text{-}1^-$ cells	30 days	↓ Nkx2.5 and GATA4 mRNAs at 4 days in SM^- cells = Nkx2.5 and GATA4 mRNAs in SM^+ cells ↓ Nkx2.5, GATA4, cMHC, and α-SA proteins in both SM^- and SM^+ cells treated with OXT	↓ cardiogenesis	[56]
OXT/OXT-GKR/OXT-G/OXT-GK: 1000 nM	m D3 ESCs	EBs treated for 5 days	↑ beating cells at 12 days in OXT- and OXT-GKR-treated cells ↑ beating cells at 5 days (only in OXT-GKR-treated cells) OXT antagonist (H-9405) inhibited OXT and OXT-GKR effects ↑ GATA4 and MLC-2v mRNAs in $OXT\text{-}GKR^+$ cells ↑ number of ventricular-like cells in $OXT\text{-}GKR^+$ cells ↑ CX43 protein in $OXT\text{-}GKR^+$ cells	↑ cardiogenesis OT-GKR stimulates cells to differentiate toward a ventricular phenotype	[67]
OXT: 100 nM	m P19 ESCs	4 days	↑ spontaneous beating ↑ cTpnl and MyoD mRNAs	↑ cardiogenesis	[61]
OXT: 100 nM	m P19Cl6 SCs	EBs or cells cultured as monolayer treated for 6 or 14 days	↑ spontaneous beating (only in EB cells) ↑ GATA4, Nkx2.5, α-cardiac actin, β-MHC, Tbx5, and Tbx20 mRNAs ↑ β-MHC and cTnI proteins	↑ cardiogenesis	[65]

Table 1. *Cont.*

Oxytocin Formulation/Dose	Species and Cell/Animal Model	Treatment	Cellular Mechanism *	Effect(s) *	Ref.
OXT/OXT-GKR/OXT-G/OXT-GK: 1000 nM	m P19 ESCs	EBs treated for 5 days	OXT-GKR is a dominant form of OXT in newborn rat hearts ↑ contracting cells in OXT-GKR-treated cells OXTR silencing inhibited OXT-GKR effects ↑ GATA4, MEF-2C, MyoG, and MyoD mRNAs ↑ DHPRα1, MLC-2v, and sarcomeric α-actinin proteins	↑ cardiogenesis	[68]
OXT: 10,000 nM	p BMSCs	1 day, then cells were cultured until day 15 without OXT	↑ eNOS and iNOS mRNAs and proteins ↑ cTnI mRNA NOS inhibitor (L-NAME) reduced cTnI and PLB mRNAs ↑ cTnT, cMHC, and cTnI proteins NOS inhibitor (L-NAME) reduced cTnT, cMHC, and cTnI proteins ↑ proliferation	↑ cardiogenesis OXT acts through NO signaling	[63]
OXT: 100 nM	h UCB-MSCs	7 days	↑ CX43, cTnI, and α-SA proteins at 7 days ↑ CX43 and cTnT mRNAs and proteins at 4 days ↑ eNOS mRNA and protein at 7 days	↑ cardiogenesis	[71]
OXT: 100 nM	h $BCRP^+/CD31^-$ CSP cells	72 h	No spontaneous beating ↑ cTnT protein ↑ α-MHC mRNA	↑ cardiogenesis	[58]
OXT-GKR: 100 nM	r $BCRP^+/CD31^-$ CSP cells	EBs treated for 5 days, then cells were cultured until day 18 without OXT	↑ cell viability ↑ formation and size of EBs ↑ synchronized contraction of cells ↑ MLC-2v, sarcomeric α-actinin, and cTnT proteins ↓ nestin protein	↑ cardiogenesis	[69]

Table 1. *Cont.*

Oxytocin Formulation/Dose	Species and Cell/Animal Model	Treatment	Cellular Mechanism *	Effect(s) *	Ref.
OXT: 100 nM	m E14Tg2 ESCs	4 days in differentiation medium, then cells were cultured until day 14 without OXT	↓ OCT4 and ↑ Mesp1 mRNAs ↑ GATA4, cTnT, MLC-2v, and α-MHC mRNAs ↑ sarcomeric α-actinin, FGF1, cTnT, and α-tubulin proteins ↑ FGF1B mRNA at the late differentiation stage Blocking FGF1 actions or its receptor impaired cardiac differentiation AKT and PKC inhibitor reduced beating cell cluster formation	↑ cardiogenesis OXT acts through FGF1B and PKC signaling pathways	[73]
OXT: 10,000 nM	m BMSCs	3 weeks	↑ GATA4, PLB, desmin, and cTnI mRNAs ↑ cTnT, cTnI, and cMHC proteins	↑ cardiogenesis	[64]
OXT: 1000, 100 or 10 nM	m ADSCs	4 days, then cells were cultured for up to 3 weeks without OXT	↑ MEF-2c, MLC-2a, MLC-2v, and CX43 mRNA, ↑ CX43, desmin, and sarcomeric α-actinin proteins	↑ cardiogenesis	[72]
OXT: 100 nM	In vitro: h EpiCs In vivo: zebrafish embryos and cardiac cryoinjured adult zebrafish	In vitro: 3 days	↑ proliferation in vitro ↑ WT1, TCF21, SNAI1, and NT5E mRNAs in vitro ↑ Ki-67, WT1, and TJP1 proteins in vitro Inhibition of OXTR impaired OXT action on Ki-67, WT1, SNAI1, and TJP1 protein expression in vitro ↑ activity of TGF-β/BMP pathway biological processes in vitro	↑ epicardial cell activation and heart regeneration	[74]

Table 1. *Cont.*

Oxytocin Formulation/Dose	Species and Cell/Animal Model	Treatment	Cellular Mechanism *	Effect(s) *	Ref.
OXT: 100 nM	In vitro: h EpiCs In vivo: zebrafish embryos and cardiac cryoinjured adult zebrafish	In vitro: 3 days	OXTR inhibition impaired the formation of the epicardium in vivo OXTR inhibition impaired the proliferation and migration of progenitor cells in vivo OXTR inhibition decreased the expression of PCNA and cTnT proteins in vivo OXTR inhibition decreased the mRNA expression of WTB1, TCF21, SNAI1a, and SNAI2 in vivo	↑ epicardial cell activation and heart regeneration	[74]

Abbreviations: OXT, oxytocin; OXT-G, OXT-Gly; OXT-GK, OXT-Gly-Lys; OXT-GKR, OXT-Gly-Lys-Arg; OXT-GKR$^+$ cells, OXT-GKR-overexpressing cells; h, human; m, mouse; p, porcine; r, rat; ADSCs, adipose-derived mesenchymal stem cells; BCRP$^+$/CD31$^-$ CSP cells, breast cancer resistance protein-positive and CD31-negative cardiac side population cells; BMSCs, bone marrow mesenchymal stem cells; D3 ESCs, D3 embryonic stem cells; E14Tg2 ESCs, E14Tg2 embryonic stem cells; EpiCs, induced pluripotent stem cell-derived epicardial cells; Sca-1$^+$/Sca-1$^-$ cells, Sca-1-positive/negative cells; P19 ESCs, P19 embryonic stem cells; P19Cl6 SCs, P19 clone 6 stem cells; Royan B1 ESCs, Royan B1 embryonic stem cells; UCB-MSCs, umbilical cord-derived mesenchymal stem cells; EB, embryonic body; α-MHC, α-myosin heavy chain; α-SA, α-sarcomeric actin; ANP, atrial natriuretic peptide; AKT, serine/threonine kinase; β-MHC, β-myosin heavy chain; BMP, bone morphogenetic protein; cMHC, cardiac myosin heavy chain; cTnI, cardiac troponin I; cTnT, cardiac troponin T; cTpnl, cardiac troponin inhibitor; CX43, connexin 43; DHPR-α1, dihydropyridine receptor-α1; eNOS, endothelial nitric oxide synthase; FGF1, fibroblast growth factor 1; FGF1B, fibroblast growth factor 1B; GATA4, GATA-binding protein 4; H-9405, [β-mercapto-β,β-cyclopentamethylene-propionyl-Tyr(Me)^2-Ile-Thr-Asn-Cys-Pro-Orn-Tyr-NH_2]; iNOS, inducible nitric oxide synthase; Ki-67, proliferation marker protein Ki-67; L-NAME, N (G)-nitro-L-arginine methyl ester; MEF-2C, myocyte enhancer factor 2C; Mesp1, mesoderm posterior 1; MLC-2a, myosin light chain 2a; MLC-2c, myosin light chain 2c; MLC-2v, myosin light chain 2v; Myf5, myogenic factor 5; MyoD, myogenic differentiation 1; MyoG, myogenin; Nkx2.5, NK2 homeobox 5; NO, nitric oxide; NOS, nitric oxide synthase; NT5E, 5′-nucleotidase ecto; OCT4, POU domain, class 5, transcription factor 1; OXTR, oxytocin receptor; PKC, protein kinase C; PCNA, proliferating cell nuclear antigen; PLB, phospholamban; sarcomeric MHC, sarcomeric myosin heavy chain; SNAI1, snail family transcriptional repressor 1; SNAI1a, snail family transcriptional repressor 1a; SNAI2, snail family transcriptional repressor 2; Tbx5, T-box transcription factor 5; Tbx20, T-box transcription factor 20; TCF21, transcription factor 21; TGF-β, transforming growth factor beta; TJP1, tight junction protein 1; vasoticin, $d(CH_2)_5{}^1$,Tyr(Me)^2,Thr-4,Orn-8,Tyr-$NH_2{}^9$; WT1, WT1 transcription factor; WT1B, WT1 transcription factor b. *Arrow and equal symbols indicate the effects of OXT on biological stem cell properties: ↑ (increase), ↓ (reduce), = (no effect).

4. Oxytocin in Stem Cell Myogenic Differentiation

It is known that OXT exerts anabolic effects on muscle tissue [20,75]. The first study to assess the role of the neurohypophysial nonapeptide arginine8-vasopressin (AVP) and its analogue OXT in myogenic differentiation was conducted in 1995 [76]. In this study, rat mononucleated L6 myoblast cells, subclone C5 (L6-C5), stimulated with OXT acquired multinucleated myotube phenotypic characteristics; the resulting myotubes were larger than those formed in the absence of OXT or AVP and had central nuclei. In addition, the myogenic differentiation capacity of OXT was confirmed at the molecular level by the increased expression of myosin protein [76].

The contribution of the OXT–OXTR pathway to myogenic differentiation was further demonstrated in myoblasts derived from human satellite cells [77]. Myoblasts treated with OXT, AVP, or [Thr(4)Gly(7)]OXT had activated OXTR signaling and showed an increase in the number of fused myoblasts and in the formation of cultured myotubes; this evidence supports the hypothesis that OXT acts in a paracrine/autocrine fashion to stimulate human myoblast fusion [77]. In 2014, the role of OXT in mice skeletal muscles and in their satellite cells was investigated [46]. The findings revealed a decrease in plasma OXT levels in old mice, while OXTR levels in skeletal muscles remained similar between young and old mice. However, a reduction in OXTR expression was observed in old satellite cells. OXT administration improved the capacity of old mice to form new muscle fibers, restoring it to levels comparable to those of young mice, suggesting that OXT administration ameliorated muscle healing in old mice. Moreover, the OXT treatment, by activating the mitogen-activated protein kinase (MAPK)/extracellular signal-regulated kinase (ERK) pathway, increased the proliferation ability of satellite cells in young mice and restored both the proliferation and myogenic differentiation abilities of satellite cells in old mice, as evidenced by the increase in the myogenic fusion index [46].

On the contrary, another study demonstrated that OXT did not influence myogenic fate [78]: murine C2C12 myoblasts were exposed to a chronic treatment with OXT, either alone or in combination with 17β-estradiol (E2), for 7 days during myocyte differentiation [78]. Neither molecule alone or in combination influenced the myotube fusion index, as well as the expression of the *myogenic regulatory factor* (*MRF*) or *MHC* genes in differentiated cells [78]. In another study, while the researchers were studying the effect of 5-azacytidine on human BMSCs cultured on polycaprolactone electrospun fibers for muscle regeneration, they observed that OXT did not affect myogenesis differentiation except when the cells were cultured in the presence of the scaffold, where a positive effect of OXT was observed [79].

In another study [80], the effects of OXT and other steroid hormones on the proliferation and differentiation abilities of bovine satellite cells (BSCs) were compared. OXT was found to increase the fusion index, reduce the number of apoptotic nuclei, enhance BSC migration ability, and upregulate the expression of both *MyoD* and *MyoG* mRNAs. Similar effects were observed in the steroid hormone-treated cells, which increased *OXT* mRNA levels, leading to the hypothesis of a key role of OXT in myogenesis [80]. The results described above are summarized in Figure 3 and Table 2.

Table 2. Effects of oxytocin on myogenic differentiation.

Oxytocin Formulation/Dose	Species and Cell/ Animal Model	Treatment	Cellular Mechanism *	Effect(s) *	Ref.
OXT: 100 nM	r L6-C5 myoblasts	5 or more days	↑ myoblast fusion index ↑ myosin protein levels	↑ myogenesis	[76]
OXT: 1000 nM	h myoblasts	20–48 h in differentiation medium	↑ myoblast fusion index	↑ myogenesis	[77]
OXT: 30 nM in vitro; 1 µg/g in vivo	m satellite cells m muscles	In vitro: 24 or 48 h In vivo: daily treatment for 4/6 days before the muscle injury and until animal sacrifice	In vitro: ↑ proliferation of old satellite cells and primary myogenic progenitors OXT acts via the MAPK/ERK signaling pathway In vivo: ↓ OXT plasma levels with age Similar OXTR protein expression in skeletal muscles between young and old mice ↓ OXTR protein expression in satellite cells with age ↑ new muscle fiber formation in OXT-treated old mice ↑ myogenic cell proliferation in OXT-treated old mice ↑ proliferation ability of satellite cells in OXT-treated old and young mice ↑ differentiation ability of satellite cells in old mice with OXT administration	↑ proliferation in vitro ↑ myogenesis in vivo	[46]
OXT: 10,000 nM	m C2C12 myoblasts	7 days during the myotube differentiation	=myoblast fusion index =MRF and MHC mRNA levels	=myogenesis	[78]

Table 2. *Cont.*

Oxytocin Formulation/Dose	Species and Cell/ Animal Model	Treatment	Cellular Mechanism *	Effect(s) *	Ref.
OXT: 10 nM	h BMSCs	28 days	=YIP-1B protein expression	=myogenesis	[79]
OXT: 31.25, 62.5, 125, 250 nM	BSCs	48 or 72 h	↑ myoblast fusion index ↓ apoptotic nuclei ↑ cell migration ↑ MyoD and MyoG mRNAs during BSC proliferation ↑ MyoG mRNA during BSC differentiation	↑ myogenesis	[80]

Abbreviations: OXT, oxytocin; h, human; m, mouse; r, rat; BMSCs, bone marrow mesenchymal stem cells; BSCs, bovine satellite cells; L6-C5 myoblasts, mononucleated L6 myoblasts, subclone C5; P19, P19 embryonic stem cell; ERK, extracellular signal-regulated kinase; MAPK, mitogen-activated protein kinase; MHC, myosin heavy chain; MRF, myogenic regulatory factor; MyoD, myogenic differentiation 1; MyoG, myogenin; OXTR, oxytocin receptor; Yip-1B, Yip1-interacting factor homolog B. * Arrow and equal symbols indicate the effects of OXT on biological stem cell properties: ↑ (increase), ↓ (reduce), = (no effect).

5. Oxytocin in Stem Cell Adipogenic, Osteogenic, and Odontogenic Differentiation

Osteoblasts and adipocytes originate from the same mesenchymal precursor cells, and an inverse relationship exists between these two lineages. In osteoporosis, bone loss is associated with an increase in bone marrow adipose tissue resulting from the production of adipocytes at the expense of osteoblasts [50]. Thus, identifying signaling pathways that promote MSC osteogenesis and reduce adipogenesis is crucial for reinforcing bone regeneration treatments.

The first study to investigate the role of OXT in these two processes [49] demonstrated that OXT promoted osteoblast differentiation and inhibited adipocyte commitment in both human ADSCs and BMSCs. The authors also speculated that ERK activation might be involved in the OXT-driven differentiation process, while ruling out the involvement of the Ras homolog family member A (RhoA) pathway. Consistent with these findings, ovariectomized (OVX) mice and rats showed a significant decrease in OXT levels, an osteoporotic phenotype, as well as an increase in bone marrow adiposity and an upregulation in the expression of *fatty acid-binding protein 4* (*FABP4*) mRNA, a marker of adipogenesis; these effects were reverted by subcutaneous OXT injections [49]. In another study [81], OVX rats received two implants at the distal femoral metaphysis; the subcutaneous injection of OXT administered after surgery resulting in an enhanced relative bone volume around the implant, an improved percent implant osseointegration, and an increased maximum push-out force and bone mass. These findings indicated that OXT promoted peri-implant bone healing and counteracted the negative effects of osteoporosis.

Consistent with the data obtained in mice, elevated levels of osteoporosis accompanied by increased adipose tissue were observed in rabbits treated with glucocorticoids [82]. OXT treatment reversed the glucocorticoid-induced marrow adiposity and prevented the osteoporosis induced by glucocorticoids.

The inverse relationship between the two lineages was also recently observed in human ADSCs [83]. OXT clearly promoted osteogenic differentiation and either had no effect or sometimes reduced adipogenic differentiation in ADSCs. Moreover, an increase in the expression of the autophagy marker genes *Beclin 1* (*BECN1*) and *Microtubule-Associated Protein 1 Light Chain 3 alpha* (*MAP1LC3A*) was observed at the onset of the osteogenesis, suggesting a role of autophagy in OXT-induced osteogenesis.

Alongside the studies evaluating the effects of OXT in both adipogenesis and osteogenesis, several studies have focused on the role of OXT in only one of these two processes. Regarding adipogenesis, it was observed that Sca-1^+ cells, previously investigated for their cardiogenic differentiation potential, were able to differentiate into adipocytes when treated with OXT [56].

On the contrary, the data reported in another study showed that OXT did not affect the adipogenic differentiation of mP19 ESCs [61].

Further insights into OXT's involvement in adipogenesis were further provided by a study focused on the evaluation of *OXTR* mRNA expression during the adipocyte differentiation process [84]. The authors found that *OXTR* mRNA levels were higher in adipocytes derived from mouse adipose tissues compared to vascular stromal cells, and they increased during 3T3-L1 adipocyte differentiation [84].

In addition, it was shown that *OXTR* mRNA expression was higher in older mice and in mice fed with a high-fat diet; moreover, OXT induced lipolysis in 3T3-L1 adipocytes, suggesting a role of OXTR in the regulation of both adipocyte differentiation and fat accumulation [84]. Notably, the effects of OXT on fat and metabolism have also been reported in in vivo models [85,86].

In a study investigating the role of OXT in osteogenesis, it was demonstrated that OXT promoted osteogenic differentiation in BMSCs derived from both cyclic adult (12 months old) and acyclic aging (24 months old) female Wistar rats cultured in osteogenic medium. OXT treatment led to an increase in the expression of both *OXT* and *OXTR*, and anticipated mineralization, and enhanced the gene expression of *bone morphogenetic protein 2* (*BMP2*), *bone sialoprotein* (*BSP*), *osteopontin* (*OPN*), and *osteocalcin* (*OCN*) in both rat populations [87]. Consistent with these findings, OXT also promoted osteogenic differentiation in human periodontal ligament SCs (PDLSCs) [51]. PDLSCs, which express OXTR, increased their migration and proliferation ability upon OXT treatment; in addition, OXT enhanced mineralized nodule formation and calcium deposition and significantly upregulated the expression of osteogenesis-related markers, such as *alkaline phosphatase* (*ALP*), *collagen I* (*Col I*), *runt-related transcription factor 2* (*RUNX2*), *OPN*, and *OCN*. Finally, the authors suggested that OXT exerts its function through the phosphorylation of ERK and AKT proteins [51].

Another study partially challenges the previously reported findings [88]: here, human dental pulp-derived stromal cells (DPSCs) were shown to express OXTR [89]. Upon blocking OXTR using specific antagonists or siRNA, the authors observed an increase in calcium deposition and in the expression of markers of osteogenic (*BMP2*, *OPN*, *OCN*, and *RUNX2*) and odontogenic processes (*dentin matrix acidic phosphoprotein 1*, *DMP1*, and *dentin sialophosphoprotein*, *DSPP*), depending on whether OXT was used together with the osteogenic induction medium or in the basal culture medium [88]. Surprisingly, treatment of the DPSCs with OXT still resulted in an increase in osteogenic differentiation, albeit to a lesser extent.

Moreover, it was found that OXTR is involved in extracellular matrix (ECM) remodeling through modulating the expression of genes related to ECM homeostasis, probably through the Yes-associated protein (YAP) signaling pathway [88]. The results on adipogenesis, osteogenesis, and odontogenesis in cells treated with OXT are summarized in Figure 3 and Table 3.

Table 3. Effects of oxytocin on adipogenic, chondrogenic, osteogenic, and odontogenic differentiation.

Oxytocin Formulation/Dose	Species and Cell/ Animal Model	Treatment	Cellular Mechanism *	Effect(s) *	Ref.
OXT: in vitro 30 nM; in vivo 1 mg/kg	in vitro: h ADSCs and h BMSCs in vivo: ovariectomized eight-week-old C57Bl/6J mice and rats	in vitro: in osteogenic or adipogenic differentiation medium; in vivo: daily treatment for 8 weeks	↑ OXTR mRNA during osteogenesis ↓ OXTR mRNA during adipogenesis ↑ mineral deposits ↑ ALP activity ↑ PDPN mRNA ↓ lipid droplet formation ↓ GPDH activity ↓ FABP4 mRNA ↑ ERK1 and ERK2 phosphorylation	↓ adipogenesis ↑ osteogenesis	[49]
OXT: 100 nM	m Sca-1$^+$ and Sca-1$^-$ cells	30 days	Sca-1$^+$ cells showed scattered cellular aggregates with an adipocytic phenotype	↑ adipogenesis	[56]
OXT: 100 nM	m P19 ESCs	20 days after aggregation in the presence of adipogenic differentiation medium	=lipid droplet formation =PPARγ mRNA	=adipogenesis	[61]
OXT: 1 nM	m 3T3-L1 cells	24 h	↑ OXTR mRNA during adipocyte differentiation; ↑ glycerol release	↑ lipolysis	[84]
OXT: 100 nM	r BMSCs	3, 7, and 14 days in the presence of osteogenic medium	↑ calcium deposits at 14 days and at 17 days in cells from 12- and 24-month-old rats, respectively ↑ OXT and OXTR mRNAs in cells from both ages ↑ ALP activity in cells from 24 month-old rats ↑ BMP2, BSP, OPN, and OCN mRNAs in cells from both ages ↑ OSX and COL1A1 mRNAs in cells from 12 month-old rats ↓ OSX and COL1A1 in cells from 24 month-old rats	↑ osteogenesis	[87]
OXT: 10, 50, 100 nM	h PDLSCs	7, 14, and 21 days in the presence of osteogenic medium	↑ calcium deposits ↑ OXTR mRNA ↑ ALP, Col I, RUNX2, OPN, and OCN mRNAs ↑ ALP, Col I, and RUNX2 proteins ↑ ERK and AKT phosphorylation ↓ PI3K phosphorylation	↑ osteogenesis ERK and AKT signaling pathways are involved in OXT-induced osteogenesis	[51]

Table 3. *Cont.*

Oxytocin Formulation/Dose	Species and Cell/ Animal Model	Treatment	Cellular Mechanism *	Effect(s) *	Ref.
OXT: 30 nM	h BMSCs and h ADSCs	2D or 3D cultures were treated for 21 days in the presence of chondrogenic differentiation medium	↑ glycosaminoglycan content ↑ ACAN, COMP, SOX9, and Col X mRNAs in 2D culture especially hADSC cultures ↓ COL1A1 mRNA in 2D culture of both cell types ↑ ACAN, SOX9, and Col X mRNAs in 3D culture of h ADSCs ↑ SOX9 and Col II proteins in 3D culture of h ADSCs	↑ chondrogenesis	[45]
OXT: 300 nM	h DPSCs	2 weeks in osteogenic medium with OXTR inhibitors or OXT	↓ OXTR mRNA during osteogenesis ↑ calcium deposits by inhibiting OXTR ↑ BMP2, OPN, OCN, and RUNX2 mRNAs with OXTR inhibitor (atosiban) ↑ DMP1 and DSPP mRNAs with OXTR inhibitor (atosiban) OXT slightly increased calcium deposits OXT slightly increased BMP2, OPN, OCN, RUNX2, and DSPP mRNAs ↑ MMP1 mRNA with OXTR inhibitor (atosiban) ↓ COL1A1 mRNA with OXTR inhibitor (atosiban) ↓ MMP1 mRNA with YAP inhibitor (verteporfin) ↑ COL1A1 mRNA with YAP inhibitor (verteporfin) ↓ YAP protein in the nucleus of cells with OXTR inhibitor (atosiban) ↑ OCN, DSPP, and DMP1 mRNAs by silencing YAP ↑ calcium deposits with YAP inhibitor (verteporfin)	↑ osteogenesis ↑ odontogenesis	[88]

Table 3. *Cont.*

Oxytocin Formulation/Dose	Species and Cell/ Animal Model	Treatment	Cellular Mechanism *	Effect(s) *	Ref.
OXT: 100, 500, 1000 nM	h ADSCs	72 h in basal medium treatment lasted for the entire differentiation protocol in the presence of adipogenic or osteogenic medium	↑ OXTR mRNA OXT alone did not affect adipogenic and osteogenic differentiation ↓ lipid droplet formation in adipogenic medium ↓ PPARγ mRNA in adipogenic medium ↑ calcium deposits in osteogenic medium ↑ OCN mRNA in osteogenic medium RUNX, OPN, fibronectin, and Col I proteins in osteogenic medium ↑ BECN1 and MAP1LC3A mRNAs in osteogenic medium	↓ adipogenesis ↑ osteogenesis	[83]

Abbreviations: OXT, oxytocin; h, human; m, mouse; r, rat; 3T3-L1, 3T3-L1 preadipocytes; ADSCs, adipose-derived mesenchymal stem cells; BMSCs, bone marrow mesenchymal stem cells; DPSCs, dental pulp stem cells; P19 ESCs, P19 embryonic stem cells; PDLSCs, periodontal ligament-derived stem cells; Sca-1^{+}/Sca-1^{-} cells, Sca-1-positive/negative cells; ACAN, aggrecan; AKT, serine/threonine kinase; ALP, alkaline phosphatase; BECN1, Beclin 1; OCN, osteocalcin; BMP2, bone morphogenetic protein 2; BSP, bone sialoprotein; ERK1, extracellular signal-regulated kinase 1; COL1A1/Col I, collagen type I alpha 1 chain;/collagen type I; Col II, collagen type II; Col X, collagen type X; COMP, cartilage oligomeric matrix protein; DMP1, dentin matrix acidic phosphoprotein 1; DSPP, dentin sialophosphoprotein; ERK1, extracellular signal-regulated kinase 1; ERK2, extracellular signal-regulated kinase 2; FABP4, fatty acid-binding protein 4; GPDH, glycerol-3-phosphate dehydrogenase; MAP1LC3A, microtubule-associated protein 1 light chain 3 alpha; MAPK, mitogen-activated protein kinase; MMP1, matrix metalloproteinase 1; OPN, osteopontin; OSX, osterix; OXTR, oxytocin receptor; PDPN, podoplanin; PI3K, phosphoinositide 3-kinase; PPARγ, peroxisome proliferator-activated receptor gamma; RUNX2, runt-related transcription factor 2; Sox9, SRY-related HMG-box gene 9; YAP, Yes-associated protein. * Arrow and equal symbols indicate the effects of OXT on biological stem cell properties: ↑ (increase), ↓ (reduce), = (no effect).

6. Oxytocin in Stem Cell Chondrogenic Differentiation

The relationship between OXT and chondrogenesis was investigated in a recent study aimed at understanding whether OXT plays a role in osteoarthritis (OA) [45].

Considering the previously reported findings that demonstrated OXTR expression in chondrocytes and its reduction in patients with OA, along with evidence that OXT treatment restored levels of collagen type II (Col II), which were diminished by OA or specific treatments [87,90], a comprehensive research study, using human ADSCs and BMSCs, was conducted to demonstrate the positive effects of OXT treatment on chondrogenesis [45]. These cells are known to express OXTR and to undergo osteogenic differentiation at the expense of adipogenic differentiation when treated with OXT. When exposed to OXT in the presence of chondrogenic medium, these cells increased the glycosaminoglycan content in the extracellular environment, and increased the expression of *aggrecan* (*ACAN*), *cartilage oligomeric matrix protein* (*COMP*), *SRY-related HMG-box gene 9* (*Sox9*), and *collagen type X* (*Col X*); in contrast, OXT treatment reduced the expression of the fibrous tissue marker *collagen type I alpha 1 chain* (*COL1A1*) [45]. Also, in a 3D cell pellet culture model used to mimic the in vivo cellular condensation process, differentiated human ADSCs expressed chondrogenic markers. Cells cultured with OXT expressed Sox9 and Col II proteins, indicating the formation of a dense filamentous matrix network surrounding the cells [45]. These findings suggested a role of OXT in chondrogenic differentiation. The results on chondrogenesis in cells treated with OXT are summarized in Figure 3 and Table 3.

7. Conclusions

In this review, we highlighted the role of OXT in SC differentiation. OXT influences SC commitment toward various mesodermal lineages, such as cardiac, adipogenic, and osteogenic lineages. The activation of specific differentiation processes mediated by OXT has not yet been fully elucidated. OXT mainly exerts its effects through OXTR, which induces several cellular responses in a cell type-dependent manner [91]. Among these, OXTR stimulates the activation of phospholipase C (PLC), which in turn increases intracellular Ca^{2+} mobilization [19]. Ca^{2+} acts as second messenger and is involved in a wide range of processes relevant for the maintenance of SC properties and differentiation [92].

For instance, Ca^{2+} regulates adipocyte differentiation and has different effects at early or late stages of the process [93], and plays a key role in cardiogenic and osteogenic differentiation [94,95]. It was postulated that Ca^{2+} can regulate SC differentiation by acting on two important processes: epigenetic processes and metabolism [90]. Both processes are controlled in a highly specific manner across various SC types and are essential for maintaining SC identity and facilitating differentiation [92]. The studies available in the literature demonstrating the effects of OXT in SC commitment provide limited insight about the molecular mechanism mediated by the OXT–OXTR pathway. Thus, deeply understanding the molecular pathways activated by the OXT–OXTR pathway in the differentiation of specific SCs could be valuable for the modulation of this process in SCs and for the development of potential therapeutic applications in regenerative medicine.

Unraveling these mechanisms is not only essential for advancing regenerative therapies, but also for understanding how OXT might influence SC differentiation during the development of organs, such as the heart. This knowledge could open new perspectives for managing neonatal complications and enable early interventions in cardiovascular diseases.

Author Contributions: Conceptualization, S.C. (Silvia Canaider) and F.F. (Federica Facchin); methodology, L.P., F.F. (Federica Forcellini) and G.P.; data curation, L.P., G.P., F.F. (Federica Forcellini) and S.C. (Sara Cruciani); writing—original draft preparation, S.C. (Silvia Canaider), P.M.A. and F.F. (Federica Facchin); writing—review and editing, S.C. (Silvia Canaider), L.P., G.P., C.V., P.M.A., F.F. (Federica Facchin) and S.C. (Sara Cruciani); supervision, S.C. (Silvia Canaider) and P.M.A.; funding acquisition, C.V. All authors have read and agreed to the published version of the manuscript.

Funding: This research was funded by the Di Bella Foundation to Guglielmo Marconi 51, 40122 Bologna, Italy.

Institutional Review Board Statement: Not applicable.

Informed Consent Statement: Not applicable.

Data Availability Statement: Not applicable.

Conflicts of Interest: The authors declare no conflicts of interest.

References

1. Du Vigneaud, V.; Ressler, C.; Trippett, S. The sequence of amino acids in oxytocin, with a proposal for the structure of oxytocin. *J. Biol. Chem.* **1953**, *205*, 949–957. [CrossRef]
2. Meyer-Lindenberg, A.; Domes, G.; Kirsch, P.; Heinrichs, M. Oxytocin and vasopressin in the human brain: Social neuropeptides for translational medicine. *Nat. Rev. Neurosci.* **2011**, *12*, 524–538. [CrossRef]
3. Carter, C.S.; Kenkel, W.M.; MacLean, E.L.; Wilson, S.R.; Perkeyble, A.M.; Yee, J.R.; Ferris, C.F.; Nazarloo, H.P.; Porges, S.W.; Davis, J.M.; et al. Is oxytocin "Nature's medicine"? *Pharmacol. Rev.* **2020**, *72*, 829–861. [CrossRef]
4. Augustine, R.A.; Seymour, A.J.; Campbell, R.E.; Grattan, D.R.; Brown, C.H. Integrative neuro-humoral regulation of oxytocin neuron activity in pregnancy and lactation. *J. Neuroendocrinol.* **2018**, *30*, e12569. [CrossRef] [PubMed]
5. Kimura, T.; Saji, F.; Nishimori, K.; Ogita, K.; Nakamura, H.; Koyama, M.; Murata, Y. Molecular regulation of the oxytocin receptor in peripheral organs. *J. Mol. Endocrinol.* **2003**, *30*, 109–115. [CrossRef] [PubMed]
6. Thibonnier, M.; Berti-Mattera, L.N.; Dulin, N.; Conarty, D.M.; Mattera, R. Signal transduction pathways of the human V1-vascular, V2-renal, V3-pituitary vasopressin and oxytocin receptors. *Prog. Brain Res.* **1998**, *119*, 147–161. [CrossRef] [PubMed]
7. Arrowsmith, S.; Wray, S. Oxytocin: Its mechanism of action and receptor signaling in the myometrium. *J. Neuroendocrinol.* **2014**, *26*, 356–369. [CrossRef]
8. Blanks, A.M.; Thornton, S. The role of oxytocin in parturition. *BJOG Int. J. Obstet. Gynaecol.* **2003**, *110*, 46–51. [CrossRef]
9. Perisic, M.; Woolcock, K.; Hering, A.; Mendel, H.; Muttenthaler, M. Oxytocin and vasopressin signaling in health and disease. *Trends Biochem. Sci.* **2004**, *49*, 361–367. [CrossRef]
10. Camerino, C. The Long Way of Oxytocin from the Uterus to the Heart in 70 Years from Its Discovery. *Int. J. Mol. Sci.* **2023**, *24*, 2556. [CrossRef]
11. Carter, C.S. Oxytocin and love: Myths, metaphors and mysteries. *Compr. Psychoneuroendocrinology* **2022**, *9*, 100107. [CrossRef] [PubMed]
12. Neumann, I.D.; Landgraf, R. Balance of Brain Oxytocin and Vasopressin: Implications for Anxiety, Depression, and Social Behaviors. *Trends Neurosci.* **2012**, *35*, 649–659. [CrossRef] [PubMed]
13. Bordt, E.A.; Smith, C.J.; Demarest, T.G.; Bilbo, S.D.; Kingsbury, M.A. Mitochondria, Oxytocin, and Vasopressin: Unfolding the Inflammatory Protein Response. *Neurotox. Res.* **2019**, *36*, 239–256. [CrossRef] [PubMed]
14. Szeto, A.; McCabe, P.M.; Nation, D.A.; Tabak, B.A.; Rossetti, M.A.; McCullough, M.E.; Schneiderman, N.; Mendez, A.J. Evaluation of enzyme immunoassay and radioimmunoassay methods for the measurement of plasma oxytocin. *Psychosom. Med.* **2011**, *73*, 393–400. [CrossRef]
15. Amri, E.Z.; Pisani, D.F. Control of bone and fat mass by oxytocin. *Horm. Mol. Biol. Clin. Investig.* **2016**, *28*, 95–104. [CrossRef]
16. Russell, G.; Lightman, S. The human stress response. *Nat. Rev. Endocrinol.* **2019**, *15*, 525–534. [CrossRef]
17. Gimpl, G.; Fahrenholz, F. The oxytocin receptor system: Structure, function, and regulation. *Physiol. Rev.* **2001**, *81*, 629–683. [CrossRef]
18. Juul, K.V.; Bichet, D.G.; Nielsen, S.; Nørgaard, J.P. The physiological and pathophysiological functions of renal and extrarenal vasopressin V2 receptors. *Am. J. Physiol. Renal Physiol.* **2014**, *306*, 931–940. [CrossRef]
19. Gutkowska, J.; Jankowski, M. Oxytocin revisited: Its role in cardiovascular regulation. *J. Neuroendocrinol.* **2012**, *24*, 599–608. [CrossRef]
20. Camerino, C. The New Frontier in Oxytocin Physiology: The Oxytonic Contraction. *Int. J. Mol. Sci.* **2020**, *21*, 5144. [CrossRef]
21. Camerino, C. Oxytocin Involvement in Body Composition Unveils the True Identity of Oxytocin. *Int. J. Mol. Sci.* **2021**, *22*, 6383. [CrossRef] [PubMed]
22. Conte, E.; Romano, A.; De Bellis, M.; de Ceglia, M.; Carratù, M.R.; Gaetani, S.; Maqoud, F.; Tricarico, D.; Camerino, C. Oxtr/TRPV1 expression and acclimation of skeletal muscle to cold-stress in male mice. *J. Endocrinol.* **2021**, *249*, 135–148. [CrossRef] [PubMed]
23. Fukushima, A.; Kataoka, N.; Nakamura, K. An oxytocinergic neural pathway that stimulates thermogenic and cardiac sympathetic outflow. *Cell Rep.* **2022**, *40*, 111380. [CrossRef] [PubMed]
24. Yamasue, H.; Domes, G. Oxytocin and Autism Spectrum Disorders. *Curr. Top. Behav. Neurosci.* **2018**, *35*, 449–465. [CrossRef]
25. Gordon, I.; Zagoory-Sharon, O.; Schneiderman, I.; Leckman, J.F.; Weller, A.; Feldman, R. Oxytocin and cortisol in romantically unattached young adults: Associations with bonding and psychological distress. *Psychophysiology* **2008**, *45*, 349–352. [CrossRef]
26. Jobst, A.; Krause, D.; Maiwald, C.; Härtl, K.; Myint, A.M.; Kästner, R.; Obermeier, M.; Padberg, F.; Brücklmeier, B.; Weidinger, E.; et al. Oxytocin course over pregnancy and postpartum period and the association with postpartum depressive symptoms. *Arch. Womens Ment. Health* **2016**, *19*, 571–579. [CrossRef]
27. Yoon, S.; Kim, Y.K. The Role of the Oxytocin System in Anxiety Disorders. *Adv. Exp. Med. Biol.* **2020**, *1191*, 103–120. [CrossRef]

28. Goh, K.K.; Chen, C.H.; Lane, H.Y. Oxytocin in Schizophrenia: Pathophysiology and Implications for Future Treatment. *Int. J. Mol. Sci.* **2021**, *22*, 2146. [CrossRef]
29. Leckman, J.F.; Goodman, W.K.; North, W.G.; Chappell, P.B.; Price, L.H.; Pauls, D.L.; Anderson, G.M.; Riddle, M.A.; McSwiggan-Hardin, M.; McDougle, C.J.; et al. Elevated cerebrospinal fluid levels of oxytocin in obsessive-compulsive disorder: Comparison with tourette's syndrome and healthy controls. *Arch. Gen. Psychiatry* **1994**, *51*, 782–792. [CrossRef]
30. Marazziti, D.; Baroni, S.; Giannaccini, G.; Catena-Dell'Osso, M.; Piccinni, A.; Massimetti, G.; Dell'Osso, L. Plasma oxytocin levels in untreated adult obsessive-compulsive disorder patients. *Neuropsychobiology* **2015**, *72*, 74–80. [CrossRef]
31. Bosch, O.J.; Meddle, S.L.; Beiderbeck, D.I.; Douglas, A.J.; Neumann, I.D. Brain oxytocin correlates with maternal aggression: Link to anxiety. *J. Neurosci.* **2005**, *25*, 6807–6815. [CrossRef] [PubMed]
32. Xu, H.; Fu, S.; Chen, Q.; Gu, M.; Zhou, J.; Liu, C.; Chen, Y.; Wang, Z. The function of oxytocin: A potential biomarker for prostate cancer diagnosis and promoter of prostate cancer. *Oncotarget* **2017**, *8*, 31215–31226. [CrossRef]
33. Guastella, A.J.; Boulton, K.A.; Whitehouse, A.J.O.; Song, Y.J.; Thapa, R.; Gregory, S.G.; Pokorski, I.; Granich, J.; DeMayo, M.M.; Ambarchi, Z.; et al. The effect of oxytocin nasal spray on social interaction in young children with autism: A randomized clinical trial. *Mol. Psychiatry* **2023**, *28*, 834–842. [CrossRef] [PubMed]
34. Zakrzewski, W.; Dobrzyński, M.; Szymonowicz, M.; Rybak, Z. Stem cells: Past, present, and future. *Stem Cell Res. Ther.* **2019**, *10*, 68. [CrossRef] [PubMed]
35. Wobus, A.M.; Boheler, K.R. Embryonic stem cells: Prospects for developmental biology and cell therapy. *Physiol. Rev.* **2005**, *85*, 635–678. [CrossRef] [PubMed]
36. Mannino, G.; Russo, C.; Maugeri, G.; Musumeci, G.; Vicario, N.; Tibullo, D.; Giuffrida, R.; Parenti, R.; Lo Furno, D. Adult stem cell niches for tissue homeostasis. *J. Cell Physiol.* **2022**, *237*, 239–257. [CrossRef]
37. Kolios, G.; Moodley, Y. Introduction to stem cells and regenerative medicine. *Respiration* **2013**, *85*, 3–10. [CrossRef]
38. Suman, S.; Domingues, A.; Ratajczak, J.; Ratajczak, M.Z. Potential Clinical Applications of Stem Cells in Regenerative Medicine. *Adv. Exp. Med. Biol.* **2019**, *1201*, 1–22. [CrossRef]
39. Amabile, G.; Meissner, A. Induced pluripotent stem cells: Current progress and potential for regenerative medicine. *Trends Mol. Med.* **2009**, *15*, 59–68. [CrossRef]
40. Sarkar, A.; Saha, S.; Paul, A.; Maji, A.; Roy, P.; Maity, T.K. Understanding stem cells and its pivotal role in regenerative medicine. *Life Sci.* **2021**, *273*, 119270. [CrossRef]
41. Takahashi, K.; Yamanaka, S. Induction of pluripotent stem cells from mouse embryonic and adult fibroblast cultures by defined factors. *Cell* **2006**, *126*, 663–676. [CrossRef] [PubMed]
42. Margiana, R.; Markov, A.; Zekiy, A.O.; Hamza, M.U.; Al-Dabbagh, K.A.; Al-Zubaidi, S.H.; Hameed, N.M.; Ahmad, I.; Sivaraman, R.; Kzar, H.H.; et al. Clinical application of mesenchymal stem cell in regenerative medicine: A narrative review. *Stem Cell Res. Ther.* **2022**, *13*, 366. [CrossRef] [PubMed]
43. Petrosyan, A.; Martins, P.N.; Solez, K.; Uygun, B.E.; Gorantla, V.S.; Orlando, G. Regenerative medicine applications: An overview of clinical trials. *Front. Bioeng. Biotechnol.* **2022**, *10*, 942750. [CrossRef] [PubMed]
44. Gutkowska, J.; Jankowski, M.; Antunes-Rodrigues, J. The role of oxytocin in cardiovascular regulation. *Braz. J. Med. Biol. Res.* **2014**, *47*, 206–214. [CrossRef] [PubMed]
45. Roux, C.H.; Pisani, D.F.; Gillet, P.; Fontas, E.; Yahia, H.B.; Djedaini, M.; Ambrosetti, D.; Michiels, J.F.; Panaia-Ferrari, P.; Breuil, V.; et al. Oxytocin Controls Chondrogenesis and Correlates with Osteoarthritis. *Int. J. Mol. Sci.* **2020**, *21*, 3966. [CrossRef]
46. Elabd, C.; Cousin, W.; Upadhyayula, P.; Chen, R.Y.; Chooljian, M.S.; Li, J.; Kung, S.; Jiang, K.P.; Conboy, I.M. Oxytocin is an age-specific circulating hormone that is necessary for muscle maintenance and regeneration. *Nat. Commun.* **2014**, *5*, 4082. [CrossRef]
47. Luo, D.; Jin, B.; Zhai, X.; Li, J.; Liu, C.; Guo, W.; Li, J. Oxytocin promotes hepatic regeneration in elderly mice. *iScience* **2021**, *24*, 102125. [CrossRef]
48. Jafarzadeh, N.; Javeri, A.; Khaleghi, M.; Taha, M.F. Oxytocin improves proliferation and neural differentiation of adipose tissue-derived stem cells. *Neurosci. Lett.* **2014**, *564*, 105–110. [CrossRef]
49. Elabd, C.; Basillais, A.; Beaupied, H.; Breuil, V.; Wagner, N.; Scheideler, M.; Zaragosi, L.E.; Massiéra, F.; Lemichez, E.; Trajanoski, Z.; et al. Oxytocin controls differentiation of human mesenchymal stem cells and reverses osteoporosis. *Stem Cells* **2008**, *26*, 2399–2407. [CrossRef]
50. Rosen, C.J.; Bouxsein, M.L. Mechanisms of disease: Is osteoporosis the obesity of bone? *Nat. Clin. Pract. Rheumatol.* **2006**, *2*, 35–43. [CrossRef]
51. Ge, B.; Liu, H.; Liang, Q.; Shang, L.; Wang, T.; Ge, S. Oxytocin facilitates the proliferation, migration and osteogenic differentiation of human periodontal stem cells in vitro. *Arch. Oral. Biol.* **2019**, *99*, 126–133. [CrossRef] [PubMed]
52. Seo, B.M.; Miura, M.; Gronthos, S.; Bartold, P.M.; Batouli, S.; Brahim, J.; Young, M.; Robey, P.G.; Wang, C.Y.; Shi, S. Investigation of multipotent postnatal stem cells from human periodontal ligament. *Lancet* **2004**, *364*, 149–155. [CrossRef] [PubMed]
53. Paquin, J.; Danalache, B.A.; Jankowski, M.; McCann, S.M.; Gutkowska, J. Oxytocin induces differentiation of P19 embryonic stem cells to cardiomyocytes. *Proc. Natl. Acad. Sci. USA* **2002**, *99*, 9550–9555. [CrossRef]
54. Jankowski, M.; Danalache, B.; Wang, D.; Bhat, P.; Hajjar, F.; Marcinkiewicz, M.; Paquin, J.; McCann, S.M.; Gutkowska, J. Oxytocin in cardiac ontogeny. *Proc. Natl. Acad. Sci. USA* **2004**, *101*, 13074–13079. [CrossRef]

55. Matsuura, K.; Nagai, T.; Nishigaki, N.; Oyama, T.; Nishi, J.; Wada, H.; Sano, M.; Toko, H.; Akazawa, H.; Sato, T.; et al. Adult cardiac Sca-1-positive cells differentiate into beating cardiomyocytes. *J. Biol. Chem.* **2004**, *279*, 11384–11391. [CrossRef] [PubMed]
56. Abdel-Latif, A.; Zuba-Surma, E.K.; Case, J.; Tiwari, S.; Hunt, G.; Ranjan, S.; Vincent, R.J.; Srour, E.F.; Bolli, R.; Dawn, B. TGF-beta1 enhances cardiomyogenic differentiation of skeletal muscle-derived adult primitive cells. *Basic. Res. Cardiol.* **2008**, *103*, 514–524. [CrossRef] [PubMed]
57. Oyama, T.; Nagai, T.; Wada, H.; Naito, A.T.; Matsuura, K.; Iwanaga, K.; Takahashi, T.; Goto, M.; Mikami, Y.; Yasuda, N.; et al. Cardiac side population cells have a potential to migrate and differentiate into cardiomyocytes in vitro and in vivo. *J. Cell Biol.* **2007**, *176*, 329–341. [CrossRef]
58. Emmert, M.Y.; Emmert, L.S.; Martens, A.; Ismail, I.; Schmidt-Richter, I.; Gawol, A.; Seifert, B.; Haverich, A.; Martin, U.; Gruh, I. Higher frequencies of BCRP+ cardiac resident cells in ischaemic human myocardium. *Eur. Heart J.* **2013**, *34*, 2830–2838. [CrossRef]
59. Gutkowska, J.; Miszkurka, M.; Danalache, B.; Gassanov, N.; Wang, D.; Jankowski, M. Functional arginine vasopressin system in early heart maturation. *Am. J. Physiol. Heart Circ. Physiol.* **2007**, *293*, H2262–H2270. [CrossRef]
60. Uchida, S.; Fuke, S.; Tsukahara, T. Upregulations of Gata4 and oxytocin receptor are important in cardiomyocyte differentiation processes of P19CL6 cells. *J. Cell Biochem.* **2007**, *100*, 629–641. [CrossRef]
61. Bouchard, F.; Paquin, J. Skeletal and cardiac myogenesis accompany adipogenesis in P19 embryonal stem cells. *Stem Cells Dev.* **2009**, *18*, 1023–1032. [CrossRef] [PubMed]
62. Danalache, B.A.; Paquin, J.; Donghao, W.; Grygorczyk, R.; Moore, J.C.; Mummery, C.L.; Gutkowska, J.; Jankowski, M. Nitric oxide signaling in oxytocin-mediated cardiomyogenesis. *Stem Cells* **2007**, *25*, 679–688. [CrossRef]
63. Ybarra, N.; del Castillo, J.R.; Troncy, E. Involvement of the nitric oxide-soluble guanylyl cyclase pathway in the oxytocin-mediated differentiation of porcine bone marrow stem cells into cardiomyocytes. *Nitric Oxide* **2011**, *24*, 25–33. [CrossRef] [PubMed]
64. Ybarra, N.; Vincent, P.; Smith, L.C.; Troncy, E. Oxytocin improves the expression of cardiac specific markers in porcine bone marrow stem cells differentiation. *Res. Vet. Sci.* **2015**, *98*, 42–50. [CrossRef]
65. Fathi, F.; Murasawa, S.; Hasegawa, S.; Asahara, T.; Kermani, A.J.; Mowla, S.J. Cardiac differentiation of P19CL6 cells by oxytocin. *Int. J. Cardiol.* **2009**, *134*, 75–81. [CrossRef]
66. Hatami, L.; Valojerdi, M.R.; Mowla, S.J. Effects of oxytocin on cardiomyocyte differentiation from mouse embryonic stem cells. *Int. J. Cardiol.* **2007**, *117*, 80–89. [CrossRef]
67. Gassanov, N.; Devost, D.; Danalache, B.; Noiseux, N.; Jankowski, M.; Zingg, H.H.; Gutkowska, J. Functional activity of the carboxyl-terminally extended oxytocin precursor Peptide during cardiac differentiation of embryonic stem cells. *Stem Cells* **2008**, *26*, 45–54. [CrossRef]
68. Danalache, B.A.; Gutkowska, J.; Slusarz, M.J.; Berezowska, I.; Jankowski, M. Oxytocin-Gly-Lys-Arg: A novel cardiomyogenic peptide. *PLoS ONE* **2010**, *5*, e13643. [CrossRef] [PubMed]
69. Danalache, B.A.; Yu, C.; Gutkowska, J.; Jankowski, M. Oxytocin-Gly-Lys-Arg stimulates cardiomyogenesis by targeting cardiac side population cells. *J. Endocrinol.* **2014**, *220*, 277–289. [CrossRef]
70. Kim, Y.S.; Kwon, J.S.; Hong, M.H.; Kim, J.; Song, C.H.; Jeong, M.H.; Cho, J.G.; Park, J.C.; Kang, J.C.; Ahn, Y. Promigratory activity of oxytocin on umbilical cord blood-derived mesenchymal stem cells. *Artif. Organs* **2010**, *34*, 453–461. [CrossRef]
71. Kim, Y.S.; Ahn, Y.; Kwon, J.S.; Cho, Y.K.; Jeong, M.H.; Cho, J.G.; Park, J.C.; Kang, J.C. Priming of mesenchymal stem cells with oxytocin enhances the cardiac repair in ischemia/reperfusion injury. *Cells Tissues Organs* **2012**, *195*, 428–442. [CrossRef] [PubMed]
72. Taha, M.F.; Javeri, A.; Karimipour, M.; Yamaghani, M.S. Priming with oxytocin and relaxin improves cardiac differentiation of adipose tissue-derived stem cells. *J. Cell Biochem.* **2019**, *120*, 5825–5834. [CrossRef] [PubMed]
73. Lin, H.Y.; Lee, D.C.; Wang, H.D.; Chi, Y.H.; Chiu, I.M. Activation of FGF1B Promoter and FGF1 Are Involved in Cardiogenesis Through the Signaling of PKC, but Not MAPK. *Stem Cells Dev.* **2015**, *24*, 2853–2863. [CrossRef] [PubMed]
74. Wasserman, A.H.; Huang, A.R.; Lewis-Israeli, Y.R.; Dooley, M.D.; Mitchell, A.L.; Venkatesan, M.; Aguirre, A. Oxytocin promotes epicardial cell activation and heart regeneration after cardiac injury. *Front. Cell Dev. Biol.* **2022**, *10*, 985298. [CrossRef]
75. Adamo, S.; Pigna, E.; Lugarà, R.; Moresi, V.; Coletti, D.; Bouché, M. Skeletal Muscle: A Significant Novel Neurohypophyseal Hormone-Secreting Organ. *Front. Physiol.* **2019**, *9*, 1885. [CrossRef]
76. Nervi, C.; Benedetti, L.; Minasi, A.; Molinaro, M.; Adamo, S. Arginine-vasopressin induces differentiation of skeletal myogenic cells and up-regulation of myogenin and Myf-5. *Cell Growth Differ.* **1995**, *6*, 81–89.
77. Breton, C.; Haenggeli, C.; Barberis, C.; Heitz, F.; Bader, C.R.; Bernheim, L.; Tribollet, E. Presence of functional oxytocin receptors in cultured human myoblasts. *J. Clin. Endocrinol. Metab.* **2002**, *87*, 1415–1418. [CrossRef] [PubMed]
78. Berio, E.; Divari, S.; Starvaggi Cucuzza, L.; Biolatti, B.; Cannizzo, F.T. 17β-estradiol upregulates oxytocin and the oxytocin receptor in C2C12 myotubes. *PeerJ* **2017**, *5*, e3124. [CrossRef]
79. Fasolino, I.; Guarino, V.; Cirillo, V.; Ambrosio, L. 5-Azacytidine-mediated hMSC behavior on electrospun scaffolds for skeletal muscle regeneration. *J. Biomed. Mater. Res. A* **2017**, *105*, 2551–2561. [CrossRef]
80. Zhang, Z.; Zhao, L.D.; Johnson, S.E.; Rhoads, M.L.; Jiang, H.; Rhoads, R.P. Oxytocin is involved in steroid hormone-stimulated bovine satellite cell proliferation and differentiation in vitro. *Domest. Anim. Endocrinol.* **2019**, *66*, 1–13. [CrossRef]
81. Wang, M.; Lan, L.; Li, T.; Li, J.; Li, Y. The effect of oxytocin on osseointegration of titanium implant in ovariectomized rats. *Connect. Tissue Res.* **2016**, *57*, 220–225. [CrossRef] [PubMed]
82. Li, B.; Jiang, Y.; Sun, J.; Liang, J.; Jin, Y. MR spectroscopy for assessing the effects of oxytocin on marrow adipogenesis induced by glucocorticoid in rabbits. *Acta Radiol.* **2016**, *57*, 701–707. [CrossRef] [PubMed]

83. Petrocelli, G.; Abruzzo, P.M.; Pampanella, L.; Tassinari, R.; Marini, S.; Zamagni, E.; Ventura, C.; Facchin, F.; Canaider, S. Oxytocin Modulates Osteogenic Commitment in Human Adipose-Derived Stem Cells. *Int. J. Mol. Sci.* **2023**, *24*, 10813. [CrossRef] [PubMed]
84. Yi, K.J.; So, K.H.; Hata, Y.; Suzuki, Y.; Kato, D.; Watanabe, K.; Aso, H.; Kasahara, Y.; Nishimori, K.; Chen, C.; et al. The regulation of oxytocin receptor gene expression during adipogenesis. *J. Neuroendocrinol.* **2015**, *27*, 335–342. [CrossRef]
85. Camerino, C. Low sympathetic tone and obese phenotype in oxytocin-deficient mice. *Obesity* **2009**, *17*, 980–984. [CrossRef]
86. Takayanagi, Y.; Kasahara, Y.; Onaka, T.; Takahashi, N.; Kawada, T.; Nishimori, K. Oxytocin receptor-deficient mice developed late-onset obesity. *Neuroreport* **2008**, *19*, 951–955. [CrossRef]
87. Santos, L.F.; Singulani, M.P.; Stringhetta-Garcia, C.T.; Oliveira, S.H.P.; Chaves-Neto, A.H.; Dornelles, R.C.M. Oxytocin effects on osteoblastic differentiation of Bone Marrow Mesenchymal Stem Cells from adult and aging female Wistar rats. *Exp. Gerontol.* **2018**, *113*, 58–63. [CrossRef]
88. Jung, J.W.; Park, S.Y.; Seo, E.J.; Jang, I.H.; Park, Y.; Lee, D.; Kim, D.; Kim, J.M. Functional expression of oxytocin receptors in pulp-dentin complex. *Biomaterials* **2023**, *293*, 121977. [CrossRef]
89. Kumagai, T.; Shindo, S.; Takeda, K.; Shiba, H. Oxytocin suppresses CXCL10 production in TNF-α-stimulated human dental pulp stem cells. *Cell Biol. Int.* **2022**, *46*, 1530–1535. [CrossRef]
90. Wu, Y.; Wu, T.; Xu, B.; Xu, X.; Chen, H.; Li, X. Oxytocin prevents cartilage matrix destruction via regulating matrix metalloproteinases. *Biochem. Biophys. Res. Commun.* **2017**, *486*, 601–606. [CrossRef]
91. McKay, E.C.; Counts, S.E. Oxytocin Receptor Signaling in Vascular Function and Stroke. *Front. Neurosci.* **2020**, *14*, 574499. [CrossRef] [PubMed]
92. Snoeck, H.W. Calcium regulation of stem cells. *EMBO Rep.* **2020**, *21*, e50028. [CrossRef] [PubMed]
93. Zhang, F.; Ye, J.; Zhu, X.; Wang, L.; Gao, P.; Shu, G.; Jiang, Q.; Wang, S. Anti-Obesity Effects of Dietary Calcium: The Evidence and Possible Mechanisms. *Int. J. Mol. Sci.* **2019**, *20*, 3072. [CrossRef] [PubMed]
94. Tyser, R.C.; Miranda, A.M.; Chen, C.M.; Davidson, S.M.; Srinivas, S.; Riley, P.R. Calcium handling precedes cardiac differentiation to initiate the first heartbeat. *eLife* **2016**, *5*, e17113. [CrossRef]
95. Barradas, A.M.; Fernandes, H.A.; Groen, N.; Chai, Y.C.; Schrooten, J.; van de Peppel, J.; van Leeuwen, J.P.; van Blitterswijk, C.A.; de Boer, J. A calcium-induced signaling cascade leading to osteogenic differentiation of human bone marrow-derived mesenchymal stromal cells. *Biomaterials* **2012**, *33*, 3205–3215. [CrossRef]

current issues in molecular biology

MDPI

Review

Is Oxytocin a Contributor to Behavioral and Metabolic Features in Prader–Willi Syndrome?

Maria Petersson [1,2,*] and Charlotte Höybye [1,2]

[1] Department of Endocrinology, Karolinska University Hospital, 171 76 Stockholm, Sweden; charlotte.hoybye@ki.se
[2] Department of Molecular Medicine and Surgery, Karolinska Institutet, 171 76 Stockholm, Sweden
* Correspondence: maria.petersson@ki.se

Abstract: Prader–Willi Syndrome (PWS) is a rare genetic disorder typically characterized by decreased social interaction, hyperphagia, poor behavioral control and temper tantrums, together with a high risk of morbid obesity unless food intake is controlled. The genetic defects that cause PWS include paternal 15q deletion (estimated in 60% of cases), chromosome 15 maternal uniparental disomy (UPD) (estimated in 35% of cases) and imprinting defects and translocations. Several studies indicate an oxytocin deficiency in PWS. Oxytocin is a hypothalamic nonapeptide with receptors located in the brain and in various other tissues in the body. It acts as a neuropeptide in several brain areas of great importance for behavioral and metabolic effects, as well as a neurohypophyseal hormone released into the circulation. Oxytocin in both rats and humans has strong and long-lasting behavioral and metabolic effects. Thus, an oxytocin deficiency might be involved in several of the behavioral and metabolic symptoms characterizing PWS. Treatment with oxytocin has, in some studies, shown improvement in psycho-social behavior and hyperphagia in individuals with PWS. This review focus on the behavioral and metabolic effects of oxytocin, the symptoms of a potential oxytocin deficiency in PWS and the effects of oxytocin treatment.

Keywords: PWS; oxytocin; metabolism; weight; behavior; social interaction; hyperphagia

1. Introduction

Prader–Willi syndrome (PWS) is a rare, genetic, neurodevelopmental disorder with a complex phenotype. It occurs in approximately 1 in 10,000 to 1 in 30,000 births and is caused by a defect in the paternal expression of imprinted genes in chromosome 15 in the region q11–13. The most common defect (60%) is paternal deletion, followed by maternal disomy (35%). The remaining defects are imprinting defects, translocations, or inversions. The majority of cases occur spontaneously, and the incidence is anticipated to be the same worldwide, although data are only available for a few countries. PWS is characterized by muscular hypotonia, motor, and cognitive developmental delays, mild to moderate intellectual disability, behavioral problems, including anxiety, compulsive and controlling behavior and temper outbursts. Hormone deficiencies are common, in particular, hypogonadism and growth hormone (GH) deficiency, leading to a short stature unless GH is administered during childhood. Most infants with PWS have feeding problems, which are switched to hyperphagia at the age of 4 to 6 years. Body composition is abnormal with more body fat than muscle mass, and the metabolic rate is low. Due to the hyperphagia, a strict diet and supervised food intake are mandatory to prevent extreme obesity and related comorbidity [1].

PWS is believed to be a hypothalamic disease, and studies of postmortem brain tissue from adults with PWS have shown a 42% decrease in number and a 54% decrease in volume of oxytocin-expressing neurons in the paraventricular nucleus (PVN) of the hypothalamus suggested to be the cause for the insatiable hunger and decreased satiety in PWS [2,3].

Very low concentrations of oxytocin in lymphoblastoid cells and in postmortem frontal cortex tissue were also demonstrated [4]. Thus, low plasma levels of oxytocin would be expected, but oxytocin is often reported to be increased in children with PWS. In a study of plasma levels of oxytocin, 23 children with PWS were compared with 18 healthy unrelated siblings matched for age and with a similar gender ratio and BMI. The children with PWS were found to have more than twice as high levels of plasma oxytocin as compared to unrelated siblings [5], and analyses of serum oxytocin in adults with PWS showed similar concentrations as in controls, but, in relation to their obesity, the concentrations were low [6]. In addition, analyses of oxytocin in cerebrospinal fluid showed higher concentrations in adolescents and adults with PWS compared to the controls [7].

The 15q11–13 region contains both protein-coding DNA genes and non-coding RNA genes. Mutations or deletions of genes in this region (Magel2 and Necdin) are in mouse models associated with hypotonia, developmental delay, hypogonadism, hyperphagia and impaired social skills. A pronounced dysregulation of the signaling pathways for oxytocin has been shown in a mouse model of PWS, and, in this model, an early postnatal administration of oxytocin improved social behavior [8].

In humans with PWS, several trials studying the intranasal administration of oxytocin have been performed. The study cohorts varied regarding age, oxytocin dose and gender, and the studies have shown varying degrees of efficacy on hyperphagia, anxiousness, distress and social behavior. In a recent systematic review and meta-analysis on the efficacy of intranasal oxytocin in individuals with PWS, four studies with 56 individuals receiving oxytocin, and 61 individuals receiving placebo, were included [9]. Pooled data from 92 individuals did not show a better effect of oxytocin on hyperphagia compared to placebo. In the same meta-analysis, pooled data from 94 individuals showed that the effect of oxytocin on weight was not different from placebo. However, in studies of children <11 years, oxytocin significantly reduced hyperphagia, especially in boys, whereas no effect was observed in individuals between 13 and 29 years suggesting a better effect in younger children [9].

Oxytocin is a neurotransmitter and neurohypophyseal hormone. Besides the classical effects on uterine contraction and milk ejection during labor and breastfeeding, oxytocin has many other effects. Oxytocin has behavioral and metabolic effects in both sexes and has been suggested to influence, for example, anxiety and compulsive and social behavior, as well as eating behavior. It also interacts with a number of other neuropeptides and hormones of importance for behavior and metabolism. Thus, a decrease or change in oxytocin synthesis or distribution could be involved in several of the symptoms in PWS described above. In the present review, similarities between low levels of oxytocin and symptoms in PWS and their possible relationships will be discussed.

2. Clinical Features and Nutritional Phases in PWS

PWS is a multisymptomatic syndrome with clinical features varying with age (Table 1) [10,11]. Typical dysmorphic features include a long narrow face, narrow bifrontal diameter, almond-shaped palpebral fissures, and a small mouth with a thin upper lip, small feet for height and age and small narrow hands with a straight ulnar border. The neonatal period is characterized by profound muscular hypotonia, a poor sucking reflex and failure to thrive. During infancy, muscular hypotonia improves, and a mild to moderate intellectual dysfunction and delayed psychomotor development become increasingly apparent along with impaired skeletal growth. Central and obstructive sleep apnea as well as excessive daytime sleepiness are common. During childhood, hyperphagia develops and a constantly supervised and restricted diet together with scheduled daily physical activities are necessary to avoid morbid obesity. Body composition is abnormal in PWS with more body fat than lean body mass. Most patients with PWS have genital hypoplasia and hypogonadism, which can be of both primary gonadal and hypothalamic origin. Delayed and incomplete pubertal development is common, and a decreased or absent growth spurt results in reduced final adult height. Typically starting in childhood are temper tantrums, a stubborn and controlling behavior, compulsive–obsessive features and a very strong preference for routines. Psychiatric and behavioral problems become more prominent with increasing age and are, together with the patient's hyperphagia and cognitive disabilities, the main limitations to independent living for adults with PWS.

Table 1. Most frequent characteristics of PWS.

Behavioral challenges	Feeding difficulties and failure to thrive in infancy
	Hyperphagia
	Poor social skills
	Temper outbursts
	Anxiety
	Compulsive behaviors
Cognition/intellectual capacity	Mild to moderate intellectual and cognitive dysfunction
Physical features	Profound hypotonia in infancy
	Short stature
	Genital hypoplasia
	Obesity
Endocrine changes	Growth hormone deficiency
	Hypogonadism

PWS is characterized by severe hypotonia and failure to thrive in infancy, which is replaced by progressive hyperphagia and a high risk of development of morbid obesity. For many years, only those two nutritional phases were believed to be present in PWS. However, in 2011, Miller et al. described seven different nutritional phases, five main phases and two sub-phases [12]. This description of the nutritional phases in PWS has greatly improved our understanding of its natural history. Briefly, phase 0 occurs in utero, with decreased fetal movements and growth restriction. In phase 1, the infant is hypotonic and not obese. In sub-phase 1a, the infant has difficulty eating with or without failure to thrive (birth to 15 months). In sub-phase 1b, the infant grows steadily, and weight is increasing at a normal rate (the median age of onset: 9 months). In phase 2, weight increases. In sub-phase 2a, the weight increases without a significant change in appetite or caloric intake (the median age of onset: 2 years), while, in sub-phase 2b, the weight gain is associated with an increased interest in food (the median age of onset: 4.5 years). Phase 3 is characterized by hyperphagia, with the typical food-seeking behavior and a lack of satiety (the median age of onset: 8 years). Some adults progress to phase 4 where the appetite is no longer insatiable. PWS is a complex model to study and understand the development of obesity and appetite regulation due to the different nutritional phases that individuals with PWS undergo, from anorexia and failure to thrive in phase 1a to hyperphagia and the lack of satiety in phase 3.

3. Oxytocin

Oxytocin is a nonapeptide produced within two hypothalamic nuclei, the supraoptical nucleus (SON) and the PVN. There is a broad spectrum of nerves connecting the neurons within the PVN with neurons in several brain areas including other parts of the hypothalamus such as the ventromedial nucleus and the arcuate nucleus and the brainstem where also oxytocin receptors are located. Oxytocin receptors are also found in many other areas within the body for example the endocrine pancreas, the adipose tissue, liver, muscles as well as the gastrointestinal tract [13–17].

In addition to being released during parturition and breastfeeding oxytocin is released in response to a number of other stimuli, induces a multitude of effects and influences several other hormones and neurotransmitters. Besides being released as a hormone from the neurohypophysis, oxytocin is as mentioned above, released into the brain and then not only from the axon into the synaptic cleft but also from the dendrites spreading oxytocin in the extracellular fluid making it possible to reach larger areas. It has even been shown that oxytocin may be produced within the dendrites [18].

Oxytocin is measurable in human plasma but of importance to remember is that concentrations in peripheral blood do not always reflect the concentrations within the CNS or the cerebrospinal fluid.

In this review we will focus on the effects of oxytocin on metabolism and behavior. Such effects were demonstrated during the 1960s–1980s, when oxytocin was reported to induce maternal and sexual behavior, to influence feeding and foraging, gluconeogenesis and glucogenolysis, as well as the release of insulin and several gastrointestinal hormones (for a review see for example [19,20].

4. Oxytocin and Metabolism

Oxytocin may, during certain circumstances, induce insulin-like effects where it stimulates glucose uptake, as well as lipogenesis, through effects on oxytocin receptors in the adipose tissue [13,21,22]. On the other hand, oxytocin has also been demonstrated to be released in response to hypoglycemia and to increase blood glucose [21,23] by stimulating both glycogenolysis and gluconeogenesis [24,25], and it may also stimulate lipolysis [26]. Interestingly, the effect on lipolysis has been suggested to be induced by a subpopulation of oxytocinergic neurons located in the adipose tissue interacting with noradrenaline release [26].

Oxytocin has also been demonstrated to release insulin through effects within the endocrine pancreas [27] and the dorsal motor nucleus (DMX) [28]. However, an opposite effect on insulin release has been demonstrated when oxytocin is injected into the dorsal vagal complex (DVC) where oxytocin decreases insulin [29]. In rats, the increase in glucose and glucagon during suckling was attenuated in response to an oxytocin antagonist [28]. It is reasonable that oxytocin may have dual effects on metabolism: one during breastfeeding and another during nonpregnant/nonbreastfeeding states, and it has been suggested that oxytocin may take part in the mobilization and transfer of energy in the production of milk during lactation [30]. Oxytocin increases the proliferation and differentiation of osteoblasts [31,32], and, in muscles, oxytocin seems to have both anabolic and protective effects [33]. An anabolic effect of oxytocin on bone and muscles during breastfeeding might serve to protect the breastfeeding mother against a too large breakdown of bone and muscles during this period with high prolactin and decreased estrogen levels. In a short-term perspective, both the stimulating and inhibiting effects of oxytocin on the release of GH and insulin-like growth factor-1 (IGF-1) have been demonstrated, but, in a long-term perspective, oxytocin seems to increase the plasma levels of IGF-1 [34,35].

Similarly, opposite effects of oxytocin have been reported on food intake and weight gain. In female rats, these effects are dependent on estrogen levels, dose and the route of administration [36,37].

In ovariectomized (OVX) rats, a decrease in weight in response to oxytocin has been demonstrated [38,39]. In humans, lower levels of oxytocin were found in postmenopausal women compared to premenopausal women and in obese women compared to women of normal weight [40]. Several recent studies have focused on oxytocin as a weight reducing hormone, changing metabolism, decreasing food intake and increasing energy expenditure [41]. However, from a physiological point of view this would not be advantageous during pregnancy and breastfeeding, but, as discussed above, oxytocin might have different effects depending on the situation and hormonal background.

Oxytocin receptors and oxytocinergic neurons projecting to other areas within the hypothalamus and the brainstem have been demonstrated, and, within these areas, oxytocin can influence metabolism through effects on other hormones and neurotransmitters, as well as the activity within the autonomic nervous system. Within the hypothalamus, oxytocinergic neurons reach other nuclei of importance for metabolism such as the ventromedial hypothalamic nucleus and the arcuate nuclei, and, in the brainstem, both vagal and sympathetic areas are reached by the oxytocinergic neurons where oxytocin receptors are also located. Within the hypothalamus, oxytocin has also been demonstrated to be released from somatodendrites spreading oxytocin within this area [14,16,42]. Oxytocin receptors are widespread in the periphery and present in, for example, the liver and the gastrointestinal tract, the endocrine part of the pancreas and muscles, as well as adipocytes [14–17]. The increase in metabolism and decrease in food intake have been suggested to be induced by an increase in activity within the sympathetic nervous system [41]. Indeed, in a mouse model deficient in oxytocin receptors, a low sympathetic tone and increased weight without an increase in food intake were seen [43]. An increase in activity within the sympathetic nervous system related to oxytocin might be true during certain circumstances, although, in contrast, oxytocin is also referred to as an anti-stress hormone since it decreases blood pressure and the activity within the hypothalamic–pituitary–adrenal (HPA)-axis. Additionally, during breastfeeding, oxytocin increases vagal nerve activity. Moreover, in rats, oxytocin decreases the activity within the locus coeruleus and increases the responsiveness of alpha 2-adrenoreceptors [44–46]. When oxytocin is administered postnatally to rats, a lifelong change with lower blood pressure and increased responsiveness of alpha 2-adrenoreceptors have been reported, while changes in weight seem to depend on sex and strain [47–49].

In hypothalamus, both the melanocyte stimulating hormone (MSH) and leptin have been demonstrated to affect oxytocin release from the PVN [50,51]. There seems also to be correlations and interactions between GLP-1 and oxytocin [52,53]. When oxytocin was administered to rats, an increase in the hunger hormone ghrelin [54] and lower levels of thyroid hormones have been reported [55]. In contrast, individuals with PWS have higher ghrelin levels and, more frequently, a central hypothyreosis compared to non-PWS individuals. Concerning the effects of oxytocin on body weight, conflicting results have been described in an overview of studies of oxytocin in humans [41]. Interestingly, in a study on obese non-PWS individuals treated with oxytocin intranasally four times a day for 8 weeks, weight decreased significantly, with the most pronounced weight loss in the most obese individuals. In individuals with a normal BMI, no weight loss was seen [56]. Furthermore, in diabetic mice, oxytocin administration decreased weight and food intake and improved glucose and fat metabolism. Meanwhile, his was not observed in non-diabetic controls [57].

Thus, as discussed above, oxytocin might have different metabolic effects depending on the physiological and experimental context. Effects may differ, for example, depending on time, gender, age, other hormones or stress, as well as if the individual is in a fed or fasting state.

In PWS, metabolism is decreased, and body composition is different, with more body fat than lean body mass, while glucose metabolism is usually normal [10]. It is likely that oxytocin, through various central and peripheral effects, plays a role for at least some of the symptoms seen in PWS, although there are several other hormones that are stronger regulators of metabolism.

5. Oxytocin and Behavior

The first behavioral effects of oxytocin that were described were the effects on maternal behavior and bonding between mother and offspring/child [58,59]. Oxytocin is released not only in response to the milk ejection reflex but also in response to the early skin-to-skin contact between mother and child, which are important for the bonding. A similar release of oxytocin in response to skin-to-skin contact has been demonstrated also in fathers [60,61].

As discussed above, in relation to the effects of oxytocin on metabolism and weight, likewise, oxytocin seems to have dual effects also in some aspects of behavior. In rats, oxytocin can induce both anxiolytic-like and sedative effects depending on dose and the route of administration but also depending on the stage of the estrous cycle [37,62], and oxytocin administered postnatally to pigs and rats induces lifelong behavioral changes [47,49,63].

In rats and mice, oxytocin induces grooming, sexual and social behavior, and it decreases aggression. Animal studies suggest that oxytocin increases social interaction by decreasing activity within the amygdala. In an animal model of depression, oxytocin has been found to induce similar effects as antidepressant medication [64].

There are many studies indicating that oxytocin plays a role for several behaviors in humans since oxytocin levels have been demonstrated to be associated with social behavior, anxiety, sexual activity, depression, trust and decision making (for reviews, see, for example [19,65]).

Oxytocin levels are lower in female patients with depression [66], and, in some but not all studies, oxytocin has been reported to improve depression [67–69].

In 2005, Kosfeld et al. showed that oxytocin administered intranasally could increase trust in humans, and, approximately ten years later, Lambert al. used an MRI to demonstrate that oxytocin affects decision making [70,71].

In another experimental setting, children exposed to a situation of social stress were tested. As a result, the children who could see or hear their mother's voice had higher levels of oxytocin and lower levels of cortisol compared to the children who had no contact with their mother [72].

Oxytocin has also been suggested to play a role in the symptoms related to autism spectrum disorders [73], and, interestingly, oxytocin was found to reduce increased activity in the amygdala in response to angry faces in participants with autism-spectrum disorders [74].

In a meta-analysis [75], lower oxytocin levels were found in children but not in adolescents with autism, and, in some studies, the administration of oxytocin was reported to improve social behavior in these individuals. In a mouse model of autism, oxytocin administered neonatally restored social behavior [76].

The above-mentioned effects might be induced by oxytocin itself, but it could also involve other hormones and neurotransmitters. Oxytocin neurons reach several brain areas of importance for behavior such as the frontal cortex, the amygdala, the hippocampus and the bed nucleus of the stria terminalis where oxytocin receptors are expressed [14,16].

For example, oxytocin increases the activity within 5HT1a/5HT2a and 5HT2c-receptors [77–79], as well as the responsiveness of alpha 2-adrenoreceptors, and decreases the activity within the HPA axis. In rats, oxytocin can affect the number of glucocorticoid receptors within for example the hippocampus [80].

In addition, oxytocin may influence dopaminergic receptors and dopaminergic transmission [81], and the effect on anxiety by oxytocin has been suggested to involve GABA-A-receptors [82].

It could be speculated that a deficiency or change in oxytocin levels or signaling might explain some of the behavioral challenges seen in PWS. Low levels of oxytocin are associated with anxiety, less social interaction, and less calmness, all behaviors common in PWS. In addition, individuals with PWS have changes in their GABA-A-receptors, a lower activity within the HPA axis and the serotoninergic transmission where especially the 5HT2C-receptors have been thought to play a role in PWS [83]. Similar changes have also been linked to oxytocin levels or oxytocin administration as discussed above.

In accordance with the effect of oxytocin administered early in life that were discussed above, oxytocin administered within 5 min after birth in a mouse model of PWS induced long-term changes since it normalized both learning and memory as well as social cognition in the mice [84]. Altogether, the behavioral alterations in PWS and the effects of oxytocin observed on those have been a great inspiration for clinical trials of oxytocin in PWS, as described below.

6. Clinical Trials in PWS

6.1. Clinical Trials on the Use of Oxytocin in PWS

The first clinical trial of oxytocin in PWS was performed by Tauber et al. in 2011. Twenty-four individuals, with a median age of 28.5 years, were included. Every individual received a single dose (24 IU) of either oxytocin or a placebo. A questionnaire was developed to assess ten social and emotional behaviors, with four questions to assess eating behavior. A first assessment of the individual's behavior was performed two days before drug administration, a second at 12 h after administration for evaluation of early effects and a third at two days after drug administration for an evaluation of late effects. There was no difference between the two groups before drug administration, but a significant increase in trust and tendencies to be less sad and conduct disruptive behavior was seen two days after oxytocin administration. No statistical difference was observed in the eating behavior scores between the groups, but there was improvement in social skills in the oxytocin-treated group [85].

In 2014, Einfeld et al. reported on a study of 30 individuals with PWS included in an 18-week double-blind, randomized, placebo-controlled trial with an intranasal use of oxytocin to examine its effect on physical, behavioral, and cognitive function. For the assessments, the Developmental Behavior Checklist (DBC), the Yale–Brown Obsessive Compulsive Scale (Y-BOCS), the Dykens Hyperphagia Questionnaire, the Reading the Mind in the Eyes Test (RMET) and the Epworth Sleepiness Scale (ESS) were used. Two different doses of oxytocin were administered for 8 weeks according to the age of the individuals. Individuals aged 16 years and above (n = 11) received 24 IU twice daily, and individuals between 13 and 15 years received 18 IU twice daily. This period was followed by a two-week washout period. For the following eight weeks, the dose was increased to 40 IU twice daily for individuals older than 16 years (n = 18) and 32 IU twice daily for individuals between 13 and 15 years. The study did not show improvement in behavioral problems or in hyperphagia in response to the oxytocin treatment. Unexpectedly, an increase in temper outbursts was noted with higher doses of oxytocin [86].

In 2016, in a randomized, double-blind, placebo-controlled, cross-over study conducted on 25 children with PWS was performed by Kuppens et al. The aim was to examine the effect of oxytocin on eating and social behavior. For the assessments, the Dykens Hyperphagia Questionnaire and the Oxytocin Study Questionnaire were used. Depending on the age, the participants received either oxytocin 24–48 IU or a placebo divided in two daily doses for four weeks. They then crossed over to the opposite treatment for another four weeks. The analyses of children <11 years compared to children >11 years did not show any differences between treatment with oxytocin or placebo. However, in children <11 years, oxytocin compared to a placebo had beneficial effects on hyperphagia and social behavior, and the children showed less anger and sadness [87].

In 2017, Miller et al. conducted a double-blind, placebo-controlled, cross-over trial in 24 children, 5 to 12 years old, with PWS. The participants received 16 IU of intranasal oxytocin or a placebo for five days, followed by a four-week washout period, and then they switched to the opposite treatment for another five days. The Aberrant Behavior Checklist, Social Responsiveness Scale (SRS-P), Repetitive Behavior Scale—Revised (RBS-R), Hyperphagia Questionnaire, and the Clinical Global Impression (CGI) scale were used to evaluate hyperphagia and other behaviors. The authors concluded that a five-day treatment with a low dose of oxytocin in children was safe, and, compared to the placebo treatment, improvements were noted in anxiety and self-injurious behaviors [88].

In 2017, Tauber et al. reported results from a single dose of intranasal oxytocin administered to infants less than six months of age. Sucking and swallowing were examined before and after the oxytocin administration by using the Neonatal Oral Motor Assessment Scale (NOMAS), video fluoroscopy of swallowing, and the Clinical Global Impression (CGI) scale. Oxytocin was well tolerated, with no side effects noted during the seven days of treatment, and oxytocin improved oral feeding and social skills [89].

In 2021, Damen et al. performed a randomized, double-blind, placebo-controlled, cross-over study on the effects of oxytocin versus the placebo treatment for three months in children with PWS to investigate changes in social behaviors and hyperphagia, with differences between males and females and between PWS molecular genetic classes. The social and eating behaviors were evaluated with the Oxytocin Study Questionnaire, the Dykens

Hyperphagia Questionnaire and the RBS-R. Forty-six children with PWS, 3 to 11 years old, were randomized to intranasal oxytocin 16–40 IU or a placebo daily divided in two doses for three months. No significant effects on hyperphagia or social behavior were seen. However, subanalyses showed a more pronounced effect of oxytocin on hyperphagia in individuals with 15q11-q13 deletions and in boys [90].

In another study the same year, Hollander et al. studied the effect of intranasal oxytocin on hyperphagia and repetitive behaviors in 23 children with PWS, 5 to 18 years of age. The hyperphagia and repetitive behaviors were evaluated with the Dykens Hyperphagia Questionnaire form and the RBS-R. The trial was randomized, double-blind, placebo-controlled and lasted eight weeks. The participants were assigned to 16 IU oxytocin or a placebo per day. Modest improvements in hyperphagia and repetitive behaviors were observed in the placebo group, while no changes were seen in the oxytocin group [91].

6.2. Clinical Trials on the Use of Carbetocin in PWS

Studies with carbetocin, an oxytocin analog with greater selectivity for oxytocin receptors and a longer half-life, have been performed and are ongoing. Carbetocin is a well-known substance, approved since 1997, to prevent uterine atony following caesarean delivery. In a 14-day phase 2 trial that was randomized, double-blind and placebo-controlled, 37 individuals with PWS, 10 to 18 years old, were assigned to carbetocin (n = 17) or a placebo (n = 29). The participants received 9.6 mg of intranasal carbetocin three times per day or a placebo. The study showed a decrease in hyperphagia and other behavioral symptoms in response to carbetocin, and it was safe and well tolerated [92]).

In a recent phase 3 trial, a total of 130 participants with PWS aged 7 to 18 years were enrolled. The participants were randomized to 9.6 mg/dose carbetocin (n = 44), 3.2 mg/dose carbetocin (n = 43) or a placebo (n = 43) three times daily during 8 weeks. During a subsequent 56-week long-term follow-up period, the placebo participants were randomly assigned to 9.6 mg (n = 64) or 3.2 mg carbetocin (n = 64), while carbetocin participants continued at their previous dose. Due to the COVID-19 pandemic, enrolment was discontinued prematurely, but the study showed that carbetocin was well tolerated and that the 3.2 mg dose was associated with clinically meaningful improvements in hyperphagia as well as anxiousness and distress behaviors [93].

In summary, beneficial effects of oxytocin treatment in PWS were seen in seven out of nine studies (Table 2). The positive effects consisted of improvements in hyperphagia and distress behaviors. However, the results of the studies cannot be compared straight away as the study designs varied, the study cohorts consisted of different age groups with different gender compositions, different doses and numbers of administrations of oxytocin were used, the duration of the studies varied and, in some studies, a limited number of participants were included. In addition, the long-term effects and risk of adverse effects in humans are not known. From animal experiments, we know that oxytocin can induce long-lasting, perhaps lifelong, behavioral and metabolic effects. The knowledge of long-term consequences are of major importance, in particular, in PWS patients due to intellectual disability, behavioral challenges and often problems describing their health and health care needs.

Table 2. Summary of clinical trials of oxytocin and carbetocin in PWS.

Dose/Duration and Study Design	Number and Gender	Age	Effect	Authors
24 IU single dose	N = 24 (16 females)	18–43 years	Trust increase	Tauber et al., 2011 [85]
18–32 IU twice daily and 24–40IU twice daily for 8 weeks Randomized, double-blind, placebo-controlled and open-label extension	N = 29 (9 females)	13–15 years and 16–29 years	No effect (increase in temper outbursts with high dose)	Einfeld et al., 2014 [86]

Table 2. *Cont.*

Dose/Duration and Study Design	Number and Gender	Age	Effect	Authors
24–48 IU for 4 weeks Randomized, double-blind, crossover	*N* = 25 (11 females)	6–14 years	Less hyperphagia, social behavior improved	Kuppens et al., 2016 [87]
16 IU for 5 days Randomized, double-blind, crossover	*N* = 24 (12 females)	5–12 years	Less anxiety and self-injury	Miller et al., 2017 [88]
7 IU for 7 days Case cohort study	*N* = 18 (8 females)	<6 months	Improved oral feeding and social skills	Tauber et al., 2017 [89]
16–40 IU for 12 weeks Randomized, double-blind, crossover	*N* = 26 (13 females)	3–11 years	Less hyperphagia in 15q-11–13 and boys	Damen et al., 2021 [90]
16 IU for 8 weeks Randomized, double-blind, placeboc-ontrolled	*N* = 23 (5 females)	5–18 years	No effect	Hollander et al., 2021 [91]
Carbetocin 9.6 mg three times daily for 2 weeks Randomized, double-blind, placebo-controlled	*N* = 37 (23 females)	10–18 years	Less hyperphagia	Dykens et al., 2021 [92]
Carbetocin 9.6 mg or 3.2 mg three times daily for 8 and 56 weeks Randomized, double-blind, placebo-controlled and open-label extension	*N* = 130 (66 females)	7–18 years	Low dose: less hyperphagia and anxiety	Roof et al., 2023 [93]

7. Conclusions

Oxytocin has a lot of effects on metabolism and behavior, including hyperphagia, but the magnitude of these effects depends on the situation and context. These symptoms are a major challenge in PWS, and, as studies have demonstrated low concentrations of oxytocin in PWS, oxytocin is, therefore, potentially an attractive treatment. In some of the clinical trials performed with oxytocin on individuals with PWS, anxiety and compulsiveness decreased. Hyperphagia could be considered as a compulsive behavior, and it also improved in some of the studies. Recent studies with the oxytocin analogue carbetocin showed similar effects with improvements in hyperphagia and behavior. Worth mentioning is that, in non-PWS individuals with obsessive compulsive disorder (OCD), changes in the oxytocin receptor have been found, and a methylation of the oxytocin receptor has even been suggested as a biomarker for the response to the treatment of OCD [94].

To our knowledge, no study has shown an effect of oxytocin on weight in PWS. However, if oxytocin is administered during a longer period, it might, at least in younger patients, have the potential to decrease weight or protect against further weight gain since oxytocin, as discussed above, in several studies decreased hyperphagia. Thus, oxytocin seems to positively affect behavior in PWS, at least during certain circumstances and especially if administered early in life.

8. Future Directions

Oxytocin is an interesting peptide that seems to modulate and balance many behavioral and metabolic effects to keep the homeostasis. Since oxytocin is influenced by and is influencing several other neurotransmitters and hormones, the effects seen in response to oxytocin may not always be caused by oxytocin itself. Oxytocin interacts directly or indirectly with GABA-receptors, 5HT-receptors, dopamine receptors, glucocorticoid receptors and the HPA axis, as well as adrenergic receptors and the autonomic nervous system. To further understand the role of oxytocin in PWS, more studies of these interactions are needed as well as studies of the effects of oxytocin during different ages and in the PWS of different genotype. For example, PWS individuals with maternal disomy 15 often have delayed diagnosis, higher verbal IQ scores with greater attention, factual knowledge and better social reasoning skills than those with the typical Type I or Type II deletions involving the 15q11-q13 region. On the other hand, patients with maternal disomy are more prone to increased episodes of psychosis and autistic behaviors [10]. Indeed, in a study by Damen et al. [90], an effect of oxytocin was only seen in the PWS with the 15q11-q13 deletion and not in the PWS with the maternal disomy, but these results need to be repeated in larger cohorts. Further studies of different ages are also of importance since oxytocin seems to have effects in particular in the youngest children. Perhaps it should be administered as early as during the neonatal period since oxytocin, when administered early in life, in animal experiments induces long-lasting and even lifelong effects, but more studies are needed. Also, the beneficial effects of oxytocin in human infants younger than 6 months need to be better established.

Other unanswered questions are what dose of oxytocin to use to be the most effective and what the long-term consequences of oxytocin treatment are. Oxytocin has a very short half-life. Different doses and administration intervals have been used in previous studies. Carbetocin has an increased half-life compared to oxytocin and crosses the blood–brain barrier more easily. It is unclear if diurnal fluctuations in oxytocin concentrations are an asset and if high concentrations can introduce a state of resistance. Future studies with varying doses and numbers of administrations will be of value to answer these questions.

Funding: This research received no external funding.

Conflicts of Interest: The authors declare no conflicts of interest.

References

1. Tauber, M.; Diene, G. Prader-Willi syndrome: Hormone therapies. *Handb. Clin. Neurol.* **2021**, *181*, 351–367. [CrossRef] [PubMed]
2. Swaab, D.F.; Purba, J.S.; Hofman, M.A. Alterations in the hypothalamic paraventricular nucleus and its oxytocin neurons (putative satiety cells) in Prader-Willi syndrome: A study of five cases. *J. Clin. Endocrinol. Metab.* **1995**, *80*, 573–579. [PubMed]
3. Swaab, D. Prader—Willi syndrome and the hypothalamus. *Acta Paediatr.* **1997**, *86*, 50–54. [CrossRef] [PubMed]
4. Bittel, D.C.; Kibiryeva, N.; Sell, S.M.; Strong, T.V.; Butler, M.G. Whole genome microarray analysis of gene expression in Prader–Willi syndrome. *Am. J. Med. Genet. Part A* **2007**, *143*, 430–442. [CrossRef] [PubMed] [PubMed Central]
5. Johnson, L.; Manzardo, A.M.; Miller, J.L.; Driscoll, D.J.; Butler, M.G. Elevated plasma oxytocin levels in children with Prader–Willi syndrome compared with healthy unrelated siblings. *Am. J. Med. Genet. Part A* **2015**, *170*, 594–601. [CrossRef] [PubMed] [PubMed Central]
6. Höybye, C.; Barkeling, B.; Espelund, U.; Petersson, M.; Thorén, M. Peptides associated with hyperphagia in adults with Prader–Willi syndrome before and during GH treatment. Growth Horm. *IGF Res.* **2003**, *13*, 322–327.
7. Martin, A.; State, M.; Anderson, G.M.; Kaye, W.M.; Hanchett, J.M.; McConaha, C.W.; North, W.G.; Leckman, J.F. Cerebrospinal fluid levels of oxytocin in Prader–Willi syndrome: A preliminary report. *Biol. Psychiatry* **1998**, *44*, 1349–1352. [CrossRef] [PubMed]
8. Meziane, H.; Schaller, F.; Bauer, S.; Villard, C.; Matarazzo, V.; Riet, F.; Guillon, G.; Lafitte, D.; Desarmenien, M.G.; Tauber, M.; et al. An Early Postnatal Oxytocin Treatment Prevents Social and Learning Deficits in Adult Mice Deficient for Magel2, a Gene Involved in Prader-Willi Syndrome and Autism. *Biol. Psychiatry* **2014**, *78*, 85–94. [CrossRef] [PubMed]
9. Shalma, N.M.; Alsharabasy, M.A.; Taha, A.M.; Alsawareah, A.; Manirambona, E.; Ahmed, S.K.; Mohamed, M.R.; Taha, N.A.; Abd-ElGawad, M. The efficacy of intranasal oxytocin in patients with Prader-Willi syndrome: A systematic review and meta-analysis. Diabetes Metab. Syndr. *Clin. Res. Rev.* **2023**, *17*, 102711. [CrossRef] [PubMed]
10. Tauber, M.; Hoybye, C. Endocrine disorders in Prader-Willi syndrome: A model to understand and treat hypothalamic dysfunction. *Lancet Diabetes Endocrinol.* **2021**, *9*, 235–246. [CrossRef] [PubMed]
11. Shaikh, M.G.; Barrett, T.; Bridges, N.; Chung, R.; Gevers, E.; Goldstone, A.P.; Holland, A.; Kanumakala, S.; Krone, R.E.; Kyriakou, A.; et al. Prader-Willi syndrome: Guidance for children and transition into adulthood. *Endocr Connect.* **2024**, *13*, e240091. [CrossRef] [PubMed]
12. Miller, J.L.; Lynn, C.H.; Driscoll, D.C.; Goldstone, A.P.; Gold, J.; Kimonis, V.; Dykens, E.; Butler, M.G.; Shuster, J.J. Nutritional phases in Prader–Willi syndrome. *Am. J. Med Genet, Part A* **2011**, *155*, 1040–1049. [CrossRef] [PubMed] [PubMed Central]
13. Bonne, D.; Cohen, P. Characterization of Oxytocin Receptors on Isolated Rat Fat Cells. *Eur. J. Biochem.* **1975**, *56*, 295–303. [CrossRef] [PubMed]
14. Tribollet, E.; Goumaz, M.; Raggenbass, M.; Dreifuss, J.J. Appearance and Transient Expression of Vasopressin and Oxytocin Receptors in the Rat Brain. *J. Recept. Res.* **1991**, *11*, 333–346. [CrossRef] [PubMed]
15. Kimura, T.; Saji, F.; Nishimori, K.; Ogita, K.; Nakamura, H.; Koyama, M.; Murata, Y. Molecular regulaton of the oxytocin receptor in peripheral organs. *J. Mol. Endocrinol.* **2003**, *30*, 109–115. [CrossRef] [PubMed]
16. Gimpl, G.; Fahrenholz, F. The Oxytocin Receptor System: Structure, Function, and Regulation. *Physiol. Rev.* **2001**, *81*, 629–683. [CrossRef]

17. Monstein, H.-J.; Grahn, N.; Truedsson, M.; Ohlsson, B. Oxytocin and oxytocin-receptor mRNA expression in the human gastrointestinal tract: A polymerase chain reaction study. *Regul. Pept.* **2004**, *119*, 39–44. [CrossRef] [PubMed]
18. Landgraf, R.; Neumann, I.D. Vasopressin and oxytocin release within the brain: A dynamic concept of multiple and variable modes of neuropeptide communication. *Front. Neuroendocr.* **2004**, *25*, 150–176. [CrossRef] [PubMed]
19. Uvnäs-Moberg, K. Oxytocin may mediate the benefits of positive social interaction and emotions. *Psychoneuroendocrinology* **1998**, *23*, 819–835. [CrossRef]
20. Uvnäs-Moberg, K. Role of efferent and afferent vagal nerve activity during reproduction: Integrating function of oxytocin on metabolism and behaviour. *Psychoneuroendocrinology* **1994**, *19*, 687–695. [CrossRef]
21. Mirsky, I.A. Effect of Oxytocin on Plasma Free Fatty Acids of Non-Diabetic and Diabetic Dogs. *Exp. Biol. Med.* **1962**, *110*, 42–44. [CrossRef] [PubMed]
22. Braun, T.; Hechter, O.; Rudinger, J. Insulin-like action of oxytocin: Evidence for separate oxytocin sensitive and insulin-sensitive systems in fat cells. *Endocrinology* **1969**, *85*, 1092–1096. [CrossRef] [PubMed]
23. Fisher, B.M.; Baylis, P.H.; Frier, B.M. Plasma oxytocin, arginine vasopressin and atrial natriuretic peptide responses to insulin-induced hypoglycemia in man. *Clin. Endocrinol.* **1987**, *26*, 179–185. [CrossRef] [PubMed]
24. Hems, D.A.; Rodrigues, L.M.; Whitton, P.D. Rapid stimulation by vasopressin, oxytocin and angiotensin II of glycogen degradation in hepatocytes suspensions. *Biochem. J.* **1978**, *172*, 311–317. [CrossRef] [PubMed]
25. Whitton, P.D.; Rodrigues, L.M.; Hems, D.A. Stimulation by vasopressin, angiotensin and oxytocin of gluconeogenesis in hepatocyte suspensions. *Biochem. J.* **1978**, *176*, 893–898. [CrossRef] [PubMed]
26. Li, E.; Wang, L.; Wang, D.; Chi, J.; Lin, Z.; Smith, G.I.; Klein, S.; Cohen, P.; Rosen, E.D. Control of lipolysis by a population of oxytocinergic sympathetic neurons. *Nature* **2024**, *625*, 175–180. [CrossRef] [PubMed]
27. Stock, S.; Fastbom, J.; Björkstrand, E.; Ungerstedt, U.; Uvnäs-Moberg, K. Effects of oxytocin on in vivo release of insulin and glucagon studied by microdialysis in the rat pancreas and autoradiographic evidence for [3H]oxytocin binding sites within the islets of Langerhans. *Regul. Pept.* **1990**, *30*, 1–13. [CrossRef] [PubMed]
28. Björkstrand, E.; Eriksson, M.; Uvnäs-Moberg, K. Evidence of a Peripheral and a Central Effect of Oxytocin on Pancreatic Hormone Release in Rats. *Neuroendocrinology* **1996**, *63*, 377–383. [CrossRef] [PubMed]
29. Siaud, P.; Puech, R.; Assenmacher, I.; Alonso, G. Microinjection of oxytocin into the dorsal vagal complex decreases pancreatic insulin secretion. *Brain Res.* **1991**, *546*, 190–194. [CrossRef]
30. Uvnäs-Moberg, K.; Prime, D.K. Oxytocin effects in mothers and infants during breastfeeding. *Infant* **2013**, *9*, 201–206.
31. Petersson, M.; Lagumdzija, A.; Stark, A.; Bucht, E. Oxytocin stimulates proliferation of human osteoblast-like cells. *Peptides* **2002**, *23*, 1121–1126. [CrossRef] [PubMed]
32. Cuscito, C.; Colaianni, G.; Tamma, R.; Greco, G.; Dell'Endice, S.; Yen, T.; Sun, L.; Zaidi, M.; Di Benedetto, A.; Zallone, A. Adrenergic stimulation decreases osteoblast oxytocin synthesis. *Ann. N. Y. Acad. Sci.* **2011**, *1237*, 53–57. [CrossRef] [PubMed]
33. Costa, D.M.; da Cruz-Filho, J.; Vasconcelos, A.B.S.; Gomes-Santos, J.V.; Reis, L.C.; de Lucca, W.; Camargo, E.A.; Lauton-Santos, S.; Zanon, N.M.; Kettelhut, D.C.; et al. Oxytocin induces anti-catabolic and anabolic effects on protein metabolism in the female rat oxidative skeletal muscle. *Life Sci.* **2021**, *279*, 119665. [CrossRef] [PubMed]
34. Sohlström, A.; Brismar, K.; Carlsson-Skwirut, C.; Bang, P.; Uvnäs-Moberg, K. Effects of oxytocin on the IGF-axis and some gastrointestinal hormones in ad libitum fed and food-restricted female rats. *Acta Physiol. Scand.* **1999**, *166*, 223–230. [CrossRef] [PubMed]
35. Petersson, M.; Lundeberg, T.; Sohlström, A.; Wiberg, U.; Uvnäs-Moberg, K. Oxytocin increases the survival of musculocutaneous flaps. *Naunyn Schmiedebergs Arch Pharmacol.* **1998**, *357*, 701–704. [CrossRef] [PubMed]
36. Uvnas-Moberg, K.; Alster, P.; Petersson, M. Dissociation of oxytocin effects on body weight in two variants of female Sprague-Dawley rats. Integr. Psychol. *Behav. Sci.* **1996**, *31*, 44–55.
37. Petersson, M.; Ahlenius, S.; Wiberg, U.; Alster, P.; Uvnäs-Moberg, K. Steroid Dependent Effects of Oxytocin on Spontaneous Motor Activity in Female Rats. *Brain Res. Bull.* **1998**, *45*, 301–305. [CrossRef] [PubMed]
38. Petersson, M.; Eklund, M.; Uvnäs-Moberg, K. Oxytocin decreases corticosterone and nociception and increases motor activity in OVX rats. *Maturitas* **2005**, *51*, 426–433. [CrossRef] [PubMed]
39. Iwasa, T.; Matsuzaki, T.; Mayila, Y.; Yanagihara, R.; Yamamoto, Y.; Kawakita, T.; Kuwahara, A.; Irahara, M. Oxytocin treatment reduced food intake and body fat and ameliorated obesity in ovariectomized female rats. *Neuropeptides* **2019**, *75*, 49–57. [CrossRef]
40. Maestrini, S.; Mele, C.; Mai, S.; Vietti, R.; Di Blasio, A.; Castello, L.; Surico, D.; Aimaretti, G.; Scacchi, M.; Marzullo, P. Plasma Oxytocin Concentration in Pre- and Postmenopausal Women: Its Relationship with Obesity, Body Composition and Metabolic Variables. *Obes. Facts* **2018**, *11*, 429–439. [CrossRef]
41. McCormack, S.E.; Blevins, J.E.; Lawson, E.A. Metabolic Effects of Oxytocin. *Endocr. Rev.* **2019**, *41*, 121–145. [CrossRef]
42. Ludwig, M.; Leng, G. Dendritic peptide release and peptide-dependent behaviours. *Nat. Rev. Neurosci.* **2006**, *7*, 126–136. [CrossRef]
43. Camerino, C. Low Sympathetic Tone and Obese Phenotype in Oxytocin-deficient Mice. *Obesity* **2009**, *17*, 980–984. [CrossRef]
44. Petersson, M.; Uvnäs-Moberg, K.; Erhardt, S.; Engberg, G. Oxytocin increases locus coeruleus alpha 2-adrenoreceptor responsiveness in rats. *Neurosci. Lett.* **1998**, *255*, 115–118. [CrossRef]
45. Petersson, M.; Hulting, A.-L.; Uvnäs-Moberg, K. Oxytocin causes a sustained decrease in plasma levels of corticosterone in rats. *Neurosci. Lett.* **1999**, *264*, 41–44. [CrossRef]

46. Petersson, M.; Diaz-Cabiale, Z.; Fuxe, K.; Uvnäs-Moberg, K. Oxytocin increases the density of high affinity alpha 2-adrenoreceptors within the hypothalamus, the amygdala and the nucleus of the solitary tract in female ovariectomized rats. *Brain Res.* **2005**, *1049*, 234–239. [CrossRef]
47. Uvnäs-Moberg, K.; Alster, P.; Petersson, M.; Sohlström, A.; Björkstrand, E. Postnatal Oxytocin Injections Cause Sustained Weight Gain and Increased Nociceptive Thresholds in Male and Female Rats. *Pediatr. Res.* **1998**, *43*, 344–348. [CrossRef]
48. Díaz-Cabiale, Z.; Olausson, H.; Sohlström, A.; Agnati, L.F.; Narváez, J.A.; Uvnäs-Moberg, K.; Fuxe, K. Long-Term Modulation By Postnatal Oxytocin of the α2-Adrenoceptor Agonist Binding Sites in Central Autonomic Regions and the Role of Prenatal Stress. *J. Neuroendocr.* **2004**, *16*, 183–190. [CrossRef]
49. Petersson, M.; Uvnäs-Moberg, K. Postnatal oxytocin treatment of spontaneously hypertensive male rats decreases blood pressure and body weight in adulthood. *Neurosci. Lett.* **2008**, *440*, 166–169. [CrossRef]
50. Blevins, J.E.; Schwartz, M.W.; Baskin, D.G. Evidence that paraventricular nucleus oxytocin neurons link hypothalamic leptin action to caudal brain stem nuclei controlling meal size. *Am. J. Physiol. Integr. Comp. Physiol.* **2004**, *287*, R87–R96. [CrossRef]
51. Sabatier, N.; Caquineau, C.; Dayanithi, G.; Bull, P.; Douglas, A.J.; Guan, X.M.; Jiang, M.; Van der Ploeg, L.; Leng, G. Alpha-melanocyte-stimulating hormone stimulates oxytocin release from the dendrites of hypothalamic neurons while inhibiting oxytocin release from their terminals in the neurohypophysis. *J. Neurosci.* **2003**, *23*, 10351–10358. [CrossRef]
52. Bojanowska, E.; Stempniak, B. Effects of centrally or systemically injected glucagon-like peptide-1 (7–36) amide on release of neurohypophysial hormones and blood pressure in the rat. *Regul. Pept.* **2000**, *91*, 75–81. [CrossRef]
53. Lee, J.; Moon, H.; Lee, H.; Oh, Y.; Kim, C.; Lee, Y.H.; Kim, M.S.; NamKoong, C.; Lee, H.W.; Kim, J.H.; et al. Antagonistic interaction between central glucagon-like Peptide-1 and oxytocin on diet-induced obesity mice. *Heliyon* **2020**, *6*, e05190. [CrossRef]
54. Szabó, R.; Ménesi, R.; Molnár, A.H.; Szalai, Z.; Daruka, L.; Tóth, G.; Gardi, J.; Gálfi, M.; Börzsei, D.; Kupai, K.; et al. New Metabolic Influencer on Oxytocin Release: The Ghrelin. *Molecules* **2019**, *24*, 735. [CrossRef]
55. Petersson, M. Oxytocin decreases plasma levels of thyroid-stimulating hormone and thyroid hormones in rats. *Regul. Pept.* **2002**, *108*, 83–87. [CrossRef]
56. Zhang, H.; Wu, C.; Chen, Q.; Chen, X.; Xu, Z.; Wu, J.; Cai, D. Treatment of Obesity and Diabetes Using Oxytocin or Analogs in Patients and Mouse Models. *PLoS ONE* **2013**, *8*, e61477. [CrossRef]
57. Plante, E.; Menaouar, A.; Danalache, B.A.; Yip, D.; Broderick, T.L.; Chiasson, J.-L.; Jankowski, M.; Gutkowska, J. Oxytocin Treatment Prevents the Cardiomyopathy Observed in Obese Diabetic Male db/db Mice. *Endocrinology* **2015**, *156*, 1416–1428. [CrossRef]
58. Kendrick, K.M.; Keverne, E.B.; Baldwin, B.A. Intracerebroventricular Oxytocin Stimulates Maternal Behaviour in the Sheep. *Neuroendocrinology* **1987**, *46*, 56–61. [CrossRef]
59. Pedersen, C.A.; Prange, A.J. Induction of maternal behavior in virgin rats after intracerebroventricular administration of oxytocin. *Proc. Natl. Acad. Sci. USA.* **1979**, *76*, 6661–6665. [CrossRef]
60. Bystrova, K.; Widström, A.M.; Matthiesen, A.S.; Ransjö-Arvidson, A.B.; Welles-Nyström, B.; Wassberg, C.; Vorontsov, I.; Uvnäs-Moberg, K. Skin-to-skin contact may reduce negative consequences of "the stress of being born": A study on temperature in newborn infants, subjected to different ward routines in St. Petersburg. *Acta Paediatr.* **2003**, *92*, 320–326. [CrossRef] [PubMed]
61. Cong, X.; Ludington-Hoe, S.M.; Hussain, N.; Cusson, R.M.; Walsh, S.; Vazquez, V.; Briere, C.-E.; Vittner, D. Parental oxytocin responses during skin-to-skin contact in pre-term infants. *Early Hum. Dev.* **2015**, *91*, 401–406. [CrossRef]
62. Uvnäs-Moberg, K.; Ahlenius, S.; Hillegaart, V.; Alster, P. High doses of oxytocin cause sedation and low doses cause an anxiolytic-like effect in male rats. *Pharmacol. Biochem. Behav.* **1994**, *49*, 101–106. [CrossRef]
63. Rault, J.-L.; Carter, C.S.; Garner, J.P.; Marchant, J.N.; Richert, B.T.; Lay, D.C. Repeated intranasal oxytocin administration in early life dysregulates the HPA axis and alters social behavior. *Physiol. Behav.* **2013**, *112*, 40–48. [CrossRef]
64. Arletti, R.; Bertolini, A. Oxytocin acts as an antidepressant in two animal models of depression. *Life Sci.* **1987**, *41*, 1725–1730. [CrossRef] [PubMed]
65. Carter, C.S. Oxytocin Pathways and the Evolution of Human Behavior. *Annu. Rev. Psychol.* **2014**, *65*, 17–39. [CrossRef] [PubMed]
66. Ozsoy, S.; Esel, E.; Kula, M. Serum oxytocin levels in patients with depression and the effects of gender and antidepressant treatment. *Psychiatry Res.* **2009**, *169*, 249–252. [CrossRef]
67. Slattery, D.A.; Neumann, I.D. Oxytocin and Major Depressive Disorder: Experimental and Clinical Evidence for Links to Aetiology and Possible Treatment. *Pharmaceuticals* **2010**, *3*, 702–724. [CrossRef]
68. Scantamburlo, G.; Ansseau, M.; Geenen, V.; Legros, J.-J. Intranasal Oxytocin as an Adjunct to Escitalopram in Major Depression. *J. Neuropsychiatry* **2011**, *23*, E5. [CrossRef]
69. McQuaid, R.J.; McInnis, O.A.; Abizaid, A.; Anisman, H. Making room for oxytocin in understanding depression. *Neurosci. Biobehav. Rev.* **2014**, *45*, 305–322. [CrossRef] [PubMed]
70. Kosfeld, M.; Heinrichs, M.; Zak, P.J.; Fischbacher, U.; Fehr, E. Oxytocin increases trust in humans. *Nature* **2005**, *435*, 673–676. [CrossRef] [PubMed]
71. Lambert, B.; Declerck, C.H.; Boone, C.; Parizel, P.M. A functional MRI study on how oxytocin affects decision making in social dilemmas: Cooperate as long as it pays off, aggress only when you think you can win. *Horm. Behav.* **2017**, *94*, 145–152. [CrossRef] [PubMed]
72. Cardoso, C.; Ellenbogen, M.A.; Orlando, M.A.; Bacon, S.L.; Joober, R. Intranasal oxytocin attenuates the cortisol response to physical stress: A dose–response study. *Psychoneuroendocrinology* **2013**, *38*, 399–407. [CrossRef] [PubMed]

73. Yoon, S.; Kim, Y.K. The Role of the Oxytocin System in Anxiety Disorders. *Adv. Exp. Med. Biol.* **2020**, *1191*, 103–120. [CrossRef] [PubMed]
74. Jeung-Maarse, H.; Schmitgen, M.M.; Schmitt, R.; Bertsch, K.; Herpertz, S.C. Oxytocin effects on amygdala reactivity to angry faces in males and females with antisocial personality disorder. *Neuropsychopharmacology* **2023**, *48*, 946–953. [CrossRef] [PubMed]
75. Moerkerke, M.; Peeters, M.; de Vries, L.; Daniels, N.; Steyaert, J.; Alaerts, K.; Boets, B. Endogenous Oxytocin Levels in Autism—A Meta-Analysis. *Brain Sci.* **2021**, *11*, 1545. [CrossRef] [PubMed]
76. Bertoni, A.; Schaller, F.; Tyzio, R.; Gaillard, S.; Santini, F.; Xolin, M.; Diabira, D.; Vaidyanathan, R.; Matarazzo, V.; Medina, I.; et al. Oxytocin administration in neonates shapes hippocampal circuitry and restores social behavior in a mouse model of autism. *Mol. Psychiatry* **2021**, *26*, 7582–7595. [CrossRef] [PubMed]
77. Yoshida, M.; Takayanagi, Y.; Inoue, K.; Kimura, T.; Young, L.J.; Onaka, T.; Nishimori, K. Evidence That Oxytocin Exerts Anxiolytic Effects via Oxytocin Receptor Expressed in Serotonergic Neurons in Mice. *J. Neurosci.* **2009**, *29*, 2259–2271. [CrossRef] [PubMed]
78. Thompson, M.; Callaghan, P.; Hunt, G.; Cornish, J.; McGregor, I. A role for oxytocin and 5-HT1A receptors in the prosocial effects of 3,4 methylenedioxymethamphetamine ("ecstasy"). *Neuroscience* **2007**, *146*, 509–514. [CrossRef] [PubMed]
79. Humble, M.B.; Uvnäs-Moberg, K.; Engström, I.; Bejerot, S. Plasma oxytocin changes and anti-obsessive response during serotonin reuptake inhibitor treatment: A placebo controlled study. *BMC Psychiatry* **2013**, *13*, 344. [CrossRef]
80. Petersson, M.; Uvnäs-Moberg, K. Systemic oxytocin treatment modulates glucocorticoid and mineralocorticoid receptor mRNA in the rat hippocampus. *Neurosci. Lett.* **2003**, *343*, 97–100. [CrossRef]
81. Love, T.M. Oxytocin, motivation and the role of dopamine. *Pharmacol. Biochem. Behav.* **2014**, *119*, 49–60. [CrossRef] [PubMed]
82. Smith, A.S.; Tabbaa, M.; Lei, K.; Eastham, P.; Butler, M.J.; Linton, L.; Altshuler, R.; Liu, Y.; Wang, Z. Local oxytocin tempers anxiety by activating GABAA receptors in the hypothalamic paraventricular nucleus. *Psychoneuroendocrinology* **2015**, *63*, 50–58. [CrossRef] [PubMed]
83. Lucignani, G.; Panzacchi, A.; Bosio, L.; Moresco, R.M.; Ravasi, L.; Coppa, I.; Chiumello, G.; Frey, K.; Koeppe, R.; Fazio, F. GABAA receptor abnormalities in Prader–Willi syndrome assessed with positron emission tomography and [11C]flumazenil. *NeuroImage* **2004**, *22*, 22–28. [CrossRef] [PubMed]
84. Schaller, F.; Watrin, F.; Sturny, R.; Massacrier, A.; Szepetowski, P.; Muscatelli, F. A single postnatal injection of oxytocin rescues the lethal feeding behaviour in mouse newborns deficient for the imprinted Magel2 gene. *Hum. Mol. Genet.* **2010**, *19*, 4895–4905. [CrossRef] [PubMed]
85. Tauber, M.; Mantoulan, C.; Copet, P.; Jauregui, J.; Demeer, G.; Diene, G.; Rogé, B.; Laurier, V.; Ehlinger, V.; Arnaud, C.; et al. Oxytocin may be useful to increase trust in others and decrease disruptive behaviours in patients with Prader-Willi syndrome: A randomised placebo-controlled trial in 24 patients. *Orphanet J. Rare Dis.* **2011**, *6*, 47. [CrossRef] [PubMed]
86. Einfeld, S.L.; Smith, E.; McGregor, I.S.; Steinbeck, K.; Taffe, J.; Rice, L.J.; Horstead, S.K.; Rogers, N.; Hodge, M.A.; Guastella, A.J. A double-blind randomized controlled trial of oxytocin nasal spray in Prader Willi syndrome. *Am. J. Med. Genet. Part A* **2014**, *164*, 2232–2239. [CrossRef] [PubMed]
87. Kuppens, R.; Donze, S.; Hokken-Koelega, A. Promising effects of oxytocin on social and food-related behaviour in young children with Prader-Willi syndrome: A randomized, double-blind, controlled crossover trial. *Clin. Endocrinol.* **2016**, *85*, 979–987. [CrossRef] [PubMed]
88. Miller, J.L.; Tamura, R.; Butler, M.G.; Kimonis, V.; Sulsona, C.; Gold, J.; Driscoll, D.J. Oxytocin treatment in children with Prader-Willi syndrome: A double-blind, placebo-controlled, crossover study. *Am. J. Med Genet. Part A* **2017**, *173*, 1243–1250. [CrossRef] [PubMed]
89. Tauber, M.; Boulanouar, K.; Diene, G.; Çabal-Berthoumieu, S.; Ehlinger, V.; Fichaux-Bourin, P.; Molinas, C.; Faye, S.; Valette, M.; Pourrinet, J.; et al. The Use of Oxytocin to Improve Feeding and Social Skills in Infants with Prader-Willi Syndrome. *Pediatrics* **2017**, *139*, e20162976. [CrossRef] [PubMed]
90. Damen, L.; Grootjen, L.N.; Juriaans, A.F.; Donze, S.H.; Huisman, T.M.; Visser, J.A.; Delhanty, P.J.D.; Hokken-Koelega, A.C.S. Oxytocin in young children with Prader-Willi syndrome: Results of a randomized, double-blind, placebo-controlled, crossover trial investigating 3 months of oxytocin. *Clin. Endocrinol.* **2021**, *94*, 774–785. [CrossRef] [PubMed] [PubMed Central]
91. Hollander, E.; Levine, K.G.; Ferretti, C.J.; Freeman, K.; Doernberg, E.; Desilva, N.; Taylor, B.P. Intranasal oxytocin versus placebo for hyperphagia and repetitive behaviors in children with Prader-Willi Syndrome: A randomized controlled pilot trial. *J. Psychiatr. Res.* **2020**, *137*, 643–651. [CrossRef] [PubMed]
92. Dykens, E.M.; Miller, J.; Angulo, M.; Roof, E.; Reidy, M.; Hatoum, H.T.; Willey, R.; Bolton, G.; Korner, P. Intranasal carbetocin reduces hyperphagia in individuals with Prader-Willi syndrome. *J. Clin. Investig.* **2018**, *3*, e98333. [CrossRef] [PubMed]
93. Roof, E.; Deal, C.L.; McCandless, S.E.; Cowan, R.L.; Miller, J.L.; Hamilton, J.K.; Roeder, E.R.; McCormack, S.E.; Lal, T.R.R.; Abdul-Latif, H.D.; et al. Intranasal Carbetocin Reduces Hyperphagia, Anxiousness, and Distress in Prader-Willi Syndrome: CARE-PWS Phase 3 Trial. *J. Clin. Endocrinol. Metab.* **2023**, *108*, 1696–1708. [CrossRef] [PubMed] [PubMed Central]
94. Bey, K.; Campos-Martin, R.; Klawohn, J.; Reuter, B.; Grutzmann, R.; Riesel, A.; Wagner, M.; Ramirez, A.; Kathmann, N. Hypermethylation of the oxytocin receptor gene (OXTR) in obsessive-compulsive disorder: Further evidence for a biomarker of disease and treatment response. *Epigenetics* **2022**, *17*, 642–652. [CrossRef] [PubMed]

current issues in molecular biology

MDPI

Review

The Role of Oxytocin in Polycystic Ovary Syndrome: A Systematic Review

Nicoletta Cera [1,2,*], Joana Pinto [1,3] and Duarte Pignatelli [3,4]

1 Faculty of Psychology and Education Sciences, University of Porto, 4099-002 Porto, Portugal; up201707095@up.pt
2 Research Unit in Medical Imaging and Radiotherapy, Cross I&D Lisbon Research Centre, Escola Superior de Saúde da Cruz Vermelha Portuguesa, 1300-125 Lisbon, Portugal
3 Faculty of Medicine, University of Porto, 4099-002 Porto, Portugal; duarte@med.up.pt
4 Department of Endocrinology, Centro Hospitalar Universitário de São João, 4200-319 Porto, Portugal
* Correspondence: cera.nicoletta@fpce.up.pt

Abstract: Polycystic Ovary Syndrome (PCOS) is the most common endocrine disorder that affects women of reproductive age, representing the primary cause of anovulatory infertility. The non-apeptide oxytocin (OT) plays an important role in cognitive, emotional, and reproductive functions in human beings. Oxytocin receptors are expressed in several body parts, including the ovaries. Despite this, the possible role played by oxytocin in symptoms of PCOS is not clear. The present systematic review aimed at understanding the presence of possible oxytocin level alterations in PCOS, the connection between alterations of OT levels and the symptoms of PCOS, and the effect of oxytocin administration in PCOS. After a systematic search in the principal databases, eight studies, five human and three animal, were included. Four human studies and one animal study highlighted the role played by oxytocin in fertility issues related to PCOS. Three human and two animal studies investigated the role of body weight and OT levels. Studies that analyzed oxytocin basal levels in women agreed that PCOS is associated with a reduction in the serum level of oxytocin. Two human studies and one animal study agreed about lower levels of oxytocin, confirming a possible implication of the dysfunction of OT in the pathogenesis of PCOS.

Keywords: oxytocin; PCOS; systematic review; fertility

1. Introduction

Polycystic Ovary Syndrome (PCOS) is the most common endocrine disorder that affects women of reproductive age [1]. Depending on the diagnostic criteria used, its worldwide prevalence ranges from 4% up to 20% [2]. PCOS is characterized by polycystic ovary morphology, androgen excess, and ovulatory dysfunction [3]. Because of its typical metabolic, reproductive, and psychological features, PCOS is a relevant public health concern [4]. In particular, if taken into consideration, about 75% of PCOS cases are estimated to be undiagnosed [5]. PCOS is associated with high levels of androgens, including dehydroepiandrosterone and androstenedione, of adrenal origin, as well as androstenedione and testosterone, of ovarian origin [6]. PCOS is also usually characterized by increased luteinizing hormone (LH) levels, elevated LH/FSH ratios [6–9], and low to normal follicle-stimulating hormone (FSH) levels [7,10]. However, other studies suggest there are no significant differences in LH/FSH ratios [11] between PCOS and control groups.

Furthermore, PCOS is the primary cause of anovulatory infertility [12], accounting for 80% of such cases [13,14]. Pregnant PCOS women, on the other hand, have a higher risk of developing gestational diabetes mellitus or suffering a first-trimester spontaneous abortion [15,16]. Interestingly, even though an elevated baseline LH/FSH ratio in PCOS was found to be related to poor ovulatory response, PCOS cases with elevated LH/FSH ratio were more likely to achieve a clinical pregnancy and live birth than women with normal LH/FSH [9]. Ovarian function and cycles are regulated by the hypothalamic–pituitary–ovarian (HPO) axis, particularly by GnRH and the Gonadotropins.

Since the discovery of its structure in the early 1950s [17], oxytocin (OT), a nine-amino acid neuropeptide [18], has captivated the attention of the scientific community for its role in several animal and human physiological functions. Immunohistochemical studies showed that OT is synthesized in the magnocellular cells of the paraventricular (PVN), supraoptic nuclei (SON), and parvocellular of PVN, along with a few neurons in the accessory nuclei of the human and animal hypothalamus, and is transported to the neurohypophysis where it is released in the blood circulation [19–22]. Moreover, OT is dendritically released by the hypothalamic neurons and then passively diffused to various target brain regions [23].

Tracing studies also showed long-range axonal projections of oxytocinergic neurons to several brain regions, such as the hippocampal C1 and C2 subregions and ventral subiculum, shell, and core of nucleus accumbens, island of Calleja, lateral globus pallidus lateral, lateral septal nucleus, medial amygdaloid nucleus, and prelimbic cortex [24,25]. The magnocellular PVN projections reach the posterior pituitary gland, where the OT is released in the bloodstream, acting as a hormone [26]. In this way, OT plays a central role as a neurotransmitter and peripherally as a hormone.

OT plays a crucial role in several behavioral and reproductive functions in human beings, such as breastfeeding [27], pregnancy, and parturition [28–30], but also in other processes, like bonding, decision-making [31], prosocial behavior [32,33], and physical activity [34]. It is also related to the pleasure associated with orgasm, both in males and females, being released in elevated quantities during this process [35,36].

OT receptors (OXTR) are expressed in several body parts, including the ovaries and prostate gland [37,38]. Specifically, OXTRs are found to be expressed in the granulosa cells and the small follicles in several mammal species, including humans [39], and OT also takes an unclear role in steroidogenesis [40].

Despite the growing interest in the role played by oxytocin in several human functions, its possible role in one or more psychological symptoms, such as depression, anxiety, and social cognitive impairments, or physical symptoms, such as metabolic dysfunctions or infertility, of PCOS [41] is still not clear or little studied. However, depression was found to be associated with OT levels [42], and animal evidence highlighted the improvement of depressive symptoms, such as immobility, after OT administration [43,44]. Moreover, increased OT serum levels were observed in patients with mood disorders [45], and were associated with depressive symptoms [42]. Intranasal oxytocin treatment improved sleep apnea and in general the sleep quality in patients affected by sleep disorders [46,47].

Obesity is one of the principal symptoms observed in PCOS women. Studies on mice with a deficiency of OT ($OT^{-/-}$) showed that OT deficiency is responsible for increased body weight and abdominal fat [48,49]. Interestingly, $OT^{-/-}$ and $OXTR^{-/-}$ normophagic mice develop late-onset obesity. Moreover, low OT levels were observed in diet-induced obese mice, and high OT levels observed in synaptotagmin-4-deficient mice that are protected against obesity [50]. Exogenous OT administration is found to decrease weight and increase muscular tone [51]. OT and its stable analog, carbetocin, have a negative modulatory effect on adipogenesis, with the promotion of osteogenesis in in human multipotent adipose-derived stem cells [52].

In the present systematic review, we aimed to disentangle the role played by oxytocin in PCOS, taking into account all animal and human studies that have been published. In particular, we aimed to understand (i) the presence of possible alterations of basal plasmatic OT level in PCOS, (ii) in which manner a possible alteration of the OT plasmatic level can be related to the symptoms of PCOS, and (iii) the presence of a possible effect of OT administration in PCOS.

2. Materials and Methods

The present systematic review (PROSPERO reg. n. 531987) followed the procedure recommended by the Preferred Reporting Items for Systematic Reviews and Meta-Analyses (PRISMA) guidelines [53]. We performed a computer-based search in the principal databases, such as PubMed, Web of Science, and Scopus, combining terms related to polycystic ovary syndrome and oxytocin (Figure 1).

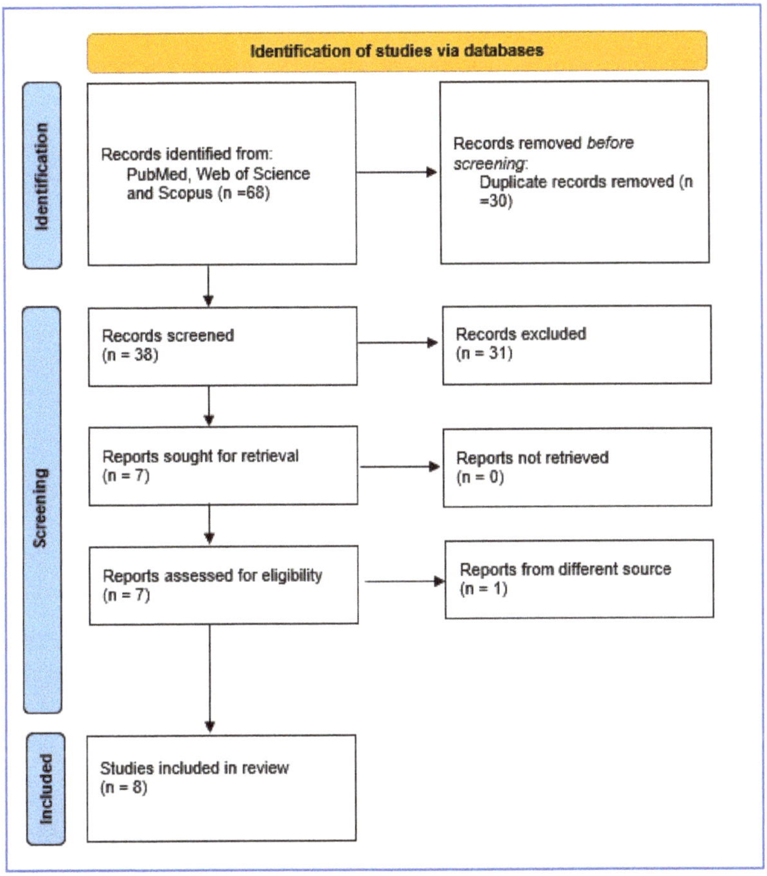

Figure 1. Flow chart of the selection process for PCOS and oxytocin. From: Page et al., 2021 [54] http://www.prisma-statement.org/ (accessed on 20 May 2024).

Moreover, our research question used the PICO strategy protocol. In particular, our research question was related to the role played by oxytocin (O—Outcome) in women or animal models with polycystic ovary syndrome (P—Population), determining the level of oxytocin in body fluids, such as blood and saliva (I—Intervention), in comparison with healthy women or animals (C—Comparison, Table 1, Supplementary Materials).

In the present systematic review, databases were selected to explore the published studies using the following keywords: "polycystic ovary syndrome" [MeSH Terms] OR ("polycystic" [All Fields] AND "ovary" [All Fields] AND "syndrome" [All Fields]) OR "polycystic ovary syndrome" [All Fields]) AND ("oxytocin" [MeSH Terms] OR "oxytocin" [All Fields] OR "oxytocin s [All Fields] OR "oxytocin [All Fields] OR "oxytocin [All Fields]", with no time limit, and using the Boolean operators AND and OR. The inclusion and exclusion criteria were determined based on the topic, study design, and population (Table 2).

Table 1. The search strategy used in the present systematic review.

Polycystic Ovary Syndrome
1. Polycystic ovary syndrome [MeSHTerms] 2. Polycystic [All Fields] 3. Ovary [All Fields] 4. Syndrome [All Fields] 5. Polycystic ovary syndrome [All Fields] OR/1-2; 4-1 AND/3-4; 1-10
Oxytocin
6. Oxytocin [MeSH Terms] 7. Oxytocin [All Fields] 8. Oxytocin s [All Fields] 9. Oxytocin [All Fields] 10. Oxytocins [All Fields] OR/6-10

Table 2. Inclusion and exclusion criteria.

Inclusion Criteria	Exclusion Criteria
✓Experimental studies ✓Randomized clinical trials ✓Clinical pilot studies ✓Case-control ✓Population genetics studies ✓Animal studies	✓Other endocrine diseases ✓Male studies ✓Reviews (scoping, narrative, and systematic) ✓Meta-analyses

In the second stage, we removed the duplicates and manually screened both the titles and abstracts to evaluate if they fulfilled the inclusion and/or exclusion criteria. After that, we retrieved the full texts of the possibly pertinent studies to verify their eligibility. Two authors independently carried out the literature search, article screening, and methodologic evaluation. Both authors discussed the results and a consensus was reached. However, a third opinion was required when a consensus was not reached.

The included studies were subsequently screened to find further articles in the reference lists related to the topic of interest. Similarly, we screened all the excluded studies to identify additional relevant bibliographic sources. To estimate the quality of the selected studies, when possible, in the current systematic review, the NOS (Newcastle–Ottawa Scale), a quality assessment scale for case-control and cohort studies (https://www.ohri.ca/programs/clinical_epidemiology/nosgen.pdf (accessed on 1 March 2024), Tables S1 and S2, Supplementary Material), was used.

Then, information associated with the characteristics of the participants and inclusion and exclusion criteria were extracted from each included article, according to the previously mentioned guidelines. The flowchart in Figure 1 depicts the steps of the selection process.

3. Results

The flowchart depicted in Figure 1 shows the selection process of the studies. We included eight published studies in the systematic review, after reaching a consensus. Moreover, we calculated Cohen's k with 87.17% (k = 0.742), indicating substantial agreement (https://idostatistics.com/cohen-kappa-free-calculator/ (accessed on 20 May 2024)) [55]. The characteristics of the studies are shown in Table 3.

Table 3. Demographics, design, assessment, and principal results as shown in the included studies.

Source	Country	Subjects	Age	Design	Assessment	Treatment	Results
Amin et al., 2023 [56]	Italy	212 women	-	Population genetics study	Single nucleotide polymorphisms (SNPs) within OXTR.	-	Out of 22 OXTR-risk variants tested, three independent variants were significantly linked to/in LD with PCOS. Three intronic variants were linked to PCOS. One intronic variant and a synonymous variant were both linked and associated with PCOS. All variants are novel and have not been previously associated with PCOS or any PCOS-related phenotype. Three of the variants were found to confer risk for PCOS, intersected with a repressed chromatin state in the ovaries.
Jahromi et al., 2018 [57]	Iran	161 women (PCOS = 80; Non-PCOS = 81)	20–35 years	Case-control	OT, AMH, BMI, LH, T, FSH, TSH, prolactin, and DHEAS. Fasting blood sugar, fasting insulin, blood sugar 2 h after 75 g glucose, insulin 2 h after 75 g glucose, and HOMA-IR.	-	The mean OT level was lower in the case group. The mean BMI, AMH, HOMA-IR, fasting insulin, and insulin 2 h after 75 g glucose were higher in the PCOS group. OT was negatively correlated to AMH when evaluated for all participants or only among controls. OT was also negatively correlated to HOMA-IR among all participants. There was not a significant relationship between OT and BMI. The calculated cutoff value for OT was 125 ng/L and for AMH was 3.6 ng/mL in the PCOS group.
Piróg et al., 2023 [58]	Poland	56 infertile women with PCOS, 18 pregnant	31.89 ± 4.59 years	Case-control	Assessment before ovarian stimulation (OS) and before hCG administration. Assessments of PNX-14, NES-1, DA, and OT serum levels were performed. Other tests: LH, FSH, estradiol, PRL, AMH, and BMI.		In the whole cohort of patients, OT levels were weakly associated with BMI ($r = 0.26$, $p = 0.048$) and FSH ($r = 0.47$, $p = 0.0002$). Pregnant group: positive correlations between baseline OT and PRL ($r = 0.47$; $p = 0.04$), as well as OT and NES-1 ($r = 0.55$; $p = 0.02$). OT level increases were associated with positive pregnancy rates. In the post-OS, in pregnant PCOS, OT was 2.7 times lower than for non-pregnant women.

Table 3. *Cont.*

Source	Country	Subjects	Age	Design	Assessment	Treatment	Results
Masrour et al., 2018 [59]	Iran	150 women	19–39 (29 ± 4.48) years	Clinical trial	OT, HCG, FSH, prolactin, follicle number, and progesterone.	The three groups at random received: 100 mg clomiphene-citrate + 8 units of OT; 100 mg clomiphene-citrate + 10,000 units of HCG; 100 mg clomiphene citrate + 8 units of OT + 10,000 units of HCG.	There was no major difference among the groups regarding the ovulation rate or the number of follicles, nor were there any significant side effects observed in any groups.
Ochsenkühn et al., 2010 [60]	Germany	86 women	18–42 (34.2 ± 4.3) years	Randomized, double-blind, placebo-controlled clinical pilot study	Follicle number, double endometrial width, estradiol, LH, and progesterone. To assess male fertility: Semen parameters (native sperm concentration, progressive motility, normal sperm morphology, semen volume, and total progressive motile sperm count).	132 homologous IUI cycles with nasal application of placebo or eight IU OT following IUI.	In 132 IUI cycles of 86 women, 17 pregnancies were achieved, accounting for a pregnancy rate of 12.9% per IUI cycle. The pregnancy rates were 13.4% per IUI cycle in the placebo group, and 12.3% per IUI cycle in the OT group. As such, the difference was not statistically significant. No relevant side effects were observed in both groups.

Table 3. *Cont.*

Source	Country	Subjects	Age	Design	Assessment	Treatment	Results
Sajadi et al., 2018 [61]	Iran	14 female rats (PCOS = 7; Control = 7)	75–95 days	Randomized clinical trial	CCh; OT.	Rats in the experimental group were subcutaneously injected with 5 m/g of free testosterone on gestational day 20; controls received solvent. The contractions of isolated uterus in offspring of both groups were recorded by the power lab system, after exposure to CCh and OT.	Uterine contractions were more irregular in PCOS rats than controls, after exposure to both contractile agonists.
Isawa et al., 2019 [62]	Japan	Female rats: Seven PCOS–OT rats: DHT-treated rats (PCOS) receiving 380 µg/day OT. Six PCOS–saline rats: without OT treatment, treated with saline solution. Seven saline rats: non-DHT-treated rats with only saline treatment.	26 days	Randomized clinical trial	Serum levels of AST, ALT, and LDH. Histological analysis of ovaries and adipocytes. Hypothalamic mRNA levels of NPY, POMC, OT, and OXTR.	At 8 weeks, seven PCOS rats were implanted with OT (380µg/day)-filled minipumps which supplied 12 µL/day for 14 days. Six PCOS rats and seven control rats were implanted with saline-filled minipumps.	Body weight changes were significant between PCOS–OT and PCOS–saline rats, with PCOS rats being lighter. The mean visceral fat weight of the PCOS–OT rats did not differ from that of the saline-control rats. No difference in the number of cystic follicles was seen between the PCOS–OT and PCOS–saline rats. No difference in hypothalamic mRNA expression of NPY, POMC, OT, and OXTR among the three groups. No difference in AST and ALT among the three groups. LDH levels were higher in PCOS–saline rats than in the other two groups.

Table 3. *Cont.*

Source	Country	Subjects	Age	Design	Assessment	Treatment	Results
Yamamoto et al., 2022 [63]	Japan	16 female rats (PCOS$_{Chronic}$ = 8; Control$_{Cronic}$ = 8; PCOS$_{Acute}$ = 8; Control$_{Acute}$ =8)	28 days	Randomized clinical trial	OT, serum level. Hypothalamic mRNA levels of NPY, POMC, OT, OXTR, prepro-orexin, and agouti-related protein. Visceral fat mRNA expression of OT and OXTR	At 10 weeks after the surgical day, all rats were injected with saline for seven consecutive days, then injected with OT (1200 µg/kg, 0.4 to 0.5 mL injection volume) for the following seven consecutive days.	The serum OT level was lower in PCOS model rats than in control rats, whereas the hypothalamic OT mRNA expression level did not differ between them. Acute intraperitoneal administration of OT during the dark phase reduced the body weight gain and food intake in PCOS model rats. However, these effects were not observed in control rats. In contrast, chronic administration of OT decreased the food intake in both the PCOS model rats and control rats.

Abbreviations: OXTR, oxytocin receptor; LD, linkage disequilibrium; PCOS, Polycystic Ovary Syndrome; OT, oxytocin; AMH, anti-mullerian hormone; BMI, Body Mass Index; LH, luteinizing hormone; T, total testosterone; FSH, follicle-stimulating hormone; TSH, thyroid-stimulating hormone; DHEAS, dehydroepiandrosterone sulfate; HOMA-IR, insulin resistance index; HCG, chorionic gonadotropin; IUI, intrauterine insemination; CCh, carbachol. PNX-14, phoenixin-14; NES-1, nesfatin-1; DA, dopamine; PRL, prolactin. AST, aspartate aminotransferase; ALT, alanine aminotransferase; LDH, lactate dehydrogenase; NPY, neuropeptide Y; POMC, proopiomelanocortin.

Both human and animal studies were included. However, the selected studies did not show homogeneity in terms of both study design and population. The five human studies included 609 women and were published between 2010 and 2023. Among these, two were randomized or pseudo-randomized clinical trials, two were case-control studies, and another one was a population genetics study.

Moreover, the principal focus of the included studies was related to the relationship between fertility and nasal oxytocin administration. Similarly, Ochsenkühn et al. included patients with PCOS as a cause of infertility together with different infertile groups [60]. Conversely, the clinical trial performed by Masrour et al. [59] included only infertile patients with PCOS [56]. During the clinical trial, the patients underwent eight units of OT, but the authors did not observe any significant changes in terms of infertility. Ochsenkühn et al., who did not observe any improvement in the fertility of the PCOS group after OT treatment [60], obtained similar results. Despite the administration of OT, these two above-mentioned studies did not assess the level of blood or salivary oxytocin in the participants [60,61]. However, since the main topic of both RCTs was to assess the effect of eight IU of intranasal OT on infertility in PCOS, the studies found the treatment not relevant to improving fertility in PCOS patients affected by infertility. As underlined by Ochsenkühn et al., the failure to detect the effect of OT on the pregnancy rate could be the result of inadequacy in dose and or mode of administration [60]. The level of OT in PCOS before a treatment can be relevant, but none of the two RCTs collected such samples in their PCOS groups. However, Jahromi et al. compared the level of oxytocin and other hormones (Table 3) in infertile women both with or without PCOS [57]. These authors found that in PCOS, the mean level of OT was inferior to the non-PCOS group, with a mean value of 124.94 ng/L compared to 207.42 ng/L ($p < 0.0001$). Moreover, since the anti-mullerian hormone usually shows high levels in PCOS, it was negatively correlated with oxytocin, and the same occurred with insulin resistance. However, the authors did not observe a significant effect of BMI on oxytocin in both groups. According to the authors, the hormonal imbalances in the hypothalamic–pituitary–ovarian (HPO) axis, namely the high LH and low FSH in the PCOS group, could be connected to the lower oxytocin levels. These low levels can in turn be implicated in chronic anovulation. Notably, this study was the first that proposed a cut-off value of the oxytocin level in women with PCOS.

Similar results were observed in a case-control study [58] assessing the hypothalamic–pituitary–ovary axis dysfunction in a sample of 56 infertile PCOS women before ovarian stimulation with 2.5 mg of letrozole and before human chorionic gonadotropin (hCG) administration. The authors assessed the serum levels of OT, dopamine (DA), phoenixin-14 (PNX-14), and nesfatin-1 (NEF-1) in groups of pregnant and non-pregnant PCOS women. Moreover, FSH, LH, AMH, TSH, and prolactin were assessed. In the whole sample, they found a weak association of OT with BMI and a stronger one with FSH ($p < 0.0002$). However, in the pregnant group, higher baseline NES-1 and OT levels (+29.2% and +44%) were observed. Similarly, OT level increases were associated with positive pregnancy rates. After OS in pregnant women, the OT levels increased compared to non-pregnant women.

Finally, Amin and colleagues [56] assessed the presence of the polymorphisms of the gene responsible for the expression of the receptor of oxytocin (OXTR) in 212 Italian PCOS patients. OXTR is widely expressed in the human body, including in the brain and ovary tissue [64]. In their genetic population study, the authors tested the hypothesis whereby the OXTR variants are in linkage disequilibrium with PCOS in Italian families. They found that five variants, out of 22, were significantly ($p < 0.05$) linked to or were in linkage disequilibrium with PCOS. However, all of these variants were not previously related to clinical manifestations of PCOS. Nonetheless, three of them (rs60345038, rs35498753, and rs237900) were found to intersect with the repressed chromatin state in the ovaries, with a negative OXTR gene expression.

In the present systematic review, animal studies were included. The three animal studies included 50 female rats and were published in 2018 and 2022, respectively. These three studies were classified as RCTs and both administered OT to PCOS rat models.

However, Sajadi et al. [61] also administered carbachol. Despite the use of OT, the main objective of the two studies was different. Sajadi and colleagues [61] studied the uterine contraction and tone in PCOS and non-PCOS rats after administration of OT or carbachol, while Yamamoto et al. [63] measured the effects of the administration of acute and chronic OT on metabolic disorders, as well as the changes in endogenous OT, in PCOS model rats.

Sajadi et al. [61] found that PCOS rats showed more irregular uterine contractions than controls, and that after being exposed to carbachol, their frequency and resting tone were significantly increased compared to controls. However, after the exposure to OT, there were no differences in frequency, resting tone, and amplitude of rhythmic contractions between both groups.

Yamamoto et al. [63] found that PCOS model rats showed lower serum OT levels than control rats. Nonetheless, the two groups did not differ in hypothalamic OT mRNA expression levels. The authors found that there were reductions in body weight gain and food intake only in PCOS model rats after acute intraperitoneal OT administration during the dark phase, whereas the chronic administration of OT decreased the food intake in both the PCOS model rats and control rats. Similar results have been observed by Isawa et al. [62]. They found no difference in the OT and OXTR mRNA expression in PCOS rats and PCOS rats treated with OT. The authors found that OT administration caused reductions in body weight, food intake, visceral fat weight, and adipocyte size in a DHT-induced rat model of PCOS. Among the included studies, one human study and two animal studies investigated the association between OT levels and BMI in humans or BW in animals (Table 4).

Table 4. BMI and body weight associations with OT levels and OT administration, as reported in the included studies.

Source	Assessment of BMI/BW	Effect of OT Administration on BMI/BW	Association of BMI/BW and OT Serum Levels
Amin et al., 2023 [56]	No	-	Risk variants in the OXTR gene pose an increased risk for obesity (BMI > 30).
Jahromi et al., 2018 [57]	Yes	-	The mean oxytocin level was lower and the mean BMI was higher in PCOS. More PCOS women had a BMI > 25 than controls. However, the relationship between BMI and oxytocin was not significant.
Piróg et al., 2023 [58]	Yes	-	BMI levels were assessed both for pregnant and non-pregnant women. No between group differences in BMI. OT levels were weakly associated with BMI ($r = 0.26$, $p = 0.048$).
Masrour et al., 2018 [59]	No	-	-
Ochsenkühn et al., 2010 [60]	No	-	-
Sajadi et al., 2018 [61]	No	-	-
Isawa et al., 2019 [62]	Yes	The PCOS–OT group was significantly lighter than the PCOS-saline group. BW changes seen in the PCOS–OT group were significantly smaller than in the PCOS–saline group. BW was not quantified in the text.	-
Yamamoto et al., 2022 [63]	Yes	Exogenous administration of OT during the dark phase reduced BW gain only in PCOS model rats, and seven days of OT administration showed no significant differences between PCOS and control rats.	-

Abbreviations: BMI, body mass index; BW, body weight.

4. Discussion

Despite the relevant role played by the nonapeptide oxytocin in several functions, such as social cognition, metabolic regulation, and reproduction, only a few recent studies have investigated its role in PCOS. Given the novelty of the topic, the present systematic review took into consideration all the studies that investigated the role played by not only serum oxytocin but also randomized clinical trials with the administration of synthetic oxytocin. The present review also took into account animal studies, since most of the current knowledge about PCOS was obtained by studies on rat models. These played a crucial role in studying and gaining insights into human pathologies during the last two centuries. Despite some similarities between animals and humans, they also show distinct characteristics. Acknowledging important disparities regarding OT between animals and humans, the insights from rat studies are nevertheless important for the comprehension of the role played by OT in PCOS.

However, despite the lack of studies about the role played in PCOS and its related symptoms or comorbidities, such as metabolic syndrome, obesity, cardiovascular risk, and mood disorders [41], several studies have been conducted disentangling the possible role of serum OT in these pathological conditions. One of the comorbidities is represented by metabolic syndrome. Among women affected by metabolic syndrome, those with prediabetes and type two diabetes showed significantly lower serum OT levels than those affected by metabolic syndrome without diabetes [65]. In obese women, lower levels of plasmatic OT than in non-obese women were observed [66]. Metabolic alterations were also studied in PCOS rats [63]. Rats to which OT was administered showed a significant decrease in weight and food intake [63]. However, this study did not quantify the lipolysis in the adipose tissue of the rats. Previous findings indicated that OT administration significantly reduced the area of adipocytes, the serum triglyceride, aspartate aminotransferase level alanine aminotransferase, and alkaline phosphatase in ovariectomized rats [67]. Moreover, according to the results reported by Yamamoto and colleagues [63], an OT-mediated mechanism was involved in obesity and food intake in the PCOS rats. However, in PCOS the authors found a decrease in serum OT levels and no difference in the hypothalamic mRNA expression of OT and OXTR after the acute systemic administration of OT. In peri- and post-menopausal rats, the administration of OT reduced the body weight and adipocyte size, without affecting the serum levels of hepatic enzymes [68]. The menopausal period is considered a risk factor for visceral adiposity and metabolic disorders. Several studies found reduced serum OT levels in pre- and postmenopausal women [69–72].

Several findings showed that oxytocin is useful for treating obesity and preventing metabolic disorders [73–78]. Among the variants observed in PCOS in the population study of Amin et al. [56], rs60345038 was also found to be relevant for type 2 diabetes [79], indicating a possible predisposition for diabetes in PCOS patients. OT values were found to be significantly lower in obese post-menopausal women compared to obese pre-menopausal women. Interestingly, significantly lower OT serum levels were observed between obese and normal-weight postmenopausal women [69]. Significant lower OT levels were also found in women with surgical menopause as compared to those with normal menopause [70]. During menopause and in T2D patients, the cardiovascular risk increases. OT and OXTR were found in the atria and ventricles of rat hearts and in the large blood vessels, indicating a possible autocrine and paracrine role played by OT, regulating the vascular tone and the atrial and ventricular load [80,81]. In ovariectomized rats with deprivation of ovarian hormones, a reduction of the OXTR mRNA levels in PVN subnuclei was observed causing autonomic dysregulation [82].

Most of the studies underestimated the relevance of the levels of OT in women or female rats with PCOS by not reporting possible OT basal level differences. Notably, only three studies found that in PCOS, the levels of OT were lower than in healthy controls. Indeed, two human studies and one animal study agreed about lower levels of PCOS, confirming a possible implication of OT in the pathogenesis of the syndrome, which could negatively affect the effect of OT administrations. Despite the complexity of the symptoms of PCOS, anovulatory infertility represents one of the most relevant issues. In PCOS, the normal follicular development is compromised and the ovaries contain an excess of small antral follicles. Estrogen is an extracellular signal that can regulate the expression of OXTR. The estrogen receptor (E2) is localized in the nucleus, cytoplasm, and mitochondria. The two E2 subtypes (E2 α and E2β) affect both OT and OXTR expression: E2 α induces OXTR expression, whereas the activated E2β induces OT transcription in rat brains [83,84].

E2 is secreted by the granulosa cells of developing antral follicles, upon which FSH controls the maturation and the selection of the dominant pre-ovulatory follicle and triggers the LH required for ovulation [85–87]. The effect of E2 is mediated by estradiol receptors α (E2 α) and β (E2 β), which showed a different expression in the granulosa cells in developing follicles.

A few studies in humans and primates have found that E2 β is the principal mediator of E2 in granulosa cells. Since E2 β is detected in both pre-antral and mature follicles. Thus, E2 is considered as a marker of follicular quality. In PCOS patients, the selection of predominant pre-ovulatory follicle is arrested, and the observed low levels of FSH, in PCOS, do not allow the stimulation of follicle maturation. Follicular fluid in PCOS women showed low levels of E2 [78,79]. However, this mechanism is still unknown [88].

A significant increase of OT and E2 has been found in the myometrium during labor in healthy women, indicating a possible paracrine OT stimulation by E2 relevant to the uterus [89]. According to Pirog and colleagues, OT can be considered a predictor of pregnancy before ovarian stimulation therapy [58]. However, a recent meta-analysis assessed the OT concentrations during the stages of the menstrual cycle in normal cycling women, finding an increase in serum OT during the follicular phase. Interestingly, the change in oxytocin concentrations from the follicular phase to ovulation was larger than the change from the follicular phase to the luteal one [90]. In the follicular fluid of infertile women without PCOS, Tachibana et al. [91] found that the OT level was extremely low, and not related to the OT serum level. OT and E2 serum levels increased during ovulatory and luteal phases [91]. However, Franik et al. [92] found that values of E2/T and E2/A indexes, relevant to homeostasis model assessment of insulin resistance (HOMA-IR), were significantly lower in the PCOS than non-PCOS subjects, but did not differ significantly between the obese and normal weight groups.

The focus of the studies that we included was only related to the infertility issues related to PCOS. In different animal models, oxytocin seems to have a role in fertility by promoting the release of PGF2α from endometrial cells. Moreover, OT is involved in the process of luteolysis [93,94]. A similar mechanism is also present in humans, but a positive role of administration of OT in folliculogenesis and increasing pregnancy rates in both humans and animals was reported in a few studies [95,96]. Ochsenkühn and colleagues [60] did not observe any increase in the pregnancy rate in couples affected by PCOS and infertility after eight IU intranasal OT administrations. The role of OT in uterine contractions has been assessed in a recent study [61] which compared PCOS and non-PCOS female rats' uteri. After administration of the oxytocin, no significant differences were observed in the amplitude, tone, and frequency in the rhythmic uterine contractions of PCOS rats. However, the increase in the dose of OT stimulated higher levels of tone, with a decrease in the contraction frequency in PCOS rats' uterine tissues. It is well-studied that in the myometrium, the number of oxytocin receptors increases during pregnancy [97], allowing the uterus to become more sensitive to oxytocin and thus affecting the pattern of contractions during pregnancy and labor. Despite this, Leonhardt et al. [98] did not find uterine morphological differences using magnetic resonance imaging, and less uterine peristaltic movement was found in PCOS assessed with transvaginal ultrasonography.

Despite the increase in the number of receptors, their genetic expression can be different in PCOS. Amin et al. [56] reported five novel genetic variants for the receptor of OT (OXTR) associated with the risk of developing PCOS in multigenerational Italian families. These variants of the OXTR gene were found to be related to the principal symptoms of

PCOS, such as anovulation or oligovulation, hyperandrogenism, polycystic ovaries, and the increased risk for metabolic alterations [79].

Some of the variants found by Amin are considered relevant for vulnerability to different disorders. The variant OXTR rs237902 found by Amin et al. [56] has been associated with schizophrenia vulnerability. Specifically, a significant association between rs237902 and negative symptoms, such as blunted affect, alogia, avolition asociality, and anhedonia, in schizophrenic patients and an overrepresentation in male aggressive children were observed [99–101].

The same variant (rs237902) was found to be related to autism spectrum disorder [102]. According to Dinsdale and Crespi [103], the relevance of oxytocin and possible alterations in the OT and OTXR system in PCOS is still not well understood. According to their review, PCOS and autism spectrum disorder (ASD) share several behavioral features that could induce speculation about a possible common role of OT in the two disorders [103]. Similarly, a recent meta-analysis [104] assessed the studies that reported the odds of PCOS women having a child with autism spectrum disorder and the risk of ASD in women with PCOS. The authors included ten studies finding that, according to the available evidence, PCOS women have increased odds of having a child with ASD. Regarding the evidence on the prevalence of ASD in PCOS women, results suggest that women with PCOS are more likely to be diagnosed with ASD. Particularly, Hergüner et al [105], after administration of the Autism-Spectrum Quotient (AQ) to a sample of PCOS women, found that patients showed higher total AQ and communication scores than age- and BMI-matched healthy women.

AQ is a 50-item self-report questionnaire that assesses social skills, communication skills, imagination abilities, attention switching, and attention to detail. AQ shows good psychometric properties.

5. Conclusions

PCOS is a multifaceted syndrome involving several symptoms affecting the patients' quality of life at different levels. The present review described the studies that analyzed the levels, or the effects of administering, OT in PCOS.

The studies reported in the present systematic review took into account only a part of the possible roles played by OT in PCOS. Most of the studies highlighted the role played by OT in fertility issues related to PCOS, and only one study found an increased pregnancy rate concomitant with high OT levels. However, studies that analyzed the basal levels of OT in PCOS women agreed that is accompanied by a reduction in the serum level of oxytocin. One could speculate that OT acts in synergy with FSH to promote follicular development towards ovulation and low levels of OT in PCOS which, together with low levels of FSH, may contribute to the anovulation that is typical of PCOS. Another possible interplay occurs between OT levels and the ovulatory LH surge, suggesting a synergy in ovulation. This also might explain why PCOS cases that have low levels of OT do not have normal rates of ovulation.

However, less still is known about possible molecular mechanisms that may also be able to affect the central tone, resulting in cognitive and behavioral alterations in normal and PCOS women. The possibility of altered OXTRs in PCOS resulting from SNP variants supports this view.

Based on previous, non-PCOS evidence, a possible OT autocrine/paracrine imbalance in the granulosa cells may be hypothesized, influenced by anovulation that is observed in the PCOS phenotype A, in which the polycystic ovary morphology is present.

Further, the partial deprivation of ovarian hormones, as shown by studies in ovariectomized rats, can affect the OTXR mRNA hypothalamic levels, resulting in behavioral alterations and obesity.

It was noticeable that different dosage regimens of OT administration or OT administration patterns were not sufficiently studied. Furthermore, despite the novelty and relevance of the topic, none of these studies analyzed the effect of OT administration on prosocial behavior or in couples' relationships and sexual satisfaction in PCOS.

6. Future Directions

Several outstanding issues need to be clarified by further studies that could disentangle the role played by OT in follicular development and ovulation/anovulation. Moreover, if the administration of OT, and in which dose and pattern of administration, can result in the improvement of ovulation in PCOS women should be investigated. Further studies are needed to clarify if OT basal serum levels are associated with metabolic disorders in PCOS. Moreover, none of the studies that were included in the present systematic review assessed the relationship between OT and social behavior and psychiatric comorbidities, such as mood and anxiety disorders, in PCOS and non-PCOS women.

Supplementary Materials: The following supporting information can be downloaded at: https://www.mdpi.com/article/10.3390/cimb46060313/s1, Ref. [106] are cited in Supplementary Materials file. Table S1: Newcastle—Ottawa quality assessment scale case control studies; Table S2: Newcastle—Ottawa quality assessment scale cohort studies.

Author Contributions: Conceptualization, N.C., D.P. and J.P.; methodology, N.C. and J.P.; investigation, N.C. and J.P.; resources, N.C. and J.P.; data curation, J.P.; writing—original draft preparation, N.C. and J.P.; writing—review and editing, N.C. and D.P.; visualization, N.C.; supervision, D.P. All authors have read and agreed to the published version of the manuscript.

Funding: This research received no external funding.

Data Availability Statement: All the data are shown in the full text.

Conflicts of Interest: The authors declare no conflicts of interest.

Abbreviations

PCOS	Polycystic Ovary Syndrome
OT	oxytocin
PVN	paraventicular nucleus
SON	supraoptic nucleus
LH	luteinizing hormone
CA	cornu ammonis
PGF2α	Prostaglandin F2α
LD	linkage disequilibrium
AMH	anti-mullerian hormone
BMI	Body Mass Index
T	total testosterone
FSH	follicle-stimulating hormone
TSH	thyroid-stimulating hormone
DHEAS	dehydroepiandrosterone sulfate
HOMA-IR	insulin resistance index
HCG	chorionic gonadotropin
IUI	intrauterine insemination
CCh	carbachol
PNX-14	phoenixin-14
NES-1	nesfatin-1
DA	dopamine
PRL	prolactin
SNPs	single nucleotides polymorphism
P4	progesterone
E2	estradiol
AST	aspartate aminotransferase
ALT	alanine aminotransferase
LDH	lactate dehydrogenase
NPY	Neuropeptide Y

POMC	proopiomelanocortin
E2/T	estradiol/testosterone ratio
E2/A	estradiol/androstenedione ratio

References

1. Witchel, S.F.; Teede, H.J.; Peña, A.S. Curtailing PCOS. *Pediatr. Res.* **2020**, *87*, 353–361. [CrossRef] [PubMed]
2. Deswal, R.; Narwal, V.; Dang, A.; Pundir, C.S. The Prevalence of Polycystic Ovary Syndrome: A Brief Systematic Review. *J. Hum. Reprod. Sci.* **2020**, *13*, 261–271. [CrossRef] [PubMed]
3. Chiaffarino, F.; Cipriani, S.; Dalmartello, M.; Ricci, E.; Esposito, G.; Fedele, F.; La Vecchia, C.; Negri, E.; Parazzini, F. Prevalence of polycystic ovary syndrome in European countries and USA: A systematic review and meta-analysis. *Eur. J. Obstet. Gynecol. Reprod. Biol.* **2022**, *279*, 159–170. [CrossRef] [PubMed]
4. Azziz, R.; Carmina, E.; Chen, Z.; Dunaif, A.; Laven, J.S.; Legro, R.S.; Lizneva, D.; Natterson-Horowtiz, B.; Teede, H.J.; Yildiz, B.O. Polycystic ovary syndrome. Nature reviews. *Dis. Primers* **2016**, *2*, 16057. [CrossRef] [PubMed]
5. Carmina, E. Diagnosis of polycystic ovary syndrome: From NIH criteria to ESHRE-ASRM guidelines. *Minerva Ginecol.* **2004**, *56*, 1–6. [PubMed]
6. Blank, S.K.; McCartney, C.R.; Chhabra, S.; Helm, K.D.; Eagleson, C.A.; Chang, R.J.; Marshall, J.C. Modulation of gonadotropin-releasing hormone pulse generator sensitivity to progesterone inhibition in hyperandrogenic adolescent girls—Implications for regulation of pubertal maturation. *J. Clin. Endocrinol. Metab.* **2009**, *94*, 2360–2366. [CrossRef] [PubMed]
7. Haqq, L.; McFarlane, J.; Dieberg, G.; Smart, N. Effect of lifestyle intervention on the reproductive endocrine profile in women with polycystic ovarian syndrome: A systematic review and meta-analysis. *Endocr. Connect.* **2014**, *3*, 36–46. [CrossRef] [PubMed]
8. Malini, N.A.; Roy George, K. Evaluation of different ranges of LH: FSH ratios in polycystic ovarian syndrome (PCOS)—Clinical based case control study. *Gen. Comp. Endocrinol.* **2018**, *260*, 51–57. [CrossRef] [PubMed]
9. Xia, Q.; Xie, L.; Wu, Q.; Cong, J.; Ma, H.; Li, J.; Cai, W.; Wu, X. Elevated baseline LH/FSH ratio is associated with poor ovulatory response but better clinical pregnancy and live birth in Chinese women with PCOS after ovulation induction. *Heliyon* **2023**, *9*, e13024. [CrossRef]
10. Kriedt, K.J.; Alchami, A.; Davies, M.C. PCOS: Diagnosis and management of related infertility. *Obstet. Gynaecol. Reprod. Med.* **2019**, *29*, 1–5. [CrossRef]
11. Cho, L.W.; Jayagopal, V.; Kilpatrick, E.S.; Holding, S.; Atkin, S.L. The LH/FSH ratio has little use in diagnosing polycystic ovarian syndrome. *Ann. Clin. Biochem.* **2006**, *43 Pt 3*, 217–219. [CrossRef] [PubMed]
12. Joham, A.E.; Norman, R.J.; Stener-Victorin, E.; Legro, R.S.; Franks, S.; Moran, L.J.; Boyle, J.; Teede, H.J. Polycystic ovary syndrome. *Lancet Diabetes Endocrinol.* **2022**, *10*, 668–680. [CrossRef] [PubMed]
13. Balen, A.H.; Morley, L.C.; Misso, M.; Franks, S.; Legro, R.S.; Wijeyaratne, C.N.; Stener-Victorin, E.; Fauser, B.C.; Norman, R.J.; Teede, H. The management of anovulatory infertility in women with polycystic ovary syndrome: An analysis of the evidence to support the development of global WHO guidance. *Hum. Reprod. Update* **2016**, *22*, 687–708. [CrossRef] [PubMed]
14. Melo, A.S.; Ferriani, R.A.; Navarro, P.A. Treatment of infertility in women with polycystic ovary syndrome: Approach to clinical practice. *Clinics* **2015**, *70*, 765–769. [CrossRef]
15. Sirmans, S.M.; Pate, K.A. Epidemiology, diagnosis, and management of polycystic ovary syndrome. *Clin. Epidemiol.* **2013**, *6*, 1–13. [CrossRef] [PubMed]
16. Salley, K.E.; Wickham, E.P.; Cheang, K.I.; Essah, P.A.; Karjane, N.W.; Nestler, J.E. Glucose intolerance in polycystic ovary syndrome—A position statement of the Androgen Excess Society. *J. Clin. Endocrinol. Metab.* **2007**, *92*, 4546–4556. [CrossRef] [PubMed]
17. Du Vigneaud, V.; Ressler, C.; Swan, J.M.; Roberts, C.W.; Katsoyannis, P.G. Oxytocin: Synthesis. *J. Am. Chem. Soc.* **1954**, *76*, 3115–3118. [CrossRef]
18. Freund-Mercier, M.J.; Moos, F.; Poulain, D.A.; Richard, P.; Rodriguez, F.; Theodosis, D.T.; Vincent, J.D. Role of central oxytocin in the control of the milk ejection reflex. *Brain Res. Bull.* **1988**, *20*, 737–741. [CrossRef] [PubMed]
19. Dierickx, K.; Vandesande, F. Immunocytochemical demonstration of separate vasopressin Neurophysin and oxytocin-neurophysin neurons in the human hypothalamus. *Cell Tissue Res.* **1979**, *196*, 203–212. [CrossRef]
20. Carter, C.S. Oxytocin pathways and the evolution of human behavior. *Annu. Rev. Psychol.* **2014**, *65*, 17–39. [CrossRef]
21. Freda, S.N.; Priest, M.F.; Badong, D.; Xiao, L.; Liu, Y.; Kozorovitskiy, Y. Brainwide input-output architecture of paraventricular oxytocin and vasopressin neurons. *bioRxiv* **2022**. [CrossRef]
22. Manjila, S.B.; Betty, R.; Kim, Y. Missing pieces in decoding the brain oxytocin puzzle: Functional insights from mouse brain wiring diagrams. *Front. Neurosci.* **2022**, *16*, 1044736. [CrossRef] [PubMed]
23. Landgraf, R.; Neumann, I.D. Vasopressin and oxytocin release within the brain: A dynamic concept of multiple and variable modes of neuropeptide communication. *Front. Neuroendocrinol.* **2004**, *25*, 150–176. [CrossRef] [PubMed]
24. Knobloch, H.S.; Charlet, A.; Hoffmann, L.C.; Eliava, M.; Khrulev, S.; Cetin, A.H.; Osten, P.; Schwarz, M.K.; Seeburg, P.H.; Stoop, R.; et al. Evoked axonal oxytocin release in the central amygdala attenuates fear response. *Neuron* **2012**, *73*, 553–566. [CrossRef] [PubMed]

25. Carcea, I.; Caraballo, N.L.; Marlin, B.J.; Ooyama, R.; Riceberg, J.S.; Mendoza Navarro, J.M.; Opendak, M.; Diaz, V.E.; Schuster, L.; Alvarado Torres, M.I.; et al. Oxytocin neurons enable social transmission of maternal behaviour. *Nature* **2021**, *596*, 553–557. [CrossRef]
26. Ludwig, M.; Leng, G. Dendritic peptide release and peptide-dependent behaviours. *Nat. Rev. Neurosci.* **2006**, *7*, 126–136. [CrossRef] [PubMed]
27. UvnäsMoberg, K.; Ekström-Bergström, A.; Buckley, S.; Massarotti, C.; Pajalic, Z.; Luegmair, K.; Kotlowska, A.; Lengler, L.; Olza, I.; Grylka-Baeschlin, S.; et al. Maternal plasma levels of oxytocin during breastfeeding—A systematic review. *PLoS ONE* **2020**, *15*, e0235806. [CrossRef] [PubMed]
28. Morris, M.; Stevens, S.W.; Adams, M.R. Plasma oxytocin during pregnancy and lactation in the cynomolgus monkey. *Biol. Reprod.* **1980**, *23*, 782–787. [CrossRef]
29. Prevost, M.; Zelkowitz, P.; Tulandi, T.; Hayton, B.; Feeley, N.; Carter, C.S.; Joseph, L.; Pournajafi-Nazarloo, H.; Ping, E.Y.; Abenhaim, H.; et al. Oxytocin in pregnancy and the postpartum: Relations to labor and its management. *Front. Public Health* **2014**, *2*, 63777. [CrossRef]
30. Levine, A.; Zagoory-Sharon, O.; Feldman, R.; Weller, A. Oxytocin during pregnancy and early postpartum: Individual patterns and maternal-fetal attachment. *Peptides* **2007**, *28*, 1162–1169. [CrossRef]
31. Bozorgmehr, A.; Alizadeh, F.; Sadeghi, B.; Shahbazi, A.; Ofogh, S.N.; Joghataei, M.T.; Razian, S.; Heydari, F.; Ghadirivasfi, M. Oxytocin moderates risky decision-making during the Iowa gambling task: A new insight based on the role of oxytocin receptor gene polymorphisms and interventional cognitive study. *Neurosci. Lett.* **2019**, *708*, 134328. [CrossRef] [PubMed]
32. Zak, P.J.; Kurzban, R.; Matzner, W.T. Oxytocin is associated with human trustworthiness. *Horm. Behav.* **2005**, *48*, 522–527. [CrossRef] [PubMed]
33. Winslow, J.T.; Insel, T.R. The social deficits of the oxytocin knockout mouse. *Neuropeptides* **2002**, *36*, 221–229. [CrossRef] [PubMed]
34. Heinrichs, M.; von Dawans, B.; Domes, G. Oxytocin, vasopressin, and human social behavior. *Front. Neuroendocrinol.* **2009**, *30*, 548–557. [CrossRef] [PubMed]
35. Cera, N.; Vargas-Cáceres, S.; Oliveira, C.; Monteiro, J.; Branco, D.; Pignatelli, D.; Rebelo, S. How relevant is the systemic oxytocin concentration for human sexual behavior? A systematic review. *Sex. Med.* **2021**, *9*, 100370. [CrossRef] [PubMed]
36. Caruso, S.; Mauro, D.; Scalia, G.; Palermo, C.I.; Rapisarda AM, C.; Cianci, A. Oxytocin plasma levels in orgasmic and anorgasmic women. *Gynecol. Endocrinol.* **2018**, *34*, 69–72. [CrossRef]
37. Uvnäs-Moberg, K. Oxytocin may mediate the benefits of positive social interaction and emotions. *Psychoneuroendocrinology* **1998**, *23*, 819–835. [CrossRef]
38. Tabak, B.A.; Leng, G.; Szeto, A.; Parker, K.J.; Verbalis, J.G.; Ziegler, T.E.; Lee, M.R.; Neumann, I.D.; Mendez, A.J. Advances in human oxytocin measurement: Challenges and proposed solutions. *Mol. Psychiatry* **2023**, *28*, 127–140. [CrossRef]
39. Yamamoto, K.; Nakano, Y.; Iwata, N.; Soejima, Y.; Suyama, A.; Hasegawa, T.; Otsuka, F. Oxytocin enhances progesterone production with upregulation of BMP-15 activity by granulosa cells. *Biochem. Biophys. Res. Commun.* **2023**, *646*, 103–109. [CrossRef]
40. Saller, S.; Kunz, L.; Dissen, G.; Stouffer, R.; Ojeda, S.; Berg, D.; Berg, U.; Mayerhofer, A. Oxytocin receptors in the primate ovary: Molecular identity and link to apoptosis in human granulosa cells. *Hum. Reprod.* **2010**, *25*, 969–976. [CrossRef]
41. Pinto, J.; Cera, N.; Pignatelli, D. Psychological symptoms and brain activity alterations in women with PCOS and their relation to the reduced quality of life: A narrative review. *J. Endocrinol. Investig.* **2024**, 1–22. [CrossRef] [PubMed]
42. Scantamburlo, G.; Hansenne, M.; Fuchs, S.; Pitchot, W.; Maréchal, P.; Pequeux, C.; Ansseau, M.; Legros, J. Plasma oxytocin levels and anxiety in patients with major depression. *Psychoneuroendocrinology* **2007**, *32*, 407–410. [CrossRef] [PubMed]
43. Meisenberg, G. Short-term behavioral effects of neurohypophyseal peptides in mice. *Peptides* **1981**, *2*, 1–8. [CrossRef] [PubMed]
44. Arletti, R.; Bertolini, A. Oxytocin acts as an antidepressant in two animal models of depression. *Life Sci.* **1987**, *41*, 1725–1730. [CrossRef] [PubMed]
45. Parker, K.J.; Kenna, H.A.; Zeitzer, J.M.; Keller, J.; Blasey, C.M.; Amico, J.A.; Schatzberg, A.F. Preliminary evidence that plasma oxytocin levels are elevated in major depression. *Psychiatry Res.* **2010**, *178*, 359–362. [CrossRef] [PubMed]
46. Jain, V.; Marbach, J.; Kimbro, S.; Andrade, D.C.; Jain, A.; Capozzi, E.; Mele, K.; Del Rio, R.; Kay, M.W.; Mendelowitz, D. Benefits of oxytocin administration in obstructive sleep apnea. *Am. J. Physiol.-Lung Cell. Mol. Physiol.* **2017**, *313*, L825–L833. [CrossRef]
47. Raymond, J.S.; Rehn, S.; Hoyos, C.M.; Bowen, M.T. The influence of oxytocin-based interventions on sleep-wake and sleep-related behaviour and neurobiology: A systematic review of preclinical and clinical studies. *Neurosci. Biobehav. Rev.* **2021**, *131*, 1005–1026. [CrossRef] [PubMed]
48. Camerino, C. Low sympathetic tone and obese phenotype in oxytocin-deficient mice. *Obesity* **2009**, *17*, 980–984. [CrossRef]
49. Takayanagi, Y.; Kasahara, Y.; Onaka, T.; Takahashi, N.; Kawada, T.; Nishimori, K. Oxytocin receptor-deficient mice developed late-onset obesity. *Neuroreport* **2008**, *19*, 951–955. [CrossRef]
50. Zhang, G.; Bai, H.; Zhang, H.; Dean, C.; Wu, Q.; Li, J.; Guariglia, S.; Meng, Q.; Cai, D. Neuropeptide exocytosis involving synaptotagmin-4 and oxytocin in hypothalamic programming of body weight and energy balance. *Neuron* **2011**, *69*, 523–535. [CrossRef]
51. Camerino, C. The New Frontier in Oxytocin Physiology: The Oxytonic Contraction. *Int. J. Mol. Sci.* **2020**, *21*, 5144. [CrossRef] [PubMed]

52. Elabd, C.; Basillais, A.; Beaupied, H.; Breuil, V.; Wagner, N.; Scheideler, M.; Zaragosi, L.-E.; Massiéra, F.; Lemichez, E.; Trajanoski, Z.; et al. Oxytocin controls differentiation of human mesenchymal stem cells and reverses osteoporosis. *Stem Cells* **2008**, *26*, 2399–2407. [CrossRef]
53. Knobloch, K.; Yoon, U.; Vogt, P.M. Preferred reporting items for systematic reviews and meta-analyses (PRISMA) statement and publication bias. *J. Cranio-Maxillofac. Surg.* **2011**, *39*, 91–92. [CrossRef]
54. Page, M.J.; McKenzie, J.E.; Bossuyt, P.M.; Boutron, I.; Hoffmann, T.C.; Mulrow, C.D.; Shamseer, L.; Tetzlaff, J.M.; Akl, E.A.; Brennan, S.E.; et al. The PRISMA 2020 statement: An updated guideline for reporting systematic reviews. *BMJ* **2021**, *29*, 372.
55. Landis, J.R.; Koch, G.G. An application of hierarchical kappa-type statistics in the assessment of majority agreement among multiple observers. *Biometrics* **1977**, *33*, 363–374. [CrossRef]
56. Amin, M.; Horst, N.; Wu, R.; Gragnoli, C. Oxytocin receptor (OXTR) is a risk gene for polycystic ovarian syndrome. *Eur. Rev. Med. Pharmacol. Sci.* **2023**, *27*, 2634–2639. [PubMed]
57. Jahromi, B.N.; Dabbaghmanesh, M.H.; Bakhshaie, P.; Parsanezhad, M.E.; Anvar, Z.; Alborzi, M.; Zarei, A.; Bakhshaei, M. Assessment of oxytocin level, glucose metabolism components and cutoff values for oxytocin and anti-mullerian hormone in infertile PCOS women. *Taiwan. J. Obstet. Gynecol.* **2018**, *57*, 555–559. [CrossRef] [PubMed]
58. Piróg, M.; Jach, R.; Ząbczyk, M.; Natorska, J. Increased Serum Levels of Phoenixin-14, Nesfatin-1 and Dopamine Are Associated with Positive Pregnancy Rate after Ovarian Stimulation. *J. Clin. Med.* **2023**, *12*, 6991. [CrossRef]
59. Masrour, M.J.; Azad, Z. A comparison of the effects of human chorionic gonadotropin and oxytocin on ovulation in PCOS patients from 2015 until 2018. *Acta Medica Mediterr.* **2018**, *34*, 1757–1763.
60. Ochsenkühn, R.; Pavlik, R.; Hecht, S.; von Schönfeldt, V.; Rogenhofer, N.; Thaler, C.J. The effect of nasal oxytocin on pregnancy rates following intrauterine insemination: Double-blind, randomized, clinical pilot study. *Arch. Gynecol. Obstet.* **2010**, *281*, 753–759. [CrossRef]
61. Sajadi, M.; Noroozzadeh, M.; Bagheripour, F.; Tehrani, F.R. Contractions in the isolated uterus of a rat model of polycystic ovary syndrome compared to controls in adulthood. *Int. J. Endocrinol. Metab.* **2018**, *16*, e63135. [CrossRef] [PubMed]
62. Iwasa, T.; Matsuzaki, T.; Mayila, Y.; Kawakita, T.; Yanagihara, R.; Irahara, M. The effects of chronic oxytocin administration on body weight and food intake in DHT-induced PCOS model rats. *Gynecol. Endocrinol.* **2020**, *36*, 55–60. [CrossRef] [PubMed]
63. Yamamoto, S.; Noguchi, H.; Takeda, A.; Arakaki, R.; Uchishiba, M.; Imaizumi, J.; Minato, S.; Kamada, S.; Kagawa, T.; Yoshida, A.; et al. Changes in endogenous oxytocin levels and the effects of exogenous oxytocin administration on body weight changes and food intake in polycystic ovary syndrome model rats. *Int. J. Mol. Sci.* **2022**, *23*, 8207. [CrossRef] [PubMed]
64. Gimpl, G.; Fahrenholz, F. The oxytocin receptor system: Structure, function, and regulation. *Physiol. Rev.* **2001**, *81*, 629–683. [CrossRef] [PubMed]
65. Akour, A.; Kasabri, V.; Bulatova, N.; Al Muhaissen, S.; Naffa, R.; Fahmawi, H.; Momani, M.; Zayed, A.; Bustanji, Y. Association of Oxytocin with Glucose Intolerance and Inflammation Biomarkers in Metabolic Syndrome Patients with and without Prediabetes. *Rev. Diabet. Stud.* **2018**, *14*, 364–371. [CrossRef] [PubMed]
66. Du, F.-M.; Kuang, H.-Y.; Duan, B.-H.; Liu, D.-N.; Yu, X.-Y. Associations of oxytocin with metabolic parameters in obese women of childbearing age. *Endokrynol. Pol.* **2019**, *70*, 417–422.
67. Iwasa, T.; Matsuzaki, T.; Mayila, Y.; Yanagihara, R.; Yamamoto, Y.; Kawakita, T.; Kuwahara, A.; Irahara, M. Oxytocin treatment reduced food intake and body fat and ameliorated obesity in ovariectomized female rats. *Neuropeptides* **2019**, *75*, 49–57. [CrossRef] [PubMed]
68. Otgontsetseg, E.; Kato, T.; Kawakita, T.; Kasai, K.; Yoshida, K.; Iwasa, T.; Irahara, M. Effects of peripheral oxytocin administration on body weight, food intake, adipocytes, and biochemical parameters in peri- and postmenopausal female rats. *Endocr. J.* **2021**, *68*, 7–16.
69. Maestrini, S.; Mele, C.; Mai, S.; Vietti, R.; Di Blasio, A.; Castello, L.; Surico, D.; Aimaretti, G.; Scacchi, M.; Marzullo, P. Plasma oxytocin concentration in pre-and postmenopausal women: Its relationship with obesity, body composition and metabolic variables. *Obes. Facts* **2018**, *11*, 429–439. [CrossRef]
70. Korkmaz, H.; Deveci, C.D.; Üstün, Y.; Pehlivanoğlu, B. Comparison of plasma oxytocin level in women with natural and surgical menopause. *Endocrine* **2023**, *82*, 209–214. [CrossRef]
71. Liu, N.; Yang, H.; Han, L.; Ma, M. Oxytocin in women's health and disease. *Front. Endocrinol.* **2022**, *13*, 786271. [CrossRef] [PubMed]
72. Breuil, V.; Panaia-Ferrari, P.; Fontas, E.; Roux, C.; Kolta, S.; Eastell, R.; Ben Yahia, H.; Faure, S.; Gossiel, F.; Benhamou, C.-L.; et al. Oxytocin, a new determinant of bone mineral density in post-menopausal women: Analysis of the OPUS cohort. *J. Clin. Endocrinol. Metab.* **2014**, *99*, E634–E641. [CrossRef] [PubMed]
73. Carr, D.B.; Utzschneider, K.M.; Hull, R.L.; Kodama, K.; Retzlaff, B.M.; Brunzell, J.D.; Shofer, J.B.; Fish, B.E.; Knopp, R.H.; Kahn, S.E. Intra-abdominal fat is a major determinant of the National Cholesterol Education Program Adult Treatment Panel III criteria for the metabolic syndrome. *Diabetes* **2004**, *53*, 2087–2094. [CrossRef] [PubMed]
74. Ferrara, C.M.; Lynch, N.A.; Nicklas, B.J.; Ryan, A.S.; Berman, D.M. Differences in adipose tissue metabolism between post-menopausal and perimenopausal women. *J. Clin. Endocrinol. Metab.* **2002**, *87*, 4166–4170. [CrossRef] [PubMed]
75. Iwasa, T.; Matsuzaki, T.; Yano, K.; Yanagihara, R.; Tungalagsuvd, A.; Munkhzaya, M.; Mayila, Y.; Kuwahara, A.; Irahara, M. The effects of chronic testosterone administration on body weight, food intake, and adipose tissue are changed by estrogen treatment in female rats. *Horm. Behav.* **2017**, *93*, 53–61. [CrossRef] [PubMed]

76. Liang, Y.Q.; Akishita, M.; Kim, S.; Ako, J.; Hashimoto, M.; Iijima, K.; Ohike, Y.; Watanabe, T.; Sudoh, N.; Toba, K.; et al. Estrogen receptor ß is involved in the anorectic action of estrogen. *Int. J. Obes.* **2002**, *26*, 1103–1109. [CrossRef] [PubMed]
77. Meli, R.; Pacilio, M.; Raso, G.M.; Esposito, E.; Coppola, A.; Nasti, A.; Di Carlo, C.; Nappi, C.; Di Carlo, R. Estrogen and raloxifene modulate leptin and its receptor in hypothalamus and adipose tissue from ovariectomized rats. *Endocrinology* **2004**, *145*, 3115–3121. [CrossRef] [PubMed]
78. Palmisano, B.T.; Zhu, L.; Stafford, J.M. Role of estrogens in the regulation of liver lipid metabolism. *Sex Gend. Factors Affect. Metab. Homeost. Diabetes Obes.* **2017**, *1043*, 227–256.
79. Chang, H.H.; Chang, W.H.; Chi, M.H.; Peng, Y.C.; Huang, C.C.; Yang, Y.K.; Chen, P.S. The OXTR polymorphism stratified the correlation of oxytocin and glucose homeostasis in non-diabetic subjects. *Diabetes Metab. Syndr. Obes.* **2019**, *12*, 2707–2713. [CrossRef]
80. Thibonnier, M.; Conarty, D.M.; Preston, J.A.; Plesnicher, C.L.; Dweik, R.A.; Erzurum, S.C. Human vascular endothelial cells express oxytocin receptors. *Endocrinology* **1999**, *140*, 1301–1309. [CrossRef]
81. Gutkowska, J.; Jankowski, M.; Antunes-Rodrigues, J. The role of oxytocin in cardiovascular regulation. *Braz. J. Med. Biol. Res.* **2014**, *47*, 206–214. [CrossRef] [PubMed]
82. De Melo, V.U.; Saldanha, R.R.; Dos Santos, C.R.; De Campos Cruz, J.; Lira, V.A.; Santana-Filho, V.J.; Michelini, L.C. Ovarian hormone deprivation reduces oxytocin expression in paraventricular nucleus preautonomic neurons and correlates with baroreflex impairment in rats. *Front. Physiol.* **2016**, *7*, 218530. [CrossRef] [PubMed]
83. Choleris, E.; Gustafsson, J.Å.; Korach, K.S.; Muglia, L.J.; Pfaff, D.W.; Ogawa, S. An estrogen-dependent four-gene micronet regulating social recognition: A study with oxytocin and estrogen receptor-α and-β knockout mice. *Proc. Natl. Acad. Sci. USA* **2003**, *100*, 6192–6197. [CrossRef]
84. Sharma, D.; Handa, R.J.; Uht, R.M. The ERβ ligand 5α-androstane, 3β, 17β-diol (3β-diol) regulates hypothalamic oxytocin (Oxt) gene expression. *Endocrinology* **2012**, *153*, 2353–2361. [CrossRef] [PubMed]
85. Hutz, R.J.; Dierschke, D.J.; Wolf, R.C. Estradiol-induced follicular atresia in rhesus monkeys is not prevented by exogenous gonadotropins. *Am. J. Primatol.* **1991**, *23*, 247–255. [CrossRef] [PubMed]
86. Agarwal, S.K.; Judd, H.L.; Magoffin, D.A. A mechanism for the suppression of estrogen production in polycystic ovary syndrome. *J. Clin. Endocrinol. Metab.* **1996**, *81*, 3686–3691. [PubMed]
87. Naessen, T.; Kushnir, M.M.; Chaika, A.; Nosenko, J.; Mogilevkina, I.; Rockwood, A.L.; Carlstrom, K.; Bergquist, J.; Kirilovas, D. Steroid profiles in ovarian follicular fluid in women with and without polycystic ovary syndrome, analyzed by liquid chromatography-tandem mass spectrometry. *Fertil. Steril.* **2010**, *94*, 2228–2233. [CrossRef] [PubMed]
88. Chauvin, S.; Cohen-Tannoudji, J.; Guigon, C.J. Estradiol Signaling at the Heart of Folliculogenesis: Its Potential Deregulation in Human Ovarian Pathologies. *Int. J. Mol. Sci.* **2022**, *23*, 512. [CrossRef] [PubMed]
89. Blanks, A.M.; Vatish, M.; Allen, M.J.; Ladds, G.; de Wit, N.C.; Slater, D.M.; Thornton, S. Paracrine oxytocin and estradiol demonstrate a spatial increase in human intrauterine tissues with labor. *J. Clin. Endocrinol. Metab.* **2003**, *88*, 3392–3400. [CrossRef]
90. Enge, S.; Klusmann, H.; Ditzen, B.; Knaevelsrud, C.; Schumacher, S. Menstrual cycle-related fluctuations in oxytocin concentrations: A systematic review and meta-analysis. *Front. Neuroendocrinol.* **2019**, *52*, 144–155. [CrossRef]
91. Tachibana, A.; Yamamoto, Y.; Noguchi, H.; Takeda, A.; Tamura, K.; Aoki, H.; Minato, S.; Uchishiba, M.; Yamamoto, S.; Kamada, S.; et al. Changes in Serum Oxytocin Levels under Physiological and Supraphysiological Gonadal Steroid Hormone Conditions in Women of Reproductive Age: A Preliminary Study. *Nutrients* **2022**, *14*, 5350. [CrossRef] [PubMed]
92. Franik, G.; Maksym, M.; Owczarek, A.J.; Chudek, J.; Madej, P.; Olszanecka-Glinianowicz, M. Estradiol/testosterone and estradiol/androstenedione indexes and nutritional status in PCOS women–A pilot study. *Eur. J. Obstet. Gynecol. Reprod. Biol.* **2019**, *242*, 166–169. [CrossRef] [PubMed]
93. Burns, P.D.; Mendes Jr, J.O.; Yemm, R.S.; Clay, C.M.; Nelson, S.E.; Hayes, S.H.; Silvia, W.J. Cellular mechanisms by which oxytocin mediates ovine endometrial prostaglandin F2α synthesis: Role of Gi proteins and mitogen-activated protein kinases. *Biol. Reprod.* **2001**, *65*, 1150–1155. [CrossRef] [PubMed]
94. Mechsner, S.; Bartley, J.; Loddenkemper, C.; Salomon, D.S.; Starzinski-Powitz, A.; Ebert, A.D. Oxytocin receptor expression in smooth muscle cells of peritoneal endometriotic lesions and ovarian endometriotic cysts. *Fertil. Steril.* **2005**, *83*, 1220–1231. [CrossRef] [PubMed]
95. Roushangar, L.; Soleimani Rad, J.; Nikpou, P.; Sayahmeli, M. Effect of oxytocin injection on folliculogenesis, ovulation and endometrial growth in mice. *Int. J. Reprod. Biomed.* **2009**, *7*, 91–95.
96. Sayyah-Melli, M.; Ouladsahebmadarek, E.; Tagavi, S.; Mostafa-Garabaghi, P.; Alizadeh, M.; Ghojazadeh, M.; Kazemi-Shishvan, M. Effect of oxytocin (OT) and OT plus human chorionic gonadotropin (hCG), in cycles induced by letrozole or clomiphene citrate (CC). *Afr. J. Pharm.* **2012**, *6*, 2112–2118.
97. Soloff, M.S.; Alexandrova, M.; Fernstrom, M.J. Oxytocin receptors: Triggers for parturition and lactation? *Science* **1979**, *204*, 1313–1315. [CrossRef] [PubMed]
98. Leonhardt, H.; Gull, B.; Kishimoto, K.; Kataoka, M.; Nilsson, L.; O Janson, P.; Stener-Victorin, E.; Hellström, M. Uterine morphology and peristalsis in women with polycystic ovary syndrome. *Acta Radiol.* **2012**, *53*, 1195–1201. [CrossRef] [PubMed]
99. Malik, A.I.; Zai, C.C.; Abu, Z.; Nowrouzi, B.; Beitchman, J.H. The role of oxytocin and oxytocin receptor gene variants in childhood-onset aggression. *Genes Brain Behav.* **2012**, *11*, 545–551. [CrossRef]

100. Montag, C.; Brockmann, E.M.; Bayerl, M.; Rujescu, D.; Müller, D.J.; Gallinat, J. Oxytocin and oxytocin receptor gene polymorphisms and risk for schizophrenia: A case-control study. *World J. Biol. Psychiatry* **2013**, *14*, 500–508. [CrossRef]
101. Correll, C.U.; Schooler, N.R. Negative Symptoms in Schizophrenia: A Review and Clinical Guide for Recognition, Assessment, and Treatment. *Neuropsychiatr. Dis. Treat.* **2020**, *16*, 519–534. [CrossRef] [PubMed] [PubMed Central]
102. Ocakoğlu, F.T.; Köse, S.; Özbaran, B.; Onay, H. The oxytocin receptor gene polymorphism -rs237902- is associated with the severity of autism spectrum disorder: A pilot study. *Asian J. Psychiatr.* **2018**, *31*, 142–149. [CrossRef] [PubMed]
103. Dinsdale, N.L.; Crespi, B.J. Revisiting the wandering womb: Oxytocin in endometriosis and bipolar disorder. *Horm. Behav.* **2017**, *96*, 69–83. [CrossRef] [PubMed]
104. Katsigianni, M.; Karageorgiou, V.; Lambrinoudaki, I.; Siristatidis, C. Maternal polycystic ovarian syndrome in autism spectrum disorder: A systematic review and meta-analysis. *Mol. Psychiatry* **2019**, *24*, 1787–1797. [CrossRef] [PubMed]
105. Hergüner, S.; Harmancı, H.; Hergüner, A.; Toy, H. Autistic traits in women with polycystic ovary syndrome. *Res. Autism Spectr. Disord.* **2012**, *6*, 1019–1022. [CrossRef]
106. Miller, S.A. *PICO Worksheet and Search Strategy*; US National Center for Dental Hygiene Research: Los Angeles, CA, USA, 2001.

MDPI AG
Grosspeteranlage 5
4052 Basel
Switzerland
Tel.: +41 61 683 77 34

Current Issues in Molecular Biology Editorial Office
E-mail: cimb@mdpi.com
www.mdpi.com/journal/cimb

www.ingramcontent.com/pod-product-compliance
Lightning Source LLC
Chambersburg PA
CBHW061557210926
51850CB00009B/2130

About the program:

AAS-IOP Astronomy ebooks is the official book program of the American Astronomical Society (AAS), and aims to share in depth the most fascinating areas of astronomy, astrophysics, solar physics and planetary science. The program includes publications in the following topics:

Books in the program range in level from short introductory texts on fast-moving areas, graduate and upper-level undergraduate textbooks, research monographs and practical handbooks.

For a complete list of published and forthcoming titles, please visit iopscience.org/books/aas.

Magnetic Fields in O, B, and A Stars